MW01343645

Science in the Ancient World

by

Dr. Jay L. Wile

Science in the Ancient World

Published by
Berean Builders
Muncie, IN
www.bereanbuilders.com

Manufactured in the United States of America
First Printing 2014

ISBN: 978-0-9890424-2-0

Printed by Courier, Inc., Kendallville, IN

Cover Photos Not in the Public Domain come from www.shutterstock.com © Stefano Panzeri, leoks, Netfalls - Remy Musser, and Panos Karas

Introduction

What is science? It is an attempt to understand the world around us and how it works. People have been doing science for a long, long time, even though it wasn't always called "science." For the majority of human history, science was known as "natural philosophy." In this course you will learn about some of the earliest natural philosophers and how they tried to understand the world around them and how it works.

As you study this course, you should be struck by two things. First, I suspect that you will be surprised at just how intelligent the ancient natural philosophers were. Nowadays, many tend to think of ancient people as ignorant and almost savage. Hopefully, this course will dispel that notion completely. It is truly amazing how much science the ancient natural philosophers were able to comprehend by observing the world around them and thinking about what they saw.

Second, once you start studying the science that was learned after Christ, you will notice that the vast majority of the natural philosophers who advanced science were Christians. That's not surprising at all. It turns out that the modern concept of science was borne out of the Christian church, because the modern view of science requires that the universe be a rational system that is operating under unchanging laws. Since Christians believe that the universe was created by a rational Lawgiver, it is not surprising that Christians were at the forefront of science. Even today, there are many Christians among the scientists who are advancing our scientific understanding of the world around us.

Of course, while you will learn a lot about the natural philosophers themselves, you will mostly be learning about the contributions they made to our scientific understanding of the world around us. As you work through the course, I hope you come to love studying the world around you like the ancient natural philosophers did!

How To Use This Book

This book contains 90 lessons split into six sections. The sections cover broad timeframes in history, and there are 15 lessons for each timeframe. However, there are two kinds of lessons. Twelve of the lessons in each timeframe are "normal" lessons, and three are "challenge" lessons. To remind you of this fact, the titles look different from one another. Normal lessons are titled in black type, while the challenge lessons are titled in red type. There are two ways to schedule these lessons so that you can finish the course in one academic year:

1. You can do a lesson every other day. This allows you to cover all 15 lessons for each timeframe in a typical academic year, which usually consists of 180 days of school.

2. You can do two lessons each week. If you follow this schedule, you will be skipping the challenge lessons. There is no harm in skipping the challenge lessons, since nothing in the normal lessons refers to what is learned in the challenge lessons.

Which do you choose? It just depends on how much science you want to do. If your children enjoy what they are doing and ask to do it a lot, you can do science every other day. If not, you can do science twice per week. Of course, if you want this book to last more than a year, you can do science even less often than twice per week. It all depends on what works best for your specific situation.

Each lesson in this book contains a hands-on activity, usually an experiment. Typically, the hands-on activity starts the lesson, but sometimes it appears later on in the lesson. In addition, each lesson contains reading. I have designed this course so that *all* elementary-age children can use it together, so the best thing to do is to have all of your elementary-age children participate in the hands-on part of the lesson, and you or the oldest child should read the lesson aloud to all of the elementary-age children. The lessons are short (less than three pages of reading each), so even the youngest student should be able to get through the lesson without *too* much discomfort.

At the end of each lesson, there is a review assignment that is split up according to the age of the student. The youngest students review the lesson by orally answering two questions. The older students have a notebooking exercise to complete, and the oldest students have a more difficult notebooking exercise to complete. How will you know which review exercise your student should do? It is really up to you. If a student is having difficulty writing, the notebooking assignments are probably not a good fit, unless you want to use them to help the student develop writing skills. Thus, non-writers should do the review exercises for the youngest students. Whether a student who can write does the exercises for the older students or the exercises for the oldest students really depends on your evaluation of the student. The exercises for the oldest students require more critical thinking and more writing than the ones for the older students.

In my opinion, all you need to do is the lesson (including the activity) and the review exercises. If you want to give your student a grade for the course, you can use the notebook or the oral questions as an evaluation tool. However, if your child is at the upper end of the elementary years, and you would like to get him used to taking tests in science, there are some tests available in the "helps and hints" book that accompanies this book. I don't think these tests are necessary, but I have made them available in case you would like to use them. In order to study for each test, the student should review his notebook. The tests assume that the student has done the notebooking exercises for the oldest student.

Question/Answer Service

One of the most important things you need to remember while using this course is that there is a way to get help if you are stuck or confused. Just go to:

http://www.askdrwile.com/

and register. Registration is free, and once you have registered, you can log in to use the website. Your information not be shared with others. The website allows you to ask questions about the material, and it also allows you to search for the answers to questions that have been asked previously. Please feel free to ask as many questions as you like!

Experiments and Activities

There are a lot of experiments and activities in this course, but most of them are purposely kept fairly simple. Ideally, the students should be doing most of the work in the experiments and activities, but ***they must be supervised by an adult***. While most of the experiments and activities are very safe, some require the use of flames, and others use sharp instruments. Thus, ***you should have the children do all of the experiments under constant adult supervision***.

Most of the experiments in the course don't take much time, but some are long-term experiments. However, you don't have to worry about looking ahead to see whether an experiment is long-term or short-term. If a long-term experiment is coming up, you will be notified of it with a note that is set off in a yellow box. It will warn you to look ahead at the upcoming long-term experiment, and it will tell you when that experiment needs to be started. Thus, you need not read ahead in the book. Everything you need to know will be presented as you need to know it.

********* **Please Do The Experiments With Common Sense And Adult Supervision**. *********

You will not find these experiments to be any more dangerous than cooking or cleaning, but that doesn't mean children can't get hurt. Supervise your children and coach them that oven burners and open flames should be avoided, and unless you are specifically instructed in the book to do so, you should *never* eat or drink anything that comes from an experiment!

Experiment Supplies

The experiments use only common household items, but of course, some items are more "common" than others. Here is a list of the things that are a bit unusual and might take some time to find. The supplies listed in red are used for the challenge lessons. If you are not doing those lessons, you don't have to worry about those supplies.

Materials That Might Take Some Time to Acquire

For the first set of lessons (Lessons 1-15):
- Hydrogen peroxide (You can get it at any drugstore or large supermarket.)
- Active dry yeast (available in the baking section of any supermarket)
- Rubbing alcohol
- A pineapple (It *cannot* be canned. It needs to be an actual pineapple.)

For the second set of lessons (Lessons 16-30):
- A hand-held, flat mirror
- Super glue

For the third set of lessons (Lessons 31-45):
- A medicine dropper

For the fourth set of lessons (Lessons 46-60):
- Two magnets (You *cannot* use refrigerator magnets for this experiment, because they are made of composite materials that behave differently from normal bar magnets.)

- Alcohol (The container needs to indicate that it is at least 90% pure. Rubbing alcohol is usually 91% pure, and denatured alcohol, which is sold in hardware stores, is usually 95% pure. Either will work.)

For the fifth set of lessons (Lessons 61-75):

- Sandpaper

For the sxith set of lessons (Lessons 76-90):

- Craft sticks or popsicle sticks (You need two if you are doing only the regular lessons, and 18 if you are doing the challenge lessons.)

Here is a list of everything you need to do the experiments, separated by the timeframe that is being studied. Remember that each timeframe is six weeks' worth of lessons, so making sure you have everything for a given timeframe ensures that you have six weeks of science supplies ready. Note that the things listed above are also contained in the list below. In addition, the materials for the challenge lessons are in red.

Supplies for Science Before Christ, Part 1 (Lessons 1-15)

- A ruler
- A measuring tape or another ruler
- A tree or tall pole that casts an easy-to-see shadow
- Four small glasses, like juice glasses
- A candle that either supports itself or is in a candle holder
- Matches or a lighter
- A medium-sized bowl
- Two rubber bands, one that is thick, and one that is thin. They need to be long enough to fit around the top and bottom of the bowl but small enough to be stretched tight when they are put there.
- A long rubber band (The longer it is, the easier it will be to see the effect.)
- An empty metal can, like the kind soup comes in
- A can opener
- Plastic wrap
- Tape
- Pepper
- Salt
- A magnifying glass
- Hydrogen peroxide (You can get it at any drugstore or large supermarket.)
- Active dry yeast (available in the baking section of any supermarket)
- Dish soap or liquid hand soap
- A 1-cup measuring cup
- A ½-cup measuring cup
- A ¼-cup measuring cup
- A measuring tablespoon
- A measuring teaspoon
- A tall glass (It is best if you can see through it, but that is not necessary.)
- A tap that can produce warm water.
- An empty sink
- Water
- Something to heat the water (A microwave will do, as will a stove and a pot.)

- A hotpad
- Two blank sheets of white paper
- Two large hard candies that are different colors (I used cinamon candies and lemon drops, but you can use whatever you like. You will need six of each.)
- A small candy (I used Skittles, but you can use whatever you like. It needs to be smaller than the other hard candies. You need eight, and they need to be all the same color.)
- Syrup, honey, or icing (Essentially, it needs to be something sweet and a bit sticky.)
- A butter knife
- A paper plate
- About 30 pennies (the duller and uglier, the better)
- A nail (It should *not* be a stainless steel or a galvinized nail.)
- White vinegar
- Rubbing alcohol
- A small amount of raw ground beef
- Two paper towels
- Jell-O, or some other form of gelatin
- A pineapple (It can't be canned. It needs to be an actual pineapple.)
- A knife to cut the pineapple
- Two spoons
- Butter (Real butter is better than margarine, but margarine will work.)
- Two tall glasses
- Elmer's glue (or any other white, all-purpose glue in a bottle)
- Pepper in a pepper shaker

Supplies for Science Before Christ, Part 2 (Lessons 16-30)

- If you can't do multiplication and division well, a calculator
- A 2-liter bottle, like the kind soda comes in
- Vinegar
- Baking soda
- A funnel
- A measuring teaspoon
- A measuring cup
- A candle
- Something you can use to light the candle
- A heavy rock (You need to be able to lift it easily.)
- A medium-weight rock (It needs to be noticeably lighter than the one above.)
- A very light rock (It needs to be a rock, but it needs to be noticeably lighter than the one above.)
- A small board or tray that you can set your rocks on and push them off the edge easily
- A ruler or sturdy stick
- A stepstool, ladder, or something that is safe to stand on so that you can reach high up
- An empty soup can with one open end
- A hammer
- A reasonably thin nail, like a finishing nail
- A long nail
- Wax paper
- Scissors

- Tape
- A dark towel or blanket
- A sunny window (If it's not sunny, you can use a dark room and a lamp that has a shape which makes it easy to recognize the top and bottom of the lamp.)
- A square or rectangle of cardboard (It needs to be larger than a paper plate)
- Six paper plates
- Crayons, colored pencils, or markers
- Some modeling clay, like Play-Doh
- A flashlight
- A hand-held, flat mirror
- A thick piece of cardboard, like the side from a cardboard box
- A white piece of paper
- A black piece of construction paper
- Thread
- Two nickels (Two of any coin will work.)
- A sink that can be filled with water
- A shovel, rake, or other tool that has a long, strong, wooden handle
- Some old hardcover books
- A heavy couch or other piece of furniture that has a small gap between its bottom and the floor.
- A dime
- A ruler
- Two balls of different size (like a baseball and a basketball)
- A baseball or tennis ball (It can be some other ball, as long as it is about the same size as a tennis ball and has a thick outer coating into which pins can be stuck without harming the ball.)
- Two straight pins
- At least six quarters and at least two pennies
- Super glue
- A plastic pen (It should be made of slick plastic.)
- A wheel (It can be from anything that rolls - a bicycle, wagon, wheelchair, etc. It works best if you can remove it from the thing to which it is attached so that there isn't a significant bump in the center.)
- String
- A sheet of cardboard
- Some Ziploc bags

Supplies for Science Soon After Christ (Lessons 31-45)

- Red grapes (You cannot use green grapes. You need the grapes that have a red peel and a whitish, fleshy interior. They are the most common grape [other than the green grapes] sold in the produce section of the supermarket.)
- Clear vinegar
- Clear ammonia (sold with the cleaners in the supermarket)
- Six small glasses, like juice glasses (The glass needs to be clear, not colored.)
- Plastic wrap
- Four bendable straws (You could use flexible plastic tubing instead, as long as it's okay to ruin the tubing in the course of the experiment.)
- A tall glass

- Tape
- A pin or needle
- Scissors
- A square of paper, about 13-cm (5-inches) by 13-cm (5-inches)
- A pencil that has never been used
- A straight pin
- A pan in which you can boil water
- Water
- Two pencils that are the same size
- A flat, hand-held mirror (It can't be a magnifying mirror.)
- A ruler or meter stick
- Some modeling clay, like Play-Doh
- A paper plate
- Crayons, colored pencils, or markers
- A small container, like a Tupperware storage container (It needs to be one you can mark up.)
- A quarter
- Two toothpicks
- A flashlight
- A marker
- Vegetable oil
- Paper towels or a dishtowel
- Glue
- A stopwatch or a watch with a second hand
- A large piece of thick cardboard (It needs to be thick enough to push a thumbtack into.)
- String
- Two pushpins or thumbtacks
- A medicine dropper
- A plastic straw (preferably one that can bend on one end)
- A spoon
- A tape measure
- A few antacid tablets (TUMS would be best, but anything that says calcium carbonate is the active ingredient will work.)
- Baking soda
- Dish soap
- A large metal spoon
- A spoon for stirring
- A paper plate or bowl
- A measuring teaspoon (If you have two measuring teaspoons, it would make things easier.)
- A ½-cup measuring cup
- Two glasses (They shouldn't be juice glasses. They should be taller than that.)
- Paper towels

Supplies for Science in the Early Middle Ages (Lessons 46-60)

- Several marbles
- A rubber band (It should be short enough to comfortably stretch between your index finger and thumb, but long enough to allow you to use it to launch a pebble into the air. It also needs to be wide enough so a pebble can fit into it.)
- A pebble that fits comfortably in the rubber band.
- An open space outside
- A bathtub
- A magnifying mirror (It is easiest if the mirror is a small, hand-held magnifying mirror. However, it can be a larger one, such as one that sits on a vanity. The larger the mirror, the more dramatic the effect.)
- Some newspaper
- Kitchen tongs or pliers
- A small candle that stands up on its own (This can be a birthday-cake candle stuck in some clay, or it can be a candle that is wide enough to stand up on its own but still not very tall.)
- A glass that is taller than the candle
- A glass jar or other glass container that is significantly larger than the glass listed above
- A glass bowl or other glass container that is larger than the glass or jar listed above
- Matches or a lighter
- Unflavored gelatin (This kind of gelatin is clear when it sets.)
- A small desert bowl or a small glass
- A round cookie cutter or a glass
- A butter knife
- A spoon
- A spatula
- A pie pan, preferably with a completely flat bottom
- Two white paper plates
- A piece of white paper that has print on it, such as a newspaper
- A flashlight
- Black construction paper
- Tape
- Scissors
- Two magnets (You cannot use refrigerator magnets for this experiment, because they are made of composite materials that behave differently from normal bar magnets.)
- Two metal sewing needles
- Two plastic one-liter bottles (Actually, anything that is transparent, tall, thin, and able to hold water will do. Glasses don't work very well because they tend to be wider at the top than the bottom, and that makes it harder to see the effect.)
- Water
- Alcohol (The container needs to indicate that it is at least 90% pure. Rubbing alcohol is usually 91% pure, and denatured alcohol, which is sold in hardware stores, is usually 95% pure. Either will work.)
- A ½-cup measuring cup (You need the kind that is used to measure out solids, because you will fill it to overflowing rather than filling it to a certain line.)
- A funnel
- A spray bottle that can produce a fine mist

- A small ball, like a baseball, tennis ball, or golf ball
- Masking tape (or another type of tape that is easy to see)
- A board that is at least 60 centimeters (2 feet) in length
- Some books
- A normal sheet of paper (writing paper, printer paper, etc.)
- Some toilet paper
- A small rock (It needs to be about the same size as a wadded-up sheet of paper.)
- An empty 12-ounce can (like the ones containing soda)
- A measuring cup
- A flat surface that can stand to get a bit wet
- Life savers mints or candies (The mints work best.)
- A small plate
- A pot for boiling water
- A large bowl
- A Styrofoam cup
- A thimble (or something else you can use to push on a needle without getting hurt)
- A birthday candle and something to hold it upright (You can just use a lump of clay.)
- A hand-held mirror that does not magnify.

Supplies for Science in the Late Middle Ages (Lessons 61-75)

- A hand-sized ball (like a baseball or tennis ball)
- A wagon or something else that you can ride on as it is pulled by another person
- A short sidewalk or driveway along which the wagon can be pulled for a little while
- A bag of M&M or Skittles candies
- Colored pencils or crayons (Optional – If you use them, you will need pencils of roughly the same colors as the candies listed above.)
- A bowl
- A graphing grid (You can get your own graph paper or use the grid that is on page A3 of your parent/teacher's *Helps and Hints* book.)
- A plastic 2-liter bottle, like the kind that soda comes in. (A plastic milk carton will also work.)
- A nail (One that is thick enough to make a hole that water can easily flow through.)
- A small container (like a juice glass or a small jar) made of clear glass
- A Ziploc bag large enough to put the glass inside and still zip shut
- A one-liter plastic bottle
- Dried beans from the supermarket
- Scissors
- A paper towel
- Water
- Legos (marbles or beads that are made of glass or plastic will work as well)
- A stopwatch or a watch with a seconds hand
- Tape
- A sheet of dark (preferably black) construction paper
- A nickel
- Several white sheets of paper
- A mirror

- Several leaves from different plants (When you pull them off the plant, make sure the stalk that connects them to the branch comes off with each leaf.)
- Finger paint (Regular paint and brushes will work as well.)
- Newspapers
- Some heavy books
- A place with several different kinds of trees and bushes to investigate
- A tree stump sitting outside. (It would actually be a bit better to find a large tree branch that has fallen on the ground and have an adult use a saw to cut it in the middle. The idea is that you need a nice, flat surface that you can sand to bring out the tree rings. It is also best to use a stump or branch from a tree that loses its leaves in the winter.)
- Sandpaper
- Some modeling clay, like Play-Doh
- Wooden matches
- A birthday candle
- Three quarters
- A tall glass
- A glass pan (or at least a pan with low edges)
- Food coloring
- A spoon for stirring
- Unflavored gelatin (This kind of gelatin is clear when it sets.)
- A round cookie cutter or a small glass
- A butter knife
- A spoon
- A spatula
- A pie pan, preferably with a completely flat bottom
- A white paper plate
- A flashlight
- Black construction paper
- Skim milk (Even 1% fat doesn't work nearly as well.)
- Vinegar
- A tablespoon
- A 1-cup measuring cup and a ½-cup measuring cup
- A saucepan
- A stove
- A funnel
- A coffee filter
- A lamp that can be pointed or tilted so its light bulb (not a compact fluorescent or LED light) can be put close to the surface of a counter. (If it's a warm, sunny day, don't worry about the lamp.)
- A plate (It can't be made of paper, foam, or plastic.)
- A cookie cutter (The more interesting the shape, the better.)

Supplies for Science in the Early Renaissance (Lessons 76-90)

- Two glasses of the same size
- A straw
- Modeling clay, like Play-Doh
- Water
- Salt
- A spoon for stirring
- A garden hose attached to a spigot
- Clothes you can get wet in
- A rock that is small enough for you to completely close your hand around it
- A pie pan or other cooking pan that has a raised edge all the way around it
- Some dirt
- Scissors
- Glue (A glue stick works best, but any glue will do.)
- Craft sticks (or Popsicle sticks): Two if you are doing only the regular lessons, 18 if you are doing the challenge lessons.
- An index card or other piece of cardboard that is sturdy but is thin enough to fold easily
- Tape
- Four spools of thread. (If the spools are empty, that's ideal, but spools with thread on them will work as well.)
- Two long rubber bands or several smaller rubber bands.
- String (It needs to be strong string. Yarn will do if you don't have any strong string.)
- A sharp knife
- A sheet of construction paper
- A working, hand-held flashlight
- Aluminum foil
- A cardboard tube from the center of a roll of paper towels
- A reasonably large funnel
- Wax paper or plastic wrap
- A stopwatch or a watch with a second hand
- A block of hard cheese that is easy to bite into (A block of cheddar cheese is ideal.)
- Two flat pieces of wood, wooden boards, or rectangular wooden blocks (They don't have to be very big.)
- A few marbles that are all the same size
- A smooth countertop that has an edge off of which you can hang something
- An empty CD case (A small box will work as well.)
- At least 40 pennies (Any coin will work, as long as you use all the same kind of coin.)
- A Ziploc bag
- Two bananas that are over ripe (They should have several dark spots on their peels.)
- A small Styrofoam cup

Science in the Ancient World
Table of Contents

Lessons 1-15: Science Before Christ, Part 1

Lessons 16-30: Science Before Christ, Part 2

Lessons 31-45: Science Soon After Christ

Lessons 46-60: Science in the Early Middle Ages

Lessons 61-75: Science in the Late Middle Ages

Johannes Guttenberg 1440

Lessons 76-90: Science in the Early Renaissance

Science Before Christ, Part 1

This is part of a temple that was built in Athens, Greece in the early 400s BC. The four sculpted women form columns that support the roof of the temple and are called caryatids (kehr ee aye' tudz).

Lessons 1-15: Science Before Christ, Part 1

Lesson 1: When Did Science Begin?

Note to the parent/teacher: This lesson needs a day that is sunny enough so that things cast an easy-to-see shadow. I wouldn't start the course until you have such a day.

Let's suppose you meet someone new. What's one of the first things you want to know about this person? You probably want to know her name, right? What's the next thing you want to know? If you are anything like me, you want to learn how old the person is. Now it's impolite to ask an adult how old she is, but it's okay for one young person to ask that of another young person. In order to get to know your new friend, then, you probably ask how old she is. Well, in this course, I want you to get to know my favorite subject: science. That's the subject's name, but how old is it? It turns out that's kind of a hard question to answer, but I will give it a try.

If you study history, you will find very, very old accounts of people treating others for diseases and injuries. More than 2,500 years before Christ was born, there are written accounts of Egyptian doctors who would try to cure a person's illness or injury. We call these doctors **ancient** because they lived a long, long time ago. These ancient doctors would have long lists of plants that would be used to cure specific diseases or at least help the person deal with the pain caused by an injury. For example, it was common for ancient people to eat something contaminated with **tapeworm** eggs. Often, those eggs would hatch inside a person's body, and the worms would grow there. Obviously, this wasn't a good thing, and ancient Egyptian doctors used to give those people roots from the pomegranate tree. It turns out that those roots contain chemicals that help the body get rid of tapeworms, so the treatment often cured the patient.

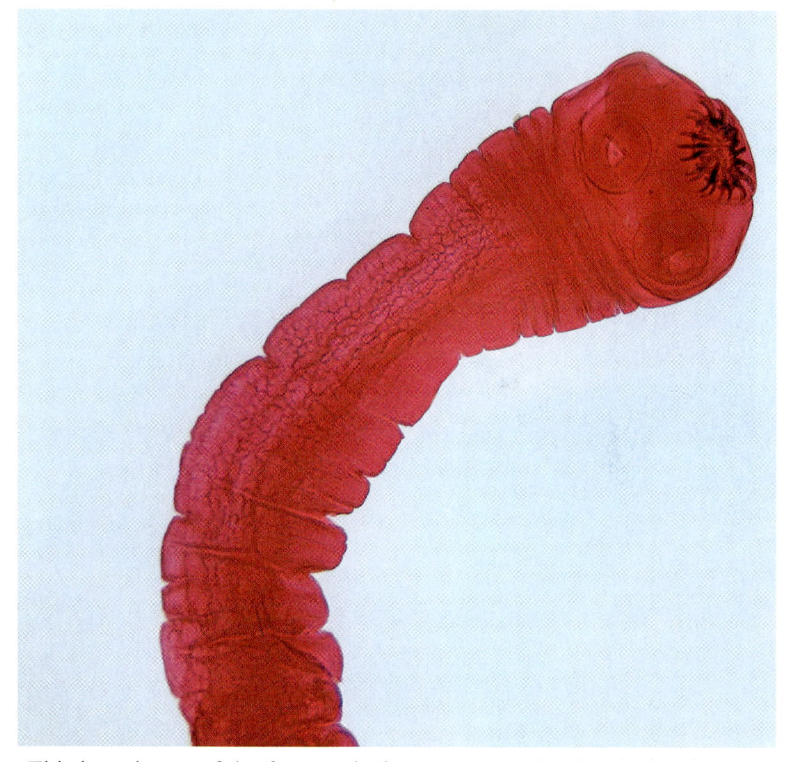

This is a picture of the front end of a tapeworm, showing its head. These worms can hatch and live inside people and animals.

Would you call that science? Some people would. However, most scientists would not. Today's doctors use science, but ancient doctors really didn't. They used **trial and error**. In other words, they would give medicine to a person (perhaps because of something they observed or heard from someone), and if the medicine didn't help, they would try something else. The hope was that eventually, they would find a medicine that would work (like the pomegranate roots), and then they would write that down so that the next time a person had the same illness, they would start with that medicine. That's not really science, however. That's just trying all sorts of things, hoping something will work.

Many scientists would say that real science started much later, only about 600 years before Christ was born. Even though there were people (especially the Egyptians) who had pretty good medicine and some impressive inventions (like paper), they didn't try to understand *why* the medicines worked or *why* the things they did produced paper. Instead, they just tried things out, and if those things produced something worthwhile, they repeated them. Otherwise, they forgot about them. About 625 years before Jesus Christ was born, however, a man named **Thales** (thay' leez) was born. As as adult, he tried to figure out *why* certain things happened. That's when many scientists think that science began.

This is a drawing of a sculpture of Thales

Since Thales was born around 625 years before Jesus, we say that was born around 625 BC (before Christ). Historians often abbreviate the word "around" with a "c," because in Latin, the word "circa" means "around." So historians would say he was born c. 625 BC. Where did he live? He lived in an ancient Greek city called Miletus (my lee' tus), so he was often called Thales of Miletus. Even though he lived there, he didn't stay there all his life. In fact, he traveled to Egypt to learn from the Egyptians. What did he learn? He learned a lot about math. The ancient Egyptians were known around the world for their excellent math skills, and Thales wanted to learn all he could from them. He obviously learned a lot, because he started making some improvements on Egyptian math.

Now wait a minute. You are supposed to be learning about science. Why are you learning about a guy who went to Egypt to learn math? Because Thales was a **philosopher** (fuh lah' suh fur), which is a person who thinks and tries to figure things out. Back then, lots of philosophers studied math, because it helped them figure things out. To see what I mean, do the following experiment, which will teach you how to figure out the height of a tree without measuring the tree itself.

Measuring Tall Things

What you will need:

- A nice, sunny day
- A ruler
- A measuring tape or another ruler
- An adult to help you
- A tree or tall pole that casts an easy-to-see shadow
- A piece of paper and pencil to write some things down

Note: An example of how to use the numbers from this experiment is given in the "Helps and Hints" book, under the "Older Students" section. The parent/teacher should feel free to help the student with the math.

What you should do:

1. Stand the ruler straight up on the ground as near to the tree as possible but so that it is still in the sun and casts an easy-to-see shadow.
2. Have an adult help you use the measuring tape or other ruler to measure how long the ruler's shadow is.
3. Write that number down.
4. Write down the length of the ruler that is casting the shadow.

5. Stand as close to the tree as possible so that you cast an easy-to-see shadow, and have the adult measure the length of your shadow from the center of your feet to the very end of your shadow.
6. Write that number down.
7. Have the adult measure your height.
8. Write that number down.
9. Have the adult help you measure the length of the tree's shadow, from the center of the tree's trunk to the end of its shadow.
10. Write that number down.
11. Go inside, and put the rulers away.

Believe it or not, you have all the information you need to figure out how high the tree is. All you have to do is use a little math. In other words, you have to use math to help you figure something out, just like an ancient philosopher!

First, take the length of the ruler (you wrote it down in step 4) and divide it by the length of the ruler's shadow (you wrote that down in step 3). We will call that number the **factor**. Now take the length of your shadow (you wrote that down in step 6) and multiply it by the factor. Compare what you got to your own height (you wrote that down in step 8). The numbers should be almost the same, because when you multiply the length of a shadow by the factor, you get the height of the thing making the shadow. Now multiply the length of the tree's shadow by the factor. That's the height of the tree! Why does this work? You'll find out in the next lesson.

LESSON REVIEW

Youngest students: Answer these questions:

1. What is the name of the philosopher who started trying to figure out *why* things happened?

2. Why did this philosopher travel to Egypt?

Older students: Remember the notebook I told you about in the introduction to the course? It is time to get it out and do some work. On the first page of your journal, write "Thales" at the top. Then, explain who Thales was and why he went to Egypt.

Oldest students: Do what the older students are doing. In addition, I want you to record the experiment you just did. To do that, start by writing out all of the measurements you made. We call them **data**. You need to label your data so that you know what they are. So for the number you wrote down in step 3, write "Length of the ruler's shadow:" and write the number after the colon. Do that with all the lengths you measured. Next, write down how you got the factor. Do that by writing "Factor =" and then, after the "=" sign, write the length of the ruler, a "÷" sign, and then the length of the ruler's shadow. On the next line, write "Factor =" and then the actual number you got when you divided the two. Next, write "My height =" and then write the length of your shadow, the "×" sign, and then the factor. On the next line, write "My height =" followed by the number you got. Then write how close it was to your actual height. Finally, write "Height of the tree =" followed by the length of the tree's shadow, the "×" sign, and the factor. On the next line, write "Height of the tree =" followed by the number you got. This is how scientists record the data from their experiments as well as the results from their experiments. (The *Helps and Hints* book has an example of how to do this.)

Lesson 2: Math and Science

In the previous lesson, you did an experiment where you measured the height of a tree by measuring the length of its shadow. Why did this work? Well, as you probably already know, the length of an object's shadow depends on how tall the object is and where the sun is in the sky. The higher it is in the sky, the shorter the shadow, and the lower it is in the sky, the longer the shadow. The sun's position affects objects in a given area the same way, however, so if one object casts a shadow that is half of its height, all objects in that area will cast a shadow that is half of their height.

In your experiment, you took something of known height (a ruler standing up straight) and measured the length of its shadow. When you divided the length of the ruler by the length of the shadow, you got a factor. That factor told you how much you needed to multiply the length of a shadow by in order to get the height of the object. So when you measured your shadow's length and then multiplied it by the factor, you ended up getting something that was very close to your measured height. In the same way, then, when you measured the length of the tree's shadow by the factor, you got something that would be very close to the measured height of the tree.

The whole reason this works is because the length of an object's shadow is determined by two things: the height of the object and the position of the sun in the sky. Since the position of the sun in the sky affects the shadows of objects in a given area the same way, if we know the height of one object, we can measure its shadow and learn how the position of the sun in the sky affects all other objects in that area. Then we can determine the height of any object just by measuring its shadow.

This is what Thales was able to do with the math he learned in Egypt. Specifically, he used it to measure the height of the pyramids in Egypt. Have you seen pictures of the pyramids? The one shown here is a good example. These impressive structures were built by the ancient Egyptians, and they are incredibly tall. The one in the picture, for example, is 136 meters (448 feet) tall. You can get a good idea of how tall it is by looking at the people I have pointed out. They look tiny compared to the pyramid, because the pyramid is so huge. The ancient Egyptians were interested in knowing how tall the pyramids were, but they had no way of measuring them. Thales used essentially the same method you used in your experiment to measure the height of the pyramids.

This is a pyramid in Egypt. Notice how small the people in the lower left of the picture look. That tells you the pyramid is *really* tall.

Do you see why Thales wanted to learn as much math as possible? Math is incredibly useful, and it helps you figure out a lot of things. In science, you have to figure out a lot of things, so math is an important part of science. In fact, for a long time, math was not even considered separate from science. In ancient times, science was known as "natural philosophy," because it was involved in figuring out how nature works. Math was a part of natural philosophy. So in ancient times, and for quite some time after that, math and science were really thought of as the same thing. It is only fairly recently that people started considering them different subjects!

Because math and science were considered the same thing back then, Thales's math education was part of his science education, and he ended up using it to figure out lots of things. For example, he used math to figure out how far ships were from shore, and he was able to predict that a total eclipse of the sun would occur in 585 BC. (If you didn't learn this already, a total eclipse of the sun happens when the moon gets directly in between the earth and the sun. This keeps the sun's light from hitting a large section of the earth, making that region dark even in the middle of the day.) Obviously, being able to predict when such an event would happen is an impressive thing, and Thales was able to do that with a combination of what we would call math and science. In his day, however, it was all called natural philosophy.

One thing that's important to remember is that scientists aren't always right. Even when they are wrong, however, their thinking can sometimes lead to an idea that is right. Thales was wrong about a lot of things, but sometimes, even his wrong thinking was on the right track. For example, Thales thought that water was the most basic form of creation. According to him, everything was made of water and, eventually, it would turn back into water. We know that this idea isn't correct, but it does contain some bits of truth. In a sense, water is in a lot of things, and if you do it right, you can make a lot of things change so that water is made. Try this experiment to see what I mean.

Making Water

What you will need:

- A small glass, like a juice glass
- A candle that either supports itself or is in a candle holder
- Matches or a lighter
- A kitchen countertop
- An adult to help you

What you should do:

1. Put the candle on a kitchen counter far from anything that can burn.
2. Have an adult use a match or lighter to light the candle.
3. Hold the glass in the air upside down over the candle, as shown in the picture on the right. Don't lower the glass too far, or the candle flame will go out.
4. As you are holding the glass over the candle, watch the glass carefully. You should notice something happening in less than a minute.
5. Pull the glass more than an arm's length away from the candle, but keep looking at it. Once again, you should notice something happening in less than a minute.
6. Repeat steps 4 and 5 to see the effect one more time.
7. Blow out the candle, put everything away, and clean up any mess you might have made.

What did you see in the experiment? You should have noticed that after just a few seconds, the glass started fogging up. However, once you pulled the glass far from the flame, the fog inside slowly went away. When you put the glass back over the flame, it fogged up again.

What explains the experiment's results? The fog you saw on the glass was actually *water*. You have probably already learned that water can exist as a gas (called "water vapor"), a solid (called "ice"), or a liquid (usually just called "water"). Well, the candle flame actually produced water vapor. However, as the water vapor moved away from the flame and hit the glass, it cooled enough to become little drops of liquid, and those drops fogged up the glass.

Why did the flame produce water? Because when you burn the wax in a candle, the chemicals in the wax interact with oxygen in the air. As a part of this interaction, the chemicals in the wax change into completely different chemicals, and one of those chemicals is water. So the *makings* of water can actually be found in wax and air! If you treat wax and air in a specific way (burn them), they can produce water.

Like burning the candle in your experiment, burning wood produces water vapor, because the wood and the oxygen in the air contain the makings of water.

In a sense, then, Thales was on the right track when he thought that things were made of water. That's not really true, but in some things (like candle wax and air), the *makings* of water can be found. If you treat these things in a specific way, they will produce water. At the same time, however, there are some things that don't contain the makings of water at all. No matter what you do to them, they can never be used to produce water.

As you learn more and more about science and its history, you will find that it is very common for even the greatest of scientists to be wrong about many, many things. Thus, no matter how great the scientist, you should not just trust his or her opinions. You should examine the evidence yourself, since even the greatest of scientists can be wrong. That's one reason I want you to do so many experiments in this course. I want you to get used to the idea of looking at the evidence yourself!

LESSON REVIEW

Youngest students: Answer these questions:

1. In the experiment for Lesson 1, you measure the height of a tree by measuring the length of its shadow. What did Thales measure in the same way?

2. What is one of the chemicals made when wax is burned?

Older students: Make a drawing of the experiment you did. Underneath the drawing, write what you saw happening to the glass and why it happened.

Oldest students: Do what the older students are doing. Also, do a bit of research and find out what other chemical (besides water) is made when wax is burned. Indicate why you didn't see it in the experiment.

Lesson 3: Pythagoras (c. 570 BC – c. 495 BC)

When you listen to music, do you ever think of science or math? Most people don't, but one famous philosopher from the past did. His name was **Pythagoras** (pih thag' ur us). He was born c. 570 BC (while Thales was still alive), and he died c. 495 BC (well after Thales died, c. 546 BC). He lived on an island called Samos (say' mahs), which was less than 100 miles from Miletus, where Thales lived. Even though travel was difficult back then, most historians think that Pythagoras visited Thales before Thales died. If nothing else, Pythagoras had heard the teachings of Thales, because he was influenced by Thales's ideas.

This is a sculpture of Pythagoras that is in Rome.

Do you remember how important math was to Thales? Because of his influence, math was very important to Pythagoras as well. In fact, when you get into high school, you will learn about the Pythagorean Theorem, which tells you how to calculate the lengths of the sides of certain triangles. For now, however, I want you to learn about Pythagoras's scientific investigations regarding music.

Like most people, Pythagoras loved music. As a natural philosopher, however, he wanted to know more about music than just how it made him feel. He wanted to learn the details about music. What made one musical sound different from another? Because math was so important to him, he wanted to know if there was any way math could be used to study music. Not surprisingly, he discovered how math and music are related. Perform the following experiment to see what he figured out.

Length and Sound

What you will need:

- A medium-sized bowl
- Two rubber bands: one that is thick and one that is thin. They need to be long enough to fit around the top and bottom of the bowl but small enough to be stretched tight when they are put there.
- Someone to help you

What you should do:

1. Stretch both rubber bands across the top and bottom of the bowl, as shown in the photo on the right. They need to be at least a few centimeters (a couple of inches) apart from one another.

2. Smooth the rubber bands out so they are not twisted. If you can't get the twists completely out, just make sure the part of each rubber band that stretches over the top of the bowl has no twists in it.
3. Pluck the thick rubber band and listen to the sound it makes. Watch the rubber band as well. What is it doing?
4. Pluck the thin rubber band and listen to the sound it makes. Watch the rubber band as well. What is it doing? Do you notice that the sound this rubber band makes is different from the sound made by the thick rubber band?
5. Have your helper pinch the thick rubber band between his thumb and forefinger right in the middle of the bowl.
6. Pluck the thick rubber band on one side of your helper's thumb and forefinger. Watch the rubber band and listen to the sound.
7. Have your helper release the thick rubber band.
8. Pluck the rubber band again, watching it and listening to it.
9. Repeat steps 5-8 with the thin rubber band. Do you hear that the rubber band makes a different sound when it is pinched? Did you notice that when your helper pinched it, the rubber band only bounced up and down on the side that you plucked? On the other side of your helper's thumb and forefinger, the rubber band stayed still.
10. Play with this a little while. Have your helper pinch the rubber bands in different places and then pluck the rubber band on each side of where your helper is pinching. Do you notice any pattern to where your helper pinches and the kind of sound that is made?
11. Put everything away.

What did you hear in the experiment? The rubber bands made different sounds when you plucked them, and when your helper pinched them, they made still different sounds. If you have ever played with a guitar, violin, harp, or other instrument that uses strings to make music, this might not have been very surprising to you. After all, we know that when a guitar player plucks the strings on his guitar, they make sounds. If you look closely at the guitar, you will see that the strings are actually different.

A guitar's strings are different, and they each make a different sound when plucked.

Look at the picture on the left. Do you see that the strings on one side of the guitar are so thin that it is hard to see them? Notice, however, that each string gets thicker so that by the time you look at the strings on the other side of the guitar, they are really easy to see. When you pluck each string, you get a different sound, don't you? What happens if you pinch one of the strings against the neck of the guitar and then pluck it? The sound changes again, doesn't it? A stringed instrument makes sounds by having its strings plucked, hit, or stroked. Each string has its own basic sound, and just like the rubber bands in your experiment, if you pinch one of the strings right, you can make it produce a sound that is different from its basic sound. Pythagoras knew all of this, because he loved music. The people of his time knew this as well, because they played stringed instruments. What Pythagoras discovered was that math could be used to study the sounds the instruments made.

Do you know what the word **pitch** means when it comes to music? It tells you how high or low a musical note is. I am not talking about how loud or soft the note is. I am talking about whether it is a high note or a low note. When people sing, women typically sing at a higher pitch than men. In other words, women usually sing higher notes than men. When you pluck a guitar string, the thinner strings make notes with a higher pitch, and the thicker strings make notes with a lower pitch.

In music, we identify the basic notes by seven letters: A, B, C, D, E, F, and G. The "A" note has the lowest pitch, and each note after that has a higher pitch. Believe it or not, those are the only basic notes in music. Even though there are a lot more *sounds* in music, there are only seven basic notes. On a guitar, the thickest string makes the note we call "E." However, the thinnest string also makes an "E." The "E" the thin string makes is noticeably higher in pitch than the "E" the thickest string makes, but they are both the same musical note. When you have two of the same note that have different pitches, we say they are an **octave** apart. The beginning of the word "octave" is "oct," which means "eight." Can you guess why we call it an octave? Remember, there are seven different musical notes. If you start at A, how many notes do you have to hit before you get back to A again? You have to hit A, B, C, D, E, F, G, and then back to A. So to repeat a note, you have to step through 8 musical notes, which is why we call it an "octave."

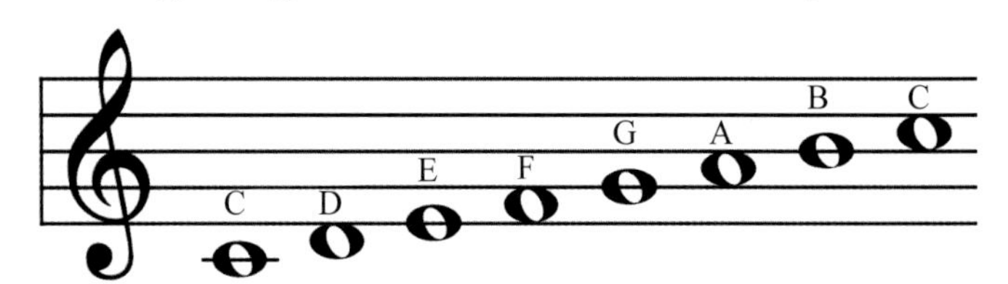

This is an octave of notes, starting and ending with "C." The "C" on the right is an octave above the "C" on the left.

Now here's what Pythagoras discovered: If you pluck a string on an instrument, it produces a given note. However, if you pinch the string exactly halfway down, it gives the same note again, *one octave higher*. What does this mean? Think about the guitar again. Its thickest string produces an "E" when it is plucked. If I pinch that string against the neck of the guitar exactly halfway down and then pluck it, guess what note it makes? It makes an "E" again, but it is an octave higher. In other words, *when pinched halfway down, the thickest string on a guitar produces the same exact pitch as the thinnest string produces if it is not pinched at all.* Pythagoras used that knowledge to actually show that there is a direct mathematical relationship between the length of a string and the note it makes when plucked. In other words, he used math to describe musical notes! Isn't that cool?

LESSON REVIEW

Youngest students: Answer these questions:

1. What does the word pitch mean when it comes to music?

2. Fill in the blank: Of the seven basic notes in music, __ has the lowest pitch and ___ has the highest pitch.

Older students: Write a description of the experiment you did, and explain the relationship between the length of the portion of rubber band that you plucked and the pitch of the sound that the rubber band made.

Oldest students: Do what the older students are doing. In addition, write (in order of their pitch) the seven letters that are used to identify the basic musical notes and explain what an octave is. Write down what Pythagoras figured out about the length of a musical string and the octave.

Lesson 4: But What Is Pitch?

It's fun to study the way that ancient philosophers learned specific things about science, but sometimes, to really understand what's going on, you have to go a little further than what people like Pythagoras actually learned. After all, Pythagoras figured out the mathematics involved in making sounds of different pitch, but he didn't really understand why the relationship worked. He just figured out that it did. I want to go a little further than Pythagoras and actually tell you *why* pinching a string makes it produce different sounds when it is plucked. Let's start with an experiment.

Sound is a Wave

What you will need:

- An empty metal can, like the kind soup comes in
- A can opener
- Plastic wrap
- Tape
- Pepper
- Someone to help you

What you should do:

1. Use the can opener to remove the top of the can (if it isn't already removed) and the bottom of the can. That way, the can is open on both ends.

2. If necessary, clean out the can.
3. Use plastic wrap to cover one end of the can. Use tape to hold the wrap to the can, and stretch the wrap tight so there are no wrinkles in the part that is covering the end of the can.
4. Shake pepper onto the plastic wrap so it is spread out over the wrap.
5. Hold the can so that the plastic wrap is on top (as shown in the picture on the left). You should hold it as high as you can while still being able to clearly see the pepper on top.
6. Have your helper get under the can so that his or her mouth is right under the opening of the can.
7. Have your helper sing "ahhh" with the lowest pitch he or she can sing. While your helper is singing, notice how the pepper behaves.
8. Have your helper sing "ahhh" at a much higher pitch, but at about the same volume he or she sang at the lower pitch.
9. Once again, notice how the pepper behaves as your helper holds the note.
10. Did you notice a difference in the pepper's behavior during the high note compared to its behavior during the low note?
11. Repeat steps 7-9 to make sure you see the difference.
12. Have your helper sing "ahhh" at any pitch he or she wants. Your helper should start singing the note with very little volume, but then he should increase the volume as time goes on. Notice what happens to the pepper as your helper increases the volume.
13. Clean up your mess and put everything away.

If you did *Science in the Beginning*, you used the can/plastic wrap/pepper device before to "see" sound. This time, however, you used it to learn a lot more about what sound is. First, when your helper sang "ahhh" under the can, the pepper bounced up and down, didn't it? That's because sound is actually a bunch of vibrations in the air. If you did *Science in the Beginning*, you knew that already. Now it's time to learn a lot more about those vibrations.

Think about what you do in order to make sound come out of your mouth. You open your mouth, and as you talk or sing, air comes out of your mouth. Now when you breathe, air is coming out of your mouth as well. However, it doesn't make a sound. What's the difference? When you breathe, air is coming out of your mouth in a smooth stream. As a result, it makes a small wind, but it doesn't really make any sound. When you talk (or sing), the air is *not* coming out of your mouth in a smooth stream. Instead, it is coming out in clumps. The molecules that make up the air crowd together, and then behind that "clump" of air, there is a lot less air. Look at the drawing on the right. The dots represent molecules in the air. Notice how they clump together in certain places and then are spread out in other places. That's what the air molecules look like as they are coming out of your mouth when you are talking or singing.

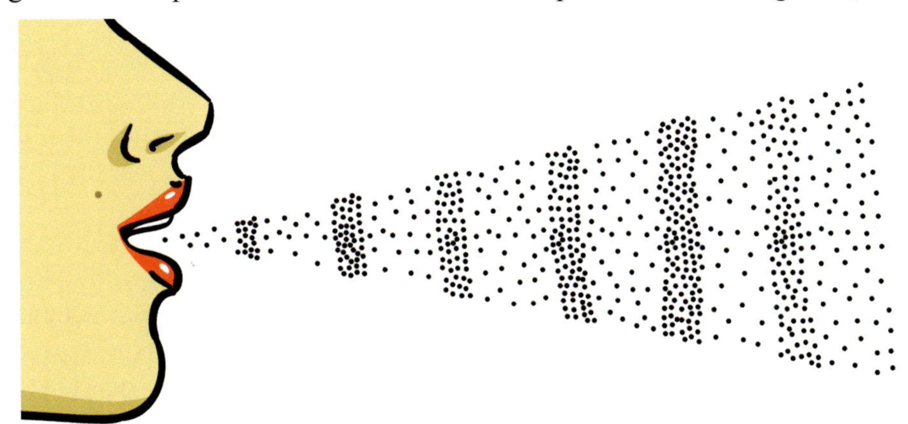

When you talk or sing, the molecules in the air clump up and spread out, forming a sound wave that travels through the air.

Can you see why the pepper on the plastic wrap bounced up and down when your helper sang under the can? What happens when a clump of air hits the plastic? The plastic bows out, because it is being hit by a clump of air. After the clump of air hits, however, there is a bunch of spread-out air that follows. This allows the plastic wrap to relax. Not long after, however, another clump of air comes and hits the plastic, and it bows out again. So the pepper bounced up and down because the plastic kept being hit by the clumps of air, causing it to bow out, but then the plastic would relax when the spread-out air reached it.

The series of clumped-up and spread-out air that makes sound is called a **sound wave**. Think about the waves that you see on the ocean. Why do you call them waves? Because there are places where the water is "clumped up" to form a "hill" of water, and behind that "hill," the water is spread out to form a little "valley." Those "hills" of water are called the **crests** of the wave, and the "valleys" are called the **troughs** (trofs) of the wave. Well, the clumps of air in sound are the crests of the sound wave, and the areas of spread-out air are the troughs.

Ocean waves have crests and troughs.

Sound, then, is really a wave that travels in air. The crests of the wave are where the air is clumped up, and the troughs are where the air is spread out. Have you ever stood in the ocean and let the waves hit you? The crests hit hard, don't they? In the same way, the crests of a sound wave hit hard as well. That's why the plastic wrap bowed out when they hit it.

Now that you know sound is a wave, you can learn why sound comes in different pitches. What happened when your helper sang a low note under the can? The pepper bounced up and down. What happened when your helper sang a high note under the can? The pepper bounced up and down, *but it bounced faster*, didn't it? Why would it bounce faster? Because the crests of the sound wave were hitting it faster. When the crests of a sound wave hit your ear slowly, your brain interprets it as a sound with a low pitch. When they hit your ear quickly, your brain interprets it as a sound with a high pitch.

Scientists have a name for how often the crests hit your ear. They call it **frequency** (free' kwuhn see), and it is a measure of how many crests hit your ear every second. The higher the frequency, the more crests hit your ear every second, and the higher the pitch. The lower the frequency, the fewer crests hit your ear every second, and the lower the pitch.

But pitch isn't the only thing that describes sound. What about volume? Well, what happened in the experiment when your helper started increasing the volume of the note he or she was singing? The pepper shouldn't have bounced up and down any faster (if your helper held the same pitch the whole time), but the pepper should have started bouncing up and down *higher*. That's because in a loud sound, there is more air in the clumped-up regions. In other words, the crests of the sound wave are higher. Just as a higher wave hits you harder when you are standing in the ocean, a sound wave with more air in its crest will hit your ear harder, making a louder sound.

When you hear a sound, then, the two main things you notice are its pitch and its volume. Since sound is a wave that travels through air, the frequency of the wave (how many crests hit your ear each second) determines the pitch, and the height of the crests in the wave determines how loud the sound is. This is more than Pythagoras knew about sound, but in the next lesson, you will see how it explains why the pitch a string makes when it is plucked changes when you pinch it.

LESSON REVIEW

Youngest students: Answer these questions:

1. What are the clumps of air in a sound wave called? What are the areas of spread-out air called?

2. Fill in the blanks: When the crests of a sound wave hit your ear quickly, you hear a ______ pitch, but when they hit your ear slowly, you hear a _____ pitch.

Older students: Draw a sound wave like the one that is drawn on page 11. Label the crests and the troughs of the wave. Below the picture, define the term "frequency," and explain how frequency relates to pitch. Also, explain how the amount of air in the crests relates to volume.

Oldest students: Do what the older students are doing. In addition, when you want to shout really loudly, you usually take in a deep breath before you shout. Using what you learned today, explain why you do that. Check your explanation and correct it if it is wrong.

Lesson 5: Why is Length Related to Pitch?

In the previous lesson, you learned that the pitch of a sound is related to the frequency of the sound waves. The more clumps of air (crests) that hit your ear every second, the higher the pitch. But how does this explain what Pythagoras discovered? Why does pinching a musical string at different points produce different pitches when the string is plucked? You are now ready to learn the answer to that question. As I usually like to do, let's start with an experiment.

Making Waves

What you will need:

- A long rubber band (The longer it is, the easier it will be to see the effect.)
- Two fixed posts across which you can stretch the rubber band so it holds tight (See the picture below to get an idea of what I mean.)

What you should do:

1. Stretch the rubber band so that it loops over both posts. You should now have a tight rubber band that stays where it is without you holding onto it. See the picture on the right.
2. Pluck the side of the rubber band that is nearest to you and watch how it vibrates back and forth. You will hear a sound when you pluck it, but you already know about that. Concentrate on how the rubber band bounces up and down.
3. Pinch that side of the rubber band with your thumb and forefinger halfway between the posts.

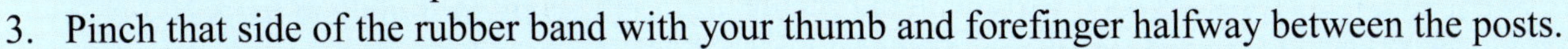

4. Pluck that side of the rubber band between one of the posts and where you are pinching it. Once again, you will hear a sound of a higher pitch, but that's not what I want you to concentrate on. Notice how the rubber band bounces up and down now.
5. Repeat steps 2-4 a couple of times to see if you can notice any difference between how the rubber band bounces when it is pinched and when it is not pinched.
6. For this lesson, don't put anything away yet. Just continue on with the lesson. You will come back to the rubber band in a moment.

Did you notice how the rubber band was bouncing up and down when you didn't have it pinched? It followed a pattern that looked like the drawing below, didn't it?

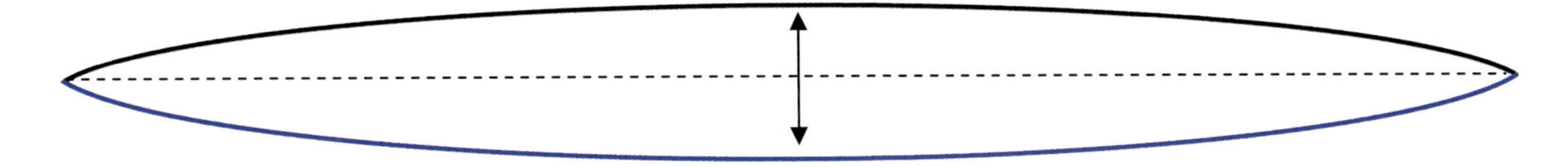

The rubber band bounced up and down so that there was a "hump" in the middle, and that hump moved above and below where the rubber band was before you plucked it (the dashed line in the drawing). Even when you pinched the rubber band, it still bounced with the same pattern, but the "hump" was half as long, right? Here's the key, however: did you see a difference in *how quickly* the hump bounced up and down? If you didn't, go back to the rubber band and repeat steps 2-4 of the experiment again. See if you can notice a difference in how quickly the hump moves up and down.

What you should have noticed is that the hump moved up and down a lot *faster* when you pinched the middle compared to when you didn't pinch the rubber band. Believe it or not, this explains why the pitch of the sound the rubber band makes is higher when it is pinched. Think about what happens when the hump of the rubber band starts moving above where the rubber band started. It pushes against the molecules in the air. What happens when those molecules get pushed? They start clumping together. When the hump starts to move down, it pushes the molecules in the air in the opposite direction. That forms a clump in the opposite direction, and it forms an area of spread-out molecules behind the first clump that it created. What do you have when you have clumps of air followed by regions of spread-out air? You have a sound wave! So the rubber band makes sound because it is pushing the air around, forming clumps of air (crests in the wave) and regions in which the air is spread out (troughs in the wave).

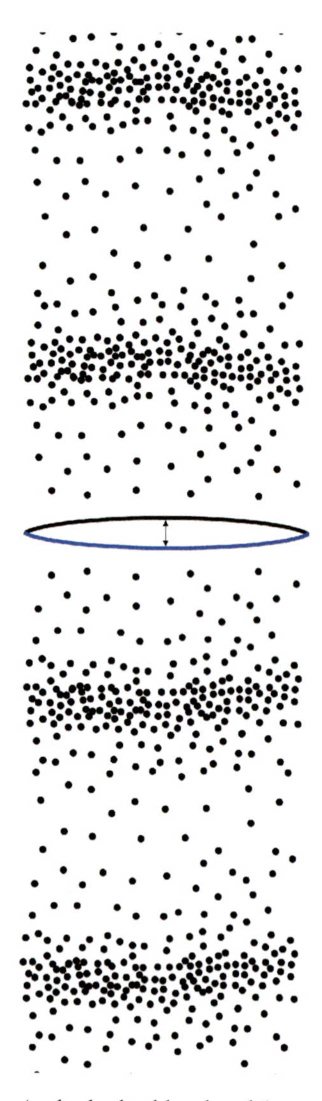

A plucked rubber band (or string) makes a sound wave by pushing air as it vibrates.

The drawing on the left shows you what I am trying to describe. As the rubber band vibrates up and down, it pushes the molecules in the air around, forming crests and troughs. That produces a sound wave, and when that sound wave hits your ears, you recognize a specific pitch, which is determined by the number of crests that hit your ear every second.

Now think about what determines how quickly the crests hit your ear. The faster the rubber band bounces up and down the more crests it will make every second, right? That means there will be more crests to hit your ear every second. Now you know why the pinched rubber band produces a sound with a higher pitch. It bounces up and down faster, making more crests each second. That means more crests hit your ear each second, which means you hear a higher pitch. So that's why pinching a rubber band (or a string on a guitar) produces a sound with a different pitch. If you shorten the length of rubber band (or string) that is vibrating, it vibrates faster. That makes more crests and troughs in the sound wave, producing a higher pitch.

Now remember, pitch isn't the only thing you notice when you hear a sound. You also notice its volume. What determines a sound's volume? The height of the crests in the wave. What do we mean by the height of a sound wave's crests? We mean the amount of air in one of the clumps. The more air that is in the clumps, the harder they hit your ear, so the louder the sound. In your experimental setup, what would determine that? Think about it and actually say an answer out loud. If you can't think of something, just guess. I want to get your mind working. Come up with *something* that you think might determine the volume of the sound that you hear. Once you have come up with something and said it out loud, go back to the rubber band. See if your guess works by testing it. Pluck the rubber band in a way that you think will produce a sound with only a little volume, and then pluck it in the way you think you need to make the sound louder. Did it work? That's not as important as actually going through the exercise of trying to think what would make a louder sound and actually testing it to see if it would.

If you weren't able to control the volume of the sound, go back to the rubber band and try again. This time, pluck the rubber band as gently as possible, just barely pulling it back before you release it. Next, pluck the rubber band harder, pulling it back much farther before you release it. You should notice that the sound is louder the harder you pluck on the string. So the distance that the hump travels when the string is plucked determines the volume of the sound. Does this make sense? It should. After all, the farther the hump travels, the more air it clumps up in the crests. That makes the crests larger, producing a louder sound.

So now you know how a vibrating string makes a sound. You also know that the shorter the string is, the faster it vibrates, which produces a sound wave that has a higher frequency and therefore a higher pitch. You also know that the distance the string's hump travels as it vibrates determines the volume of the sound, because that determines how much air is clumped up into the crests. Obviously, this reasoning applies to any stringed instrument. A guitar player changes the pitch of the strings by changing the length that vibrates, and he controls the volume of the sound by how hard he plucks the strings. A violin player does the same thing. The bow on a violin has tiny "hooks" that pluck the strings as the violin player pulls it over the strings. She controls the pitch of the strings by pinching them to change the length that vibrates, and she controls the volume by pressing down harder with the bow or pulling the bow faster, either of which is equivalent to plucking the strings harder.

This is the inside of a piano, showing the hammers and strings that produce the sounds the piano makes.

What you might not realize is that this reasoning applies to instruments that don't look like stringed instruments at first glance. Consider a piano. How does it make music? When you hit a key, the key causes a hammer to strike a string. Actually, it's one or two strings, depending on the actual note. The strings for a given key are all the same length, so they vibrate at a specific frequency, producing a specific pitch. You control the volume by how hard you hit the keys. The harder you hit, the harder the hammer strikes the strings, which is the same as plucking the strings harder. So the science you just learned behind music applies to many different instruments!

LESSON REVIEW

Youngest students: Answer these questions:

1. When a string vibrates quickly, does it produce a sound with a high pitch or a low pitch?

2. When you pluck a string gently, does it make a loud sound or a soft sound?

Older students: Draw a picture like the one on page 14, and use it to explain how a vibrating string makes a sound wave. Describe how the speed of the vibration affects the pitch and the distance over which the string vibrates determines volume.

Oldest students: Do what the older students are doing. In addition, try to find at least one other property of a string (besides its length) that affects how quickly it vibrates. Hint: Think about how a stringed instrument is tuned.

Lesson 6: Democritus (c. 460 BC – c. 370 BC)

This is what Dutch painter Hendrick ter Brugghen thought Democritus may have looked like.

About 35 years after Pythagoras is thought to have died, another famous Greek philosopher was born. His name was **Democritus** (dih mah' kruh tus), which means "chosen of the people." His father was incredibly rich, and Democritus chose to spend that wealth on travel and books so that he could learn as much as possible. In his travels, he met a philosopher named Leucippus (lew sip' us), who thought that all of matter was composed of tiny particles called **atoms**. Democritus really liked that idea and expanded on it quite a bit. As a result, even though Leucippus was the first Greek philosopher we know of to really talk about atoms, Democritus is usually the one credited with the idea. He was often called "the laughing philosopher." Unfortunately, that's not because he was a jolly fellow. It's because he often laughed *at* other people, not *with* them.

Now you probably know something about atoms already, but you need to first understand how revolutionary the idea was during this time in history. Why would anyone believe in atoms? After all, if you look at this page of paper, it looks smooth, doesn't it? It doesn't look like it is made up of a bunch of tiny particles. Democritus, however, knew that looks can be deceiving. As an example, he talked about walking towards the sea shore. When you are far away, it looks like the water ends at this smooth strip of yellow that we call "the beach." As you get closer to the beach, you realize it is not a smooth strip of yellow. Instead, it is made up of a bunch of tiny grains of sand. Those grains are so small, however, that you don't notice them until you are very close to the sand.

Democritus suggested that everything we see is really like the sand on the beach. From a distance, it all looks smooth and continuous. However, if you were able to observe any particular thing closely enough, you would see that it is made of little "grains" called "atoms." So even though the water hitting the beach looks smooth and continuous when you collect it in a bowl, if you were able to look closely enough, you would see that it is made of tiny atoms, just as the beach is made of tiny grains of sand. While a lot of what Democritus thought about atoms turned out to be wrong, at least he got that part right. I want you to see how profoundly correct Democritus was by doing the following experiment.

Lots of Dots

What you will need:
- This book and another book that has pictures in it
- A magnifying glass

What you should do:

1. Look closely at the picture of Democritus on the previous page. Can you see any lines or dots in the picture? Probably not. The picture looks smooth, doesn't it?
2. Now look at the picture with the magnifying glass. Can you see dots now? You should be able to see them. Even though it doesn't look like it, the picture of Democritus is actually made of a bunch of dots.
3. Look at some other pictures from this book and the other book. First look at them without the magnifying glass and then look at them with the magnifying glass. Notice that with the magnifying glass, you can see they are all made up of tiny dots.
4. Look at the letters in the words of a page without using the magnifying glass. Can you see any dots in the letters? Probably not.
5. Use the magnifying glass to look at some of the letters on the page. Do you see any dots? Probably not, at least not unless you have a pretty powerful magnifying glass.
6. Put the magnifying glass and the other book away.

Why are there dots in the pictures you examined? Because that's how a printing press (and a printer hooked up to a computer) draws things. It actually just puts a bunch of dots on the page, and those dots "add up" to what you see on the page. Since the dots are tiny, you can't see them without something that magnifies them. Instead, you just see the overall picture the dots make. However, with the magnifying glass, you were able to look at the pictures more closely. As a result, you saw that what appears to be a smooth picture is, in fact, made up of a bunch of little dots.

But wait a minute. Why didn't you see dots in the letters of the words on the page? Those letters were made with the same printing press. If the press just makes dots on the page, where are they in the letters of the words? Well, they are there, but they are *a lot* smaller than the dots that make up the pictures. When printers make pictures, they generally use about 300 dots for every inch of paper. That makes the dots really tiny, but with almost any decent magnifying glass, you can see them. However, most printing presses use 2,400 dots for every inch of paper when they are making letters. So the dots in the letters you are reading are about *eight times smaller* than the dots in the pictures you are seeing. As a result, your magnifying glass would need to be eight times as good in order for you to see them. If you got a good enough magnifying glass, or a microscope, you would be able to see them!

Your experiment really demonstrates what Democritus meant with his example of walking towards a beach. Looks can be deceiving. When you look at the pictures in this book, the ink looks like it has been put on the page very smoothly. However, when you magnify the pictures, you can see the dots used by the printer. On the other hand, even with the magnifying glass, you couldn't see any dots in the text. That's not because the dots aren't there. It's just because the dots are too small to see with the magnifying glass you used.

In Democritus's mind, everything around you is like that. It doesn't matter how smooth an object looks. If you were able to magnify it enough, you would eventually see that it is made of individual particles called atoms. Modern scientists agree with Democritus. We can't really see atoms, because even the most powerful microscopes that use light cannot magnify things enough for us to see them. However, we are able to do experiments that produce evidence that tells us they exist. So even though we can't see them, we have a lot of evidence to tell us that they are real.

This actually brings up a very important point about science. Some people say that "seeing is believing." However, that's not really true. First, if you have ever seen a magician perform on stage, you know that sometimes, you can't believe what you see. A magician can make it appear like he is levitating a person or making a rabbit magically come out of his hat, but those things aren't really happening. In the same way, sometimes you must believe in things you cannot see, such as atoms. So in the end, scientists have to be skeptical of what they see, because sometimes, looks can be deceiving. At the same time, however, sometimes they have to believe in things they cannot see. In the end, the reason a scientist believes in something is because there is sufficient evidence to back up the belief. That evidence isn't always what the scientist sees with his eyes, however, since Democritus showed us more than 2,000 years ago that looks can be very deceiving!

This Georges Seurat painting is a pointillist painting. It is made of individual dots that combine to make an image.

You will learn more about atoms in the lessons ahead, but before I end this lesson, I want you to know that long before printing presses and computer printers were making pictures using dots, there were some painters who made pictures this way. In 1886, a famous French artist named Georges (jorzh) Seurat (suh rah') developed a method of painting in which he put tiny dots on a canvas so that when you stand back from the painting, a smooth picture is seen. Up close, however, you can easily see the dots that make up the painting. One of his paintings, called *Young Woman Powdering Herself* is shown on the left. This style of painting is called Pointillism (poyn' tuh lih' zuhm). It was never really popular, and many art critics made fun of it at the time. Nevertheless, some famous artists, such as Vincent van Gogh (go), did use it.

LESSON REVIEW

Youngest students: Answer these questions:

1. Should you believe something is real just because you see it?

2. Why do scientists think that atoms are real, even though we can't see them?

Older students: If you feel artistic, try to make a pointillist drawing. Use a pen or marker with a fine point to make individual dots that end up producing some kind of recognizable drawing. If you don't want to do that, find an example of a pointillist painting (or photocopy the one above – you have my permission to do that) and paste it in your notebook. Then explain how a pointillist painting illustrates the concept of atoms. Also, explain why scientists think atoms exist, even though we can't see them.

Oldest students: Do what the older students are doing. In addition, answer this question in your notebook: "If a printer makes a picture using 300 dots per inch, what is the widest each dot can be?" Check your answer and correct it if it is wrong.

Lesson 7: Some Incorrect Ideas about Atoms

Even though Democritus had the basic idea of atoms correct, he had a lot of wrong ideas about what atoms are like. For example, Democritus thought that the differences between atoms were mostly based on shape. Iron was very hard, for example, because iron atoms had tiny hooks that allowed them to link together, making it very hard to separate the atoms. The atoms that made up water on the other hand, were smooth. This made them slippery, which allowed water to be poured. The atoms that made up salt were sharp, which accounted for salt's sharp taste on the tongue. We now know that atoms are all basically the same shape, and you will learn about that in a later lesson.

Another very interesting idea that Democritus had was that atoms were **indivisible** (in' duh vih' zuh buhl). Now don't get this word confused with "invisible." Something that is invisible cannot be seen. However, something that is indivisible is something that cannot be broken down into smaller parts. Think about a piece of paper, for example. It is easy for you to rip the paper in half. This means that the paper is divisible. It can be broken down into smaller pieces of paper. What about a sheet of metal? At first, you might think it is indivisible, because *you* can't tear it in half. However, if you have the right equipment (like a saw that cuts metal), you can easily break it down into smaller pieces of metal.

Democritus thought that once you started breaking something into smaller and smaller bits, you would eventually get to the atoms that make it up. Once you got down to a single atom, however, no matter what you used or how hard you tried, you could not break that atom down into anything smaller. As a result, while all *objects* could be broken down into smaller parts, atoms were as small as anything got. This, of course, meant that atoms could not be broken down into smaller parts and were therefore considered to be indivisible.

We now know that this is wrong as well. For example, let's suppose you took a grain of salt and started breaking it down into smaller and smaller bits of salt. If you could look closely enough at those bits of salt, you would find that they are actually made up of two different things: sodium **ions** (eye' uhnz) and chloride ions. If you don't recognize the term "ion," don't worry about it. You will learn what an ion is in detail in a few lessons. For right now, think of an ion as a kind of atom.

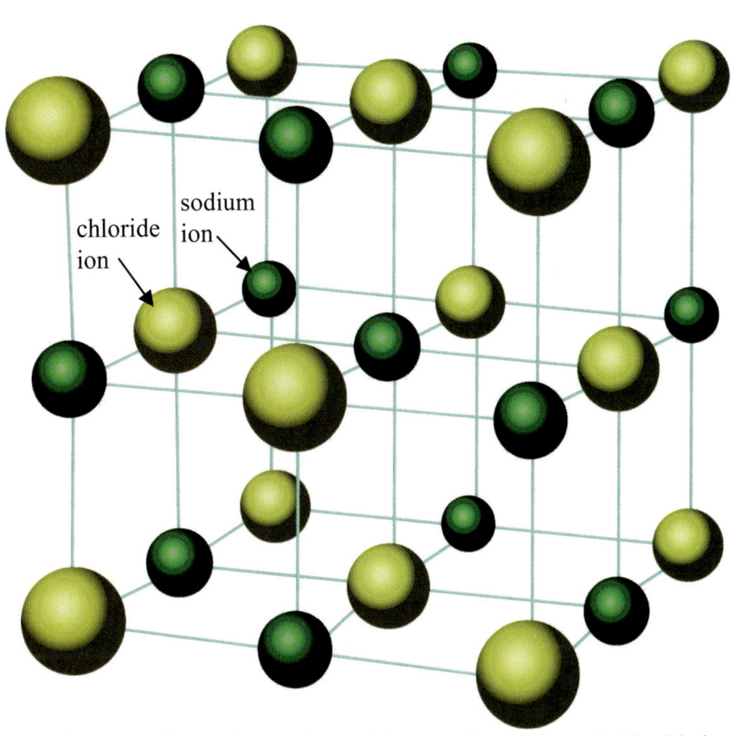

Salt is actually made up of two things: sodium ions and chloride ions. This is a drawing of what salt would look like if you could see those ions.

Salt, then, is not made up of salt atoms. It is made up of two different kinds of ions in a special arrangement. If you separate the sodium ions from the chloride ions, you no longer have salt. In order to have something that looks, tastes, and behaves like salt, you need *both* sodium ions and chloride ions. If you have just one and not the other, you no longer have salt.

So in the end, you can break salt down into parts that are so small that they are no longer salt anymore. This is because salt is not made of salt atoms, as Democritus thought. Instead, it is made of salt ***molecules***. It turns out that atoms (and ions) can join together to make things called molecules, and most of what we see around us is made of molecules, not atoms. If you are clever enough, you can break apart those molecules, making something entirely different. Perform the following experiment to see what I mean.

Decomposing Hydrogen Peroxide...in an Interesting Way

What you will need:
- Hydrogen peroxide (You can get it at any drugstore or large supermarket.)
- Active dry yeast (available in the baking section of any supermarket)
- Dish soap or liquid hand soap
- A 1-cup measuring cup
- A ¼-cup measuring cup
- A measuring tablespoon
- A tall glass (It is best if you can see through it, but that is not necessary.)
- A small glass, like a juice glass
- A tap that can produce warm water
- An empty sink

What you should do:
1. Run water from the tap until the water is warm but not hot.
2. Fill the ¼-cup measuring cup with warm water and pour the water into the small glass.
3. Add 1 tablespoon of active dry yeast to the warm water and stir.
4. Add 1 cup of hydrogen peroxide to the tall glass.
5. Add ¼ cup of dish soap to the tall glass as well.
6. Swirl the tall glass so the hydrogen peroxide and the dish soap mix well.
7. Set the tall glass in the middle of the empty sink. Observe it. Nothing exciting is happening, right?
8. Wait for two minutes.
9. Add the contents of the small glass to the tall glass.
10. Watch what happens.
11. Once it's all over, clean up your mess and put everything away.

What did you see in the experiment? Once you added the yeast/water mixture to the tall glass, suds should have immediately started forming. They should have climbed up the inside of the glass and eventually spilled over into the sink, which is why I told you to put the tall glass in the sink to begin with. Why did this happen? Because hydrogen peroxide is made up of molecules that contain two hydrogen atoms and two oxygen atoms. This arrangement of atoms is not very stable, however, so hydrogen peroxide molecules try to lose one of their oxygen atoms. When they do that, they turn into water, which is very stable. In other words, hydrogen peroxide molecules like to break apart into a water molecule and an oxygen atom. This normally happens very slowly, but yeast cause the process to speed up.

In your experiment, then, the hydrogen peroxide was breaking down into water and oxygen. Scientists call this **decomposition** (dee kom' puh zish' uhn), which means to break down into smaller things. So the hydrogen peroxide decomposed into water and oxygen. Remember, however, that oxygen is a gas, so it formed bubbles as it escaped the solution. Those bubbles caused the dish soap to foam up, and that's why so much foam was formed. In the end, then, the dish soap was used in the experiment so that you could see that a gas was being formed when hydrogen peroxide decomposed.

Now if Democritus's view of atoms were correct, such a thing could not happen. According to him, a substance like hydrogen peroxide was made of indivisible hydrogen peroxide atoms. Those atoms shouldn't be able to break down into smaller things. Nevertheless, experiments show that they can. That's because rather than being made of *atoms*, hydrogen peroxide is made of *molecules*. But what about the atoms that make up those molecules? Sure, molecules can be broken down into atoms, but can atoms be broken down into something smaller?

For a long time, scientists thought that the answer to that question was "no." That's because they couldn't figure out a way to do it. For example, scientists could break down hydrogen peroxide into two smaller things (water and oxygen). They could even break the water down into two smaller things (hydrogen and oxygen). Try as they might, however, they could not get the hydrogen to break down into smaller things, and they could not get the oxygen to break down into smaller things. Because of this, they called hydrogen and oxygen **elements** (el' uh muhntz), and they thought that elements were what Democritus called atoms. They thought that elements were indivisible.

It turns out that scientists were wrong about that as well. We now know that if you use enough energy in the right way, you can even break down the hydrogen and oxygen atoms that make up the hydrogen peroxide molecules. So unlike Democritus thought, atoms are not indivisible. They can, indeed, be broken down. Interestingly enough, scientists have only known this since the very late 1800s. Until then, scientists thought that atoms were indivisible. In other words, Democritus's incorrect idea about atoms was retained by scientists for over 2,000 years!

Now you might think that I am being hard on Democritus by discussing his incorrect ideas about atoms. I assure you, I am not. I am just pointing out how science works. Even incorrect ideas can lead scientists to correct ideas, as long as scientists work diligently enough. In addition, Democritus had some correct ideas about atoms as well. You will learn about those in the next lesson.

LESSON REVIEW

Youngest students: Answer these questions:

1. When atoms join together, what do they make?

2. What do scientists call the process in which a molecule breaks down into smaller things?

Older students: Make two drawings that illustrate what happened in the experiment. First, draw the tall glass with the hydrogen peroxide and dish soap in it, and draw the smaller glass with the yeast and water as if it is about to be poured into the tall glass. Then draw the tall glass with lots of foam coming out of it. Explain what made the foam, using the term "decomposition" or "decomposed" in your explanation.

Oldest students: Do what the older students are doing. In addition, explain the difference between atoms and molecules and indicate that even atoms are not indivisible.

Lesson 8: Some Correct Ideas about Atoms

As I already mentioned, Democritus had some incorrect ideas when it came to atoms. However, he had some correct ideas as well. One of the most important correct ideas he had about atoms is that atoms are always in motion. Today we understand that while some things are made up of atoms (a bar of iron is made up of atoms), most things are made up of molecules. If we modify Democritus's idea to say that atoms and molecules are always in motion, we come up with something that modern scientists still think is true.

Now you might have already learned that atoms and molecules are in constant motion, but it might be a bit hard for you to believe. After all, this page that you are reading is made up of molecules, and the page is not moving. If the molecules that make up the page are moving, wouldn't we see the page moving? If nothing else, wouldn't we *feel* that motion when we touch the paper? Well, let's do a quick experiment which allows you to see that even in something that is not moving, the molecules which make it up are moving like crazy.

The Motion of Water Molecules

What you will need:

- Two small glasses, like juice glasses (At least one of them should be able to withstand boiling-hot water.)
- Water
- Something to heat the water (A microwave will do, as will a stove and a pot.)
- A hotpad
- Pepper
- A blank sheet of white paper

What you should do:

1. Fill one of the glasses ¾ of the way with water that is room temperature or cooler.
2. Fill the other glass ¾ of the way with water that is near boiling. You can do this by putting water in the glass and then heating it in the microwave oven until you see it boiling, or you can heat water to boiling in a pot on the stove and then pour it into the glass.
3. Using the hotpad to handle the glass with hot water in it, put the glasses right next to each other on a counter or tabletop. If you are using a table, put the glasses on coasters.
4. Put the white sheet of paper behind the glasses.
5. Position yourself so that your eyes are level with the center of the glasses and observe the water. You should be seeing the white paper in the background. Is the water moving? Depending on how you held and moved the glasses, there might be some motion in the water, but that motion should quickly die out. In the end, you should see nothing moving in the glasses.
6. Shake pepper on the surface of the water in both glasses. You want to completely cover the surface of the water in each glass with a lot of pepper, so don't be stingy!
7. Put the white sheet of paper behind the glasses again.
8. Position yourself so that your eyes are level with the center of the glasses and observe the water now.
9. If there is no pepper falling down into the water from the surface, gently tap the glasses. You should see pepper slowly falling down from the surface of the room-temperature water to the bottom of the glass. However, what happens in the glass that holds hot water should be quite different. What do you see happening there?

10. Continue your observations for a while, gently tapping the glasses from time to time to encourage pepper to fall from the surface of the water.
11. Clean up your mess and put everything away.

What did you see in your experiment? Hopefully, you saw that while the pepper fell gently down to the bottom of the glass that contained room-temperature water, the pepper didn't do that in the glass with hot water. Instead, the pepper started to fall, but it *got pushed back up to the surface*! Why did that happen?

As you probably already know, something sinks because it weighs more than an equal volume of water. Since most of the pepper weighs more than a pepper-sized sample of water, the pepper starts to sink. However, in order to sink, what must it do? It must push the water molecules out of the way so that it can fall down into the water. That's why it has to be heavier than an equal volume of water. If it isn't heavier than an equal volume of water, it can't push the water molecules out of the way and fall between them. In the glass that contained room-temperature water, then, the pepper successfully pushed water molecules out of the way and fell between them, sinking to the bottom of the glass.

In the hot water, the pepper *started* to do that. You saw pepper falling from the surface, but rather quickly, the pepper started *moving up to the surface again*! What pushed the pepper up to the surface of the water? The motion of the water molecules did! Even though the water looked like it wasn't moving, the water molecules were moving like crazy. Water molecules near the surface of the glass were cooling down, and that made them sink. Hotter water molecules were rising from the bottom of the glass, and those rising, hot water molecules collided with the pepper, pushing the pepper back up to the surface. So even though it didn't look like the hot water was moving, its molecules were!

But wait a minute. If water molecules are always in motion, why didn't the water molecules in the room-temperature water push the pepper back up to the surface? Because the water molecules in the room-temperature water were moving *randomly*. Some of the molecules were moving one way, and some were moving the other. As a result, some water molecules pushed the pepper up, but other water molecules pushed it down. Others pushed it to the right, and still others pushed it to the left. Since water molecules were pushing in all directions, they didn't have any overall effect on the pepper, and it just sank. In the hot water, the hot water molecules were moving up, and that forced the pepper up as well.

But wait another minute. In the hot water, there were cooler molecules falling down. Why didn't they push the pepper down, allowing it to sink? Because ***the hot water molecules had more energy than the cooler water molecules***. This is the most important lesson from the experiment. When something is hot, the molecules or atoms that make it up move with a lot more energy. When something is cold, the molecules move around with less energy. Because hot molecules move more quickly than cold molecules, the molecules that were rising to the surface in the glass of hot water hit the pepper harder than the water molecules that were sinking to the bottom of the glass. As a result, the pepper was pushed back to the surface and could not sink very easily.

Now even though Democritus did give us the idea that atoms are in constant motion, he didn't know that the hotter the molecules were, the faster they moved. That's something scientists figured out much later. Nevertheless, in order for them to have figured this out, they needed the idea that atoms were in constant motion, and Democritus (or his teacher, Leucippus – we aren't sure which) gave us this initial idea, which leads to our current understanding of how temperature affects the motion of molecules and atoms.

What's the practical significance of all this? Well, you have probably already learned that matter comes in three basic phases: solid, liquid, and gas. Water, for example, can be solid (what we call ice), liquid (what we usually call water), and gas (what we call water vapor). What's the difference between water in these three phases? The only significant difference is how much the water molecules move. When water is a solid, the water molecules are held in a specific arrangement and do nothing but vibrate back and forth (like someone shivering). If you add enough energy in the form of heat, however, the temperature rises, and the ice melts. It melts because the water molecules move around with enough energy to break apart from their specific arrangement. That turns the solid (ice) into a liquid (water). If you keep heating the water, the molecules move faster and faster until they move so quickly that they can escape the confines of the container that the water is being heated in. At that point, they become a gas (water vapor).

So the thing that really separates solids, liquids, and gases is how much their molecules or atoms are moving. In solids, there is some motion, but it is mostly just a little bit of vibration. In liquids, there is more motion, but the molecules or atoms still stay relatively close to one another. In a gas, the molecules or atoms are moving so quickly that they fly far apart from one another. This is why it is so important to understand that molecules and atoms are in constant motion. This motion is really what determines the phase of the object. Democritus didn't know that, of course, but his idea of atoms being in constant motion helped others figure it out.

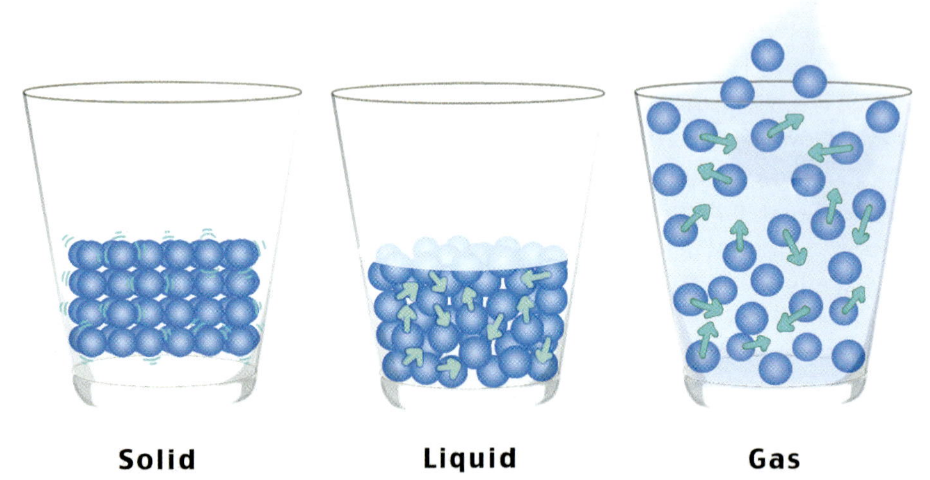

When something is solid (left), its molecules or atoms are in a specific arrangement and only vibrate. When it is a liquid (middle), its molecules or atoms are no longer in a specific arrangement and move around a bit. When it is a gas (right), its molecules or atoms are far from one another and move around a lot.

LESSON REVIEW

Youngest students: Answer these questions:

1. Which have more energy: the molecules in hot water or the molecules in cold water?

2. Which has the most motion in its molecules: a liquid, a solid, or a gas?

Older students: Draw a picture like the one above to illustrate the difference between solids, liquids, and gases. Write an explanation of the picture in your own words.

Oldest students: Do what the older students are doing. In addition, explain why increasing the temperature of something can change it from solid to liquid to gas.

Lesson 9: An Introduction to Atoms

Scientists have learned a lot about atoms since the time of Democritus, and I want you to learn a little bit about what we currently know. This is just a brief introduction, but it will allow you to have at least a basic understanding of what an atom is. Along the way, you will also learn what an ion is.

Today, we know that even though atoms are really small, they are actually made up of even smaller things! An atom has three basic parts: **protons** (pro' tahnz), **neutrons** (new' trahnz), and **electrons** (ih lek' trahnz). The protons and neutrons are packed together in the center of the atom, which we call the **nucleus** (new' klee us). The electrons are not with the protons and neutrons. Instead, they whirl around the nucleus. The illustration on the right gives you an idea of what scientists think an atom looks like. The actual picture is more complicated, but the illustration on the right is a good place to start.

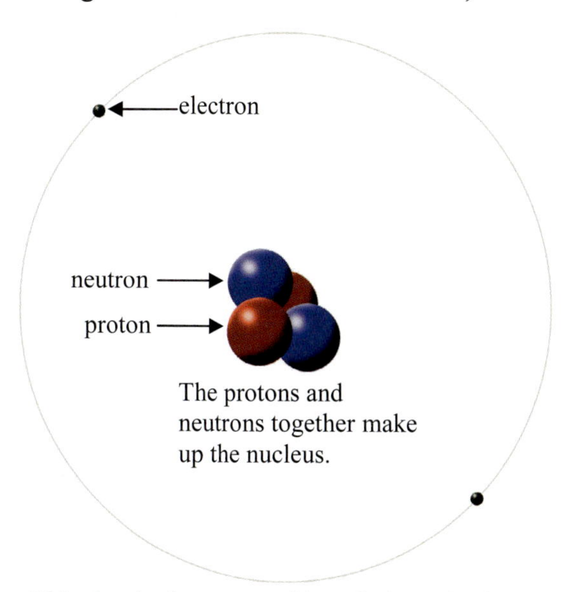

This sketch gives you an idea of what scientists think a helium atom looks like.

The particular atom in the illustration is helium. When you buy a balloon that floats, it is filled with lots of tiny atoms that look something like what you see on the right. Notice that it has two protons (the red balls) and two neutrons (the blue balls). The protons and neutrons together make up the nucleus of the atom. Outside the nucleus, there are two electrons that whirl around it. The gray circle in the drawing shows you a path the electrons can take as they travel. In other words, the gray circle is not a part of the atom. It is just showing you how the electrons move around the nucleus.

So now you see why Democritus was wrong about atoms being indivisible. In fact, if we do it right, we can break atoms down into three smaller particles: protons, neutrons, and electrons. Also, the difference between one atom and another is simply the number of protons, neutrons, and electrons in each. For example, as you just read, a helium atom has two protons, two electrons, and usually two neutrons. A carbon atom, however, has six protons, six electrons, and usually six neutrons. Even though the two atoms are made up of the same basic things (protons, neutrons, and electrons), they end up being *very, very* different. Helium is a gas that allows balloons to float in the air, and it cannot burn. Carbon is a black solid, and it burns very nicely. In fact, we make a lot of electricity by burning the carbon that is in coal.

So now you see why Democritus was wrong about different atoms having different shapes. All atoms have roughly the same overall shape – they are all like tiny balls. What makes one atom different from another atom is not the shape, but the number of protons, neutrons, and electrons each has. Adding just one proton to a helium atom, for example, changes it from a lighter-than-air gas into lithium, which is a flexible, shiny solid that will pop and fizz if you put it in water. Interestingly enough, modern scientists can actually do that kind of transformation in the lab, if they have access to something called a particle accelerator!

What's the difference between protons, neutrons, and electrons? Well, one thing is their size. Protons and neutrons are roughly the same size (as shown in the illustration above), but electrons are smaller. In fact, electrons are *a lot* smaller than protons and neutrons. If I tried to make a tiny dot that would represent the proper size of an electron in the illustration, you wouldn't be able to see it without a magnifying glass! So compared to protons and neutrons, electrons are incredibly tiny.

The other difference is that protons and electrons have electrical charge, but neutrons do not. In fact, the name "neutron" comes from the word "neutral," which means no electrical charge. Protons have positive electrical charge, and electrons have negative electrical charge. What happens when you turn on a light switch in your home? The light comes on, right? Why does the light come on? Because, when the switch is on, there is electricity running the light. Do you know what the electricity that runs the light really is? It's a bunch of electrons. When electrons flow through a wire, you have electricity, which can be used to run all sorts of great things, like light bulbs, televisions, and computers.

The fact that protons and electrons have electrical charge is very important. Look at the helium atom on the previous page. It has two protons. Each proton has a positive electrical charge. A helium atom also has two electrons. Each electron has a negative electrical charge. So a helium atom has two positive electrical charges and two negative electrical charges. Guess what that means? It means that overall, helium has no electrical charge. The two positive protons cancel out the two negative electrons, so that in the end, the helium atom has no total electrical charge.

Now what would happen if I pulled an electron away from the helium atom? It would then have two protons (two positive charges) but only one electron (one negative charge). What would that mean? It would mean that suddenly, the helium would have an electrical charge. The electron would cancel out one of the protons, but the other proton would have nothing to cancel it out. As a result, the helium atom would have an overall positive electrical charge. Guess what we would call it then? We wouldn't call it an atom. We would call it an **ion**! When an atom has more protons than electrons, then, it is no longer an atom. It is a positively charged ion. When an atom has more electrons than protons, the protons can't cancel out all the electrons, so the atom is a negatively charged ion. So that you are sure you understand this, do the following activity.

Sweet Atoms

What you will need:

- Two large hard candies that are different colors (I used cinnamon candies and lemon drops, but you can use whatever you like. You will need six of each.)
- A small candy (I used Skittles, but you can use whatever you like. It needs to be smaller than the other hard candies. You need eight, and they need to be all the same color.)
- Syrup, honey, or icing (Essentially, it needs to be something sweet and a bit sticky.)
- A butter knife
- A paper plate

What you should do:

1. Put the paper plate on a clean counter or table.
2. Choose one of the colors of large, hard candy to represent protons. The other color will represent neutrons.
3. Put a single hard candy that represents a proton at the very center of the paper plate.
4. Most paper plates have a circle that is near the edge of the flat part of the plate. If yours doesn't, just imagine such a circle. In the picture on the next page, the first two Skittles are on the circle I am talking about. Dip the knife into the syrup, honey, or icing and put a tiny dot of it somewhere on that circle.

5. Put your small candy right on top of the drop, so the drop holds it there. The small candy represents an electron. Now look what you have. You have a "proton" in the center of the plate, and a distance away, you have an "electron." You can imagine the electron whirling around the proton in a circle. This represents the simplest atom in creation: a **hydrogen** atom. It has one proton in its nucleus, and it has one electron whirling around it.
6. Put another hard candy that represents a proton next to the one that is already there on the plate.
7. Put two of the hard candies that represent neutrons in the center of the plate with the protons.
8. Use the knife to make another drop of syrup, honey, or icing somewhere on the same circle that the small candy is on. Put a small candy on the drop. You now have a **helium** atom. It has two protons and two neutrons in its nucleus, and it has two electrons whirling around that nucleus.
9. Add four more protons to the center of the plate so that you have a total of six.
10. Add four more neutrons to the center of the plate so that you have a total of six.
11. You need to also add four more electrons, but they can't go on the same circle you have been using. There might be room for them on the plate, but there isn't room for them there in an actual atom. Only two electrons can whirl around an actual nucleus when they are that close to it. Add the four electrons to the outside edge of the plate. Use the syrup, honey, or icing like you did before to "glue" the electrons to the outside edge of the plate. You should now have something that looks like the picture on the right. This is a model of a **carbon** atom. It has six protons, six neutrons, and six electrons. Two of the electrons are close to the nucleus, and four are farther away from it.

12. Add one more electron to the outer edge. What do you have now? You have something that has six protons but seven electrons. What is that? It is a negative carbon ion. Its six protons can cancel out the electrical charge of six electrons, but there is nothing to cancel out the negative charge of the seventh electron. That means it is a negatively charged ion.
13. Pull off and eat two of the electrons. What do you have now? You have something that has six protons and five electrons. The five electrons can cancel out five of the protons, but they can't cancel out the sixth. That means there is an extra positive charge, so this is a positively charged carbon ion.
14. Clean up your mess, and eat the parts of the atom that your parent says are okay to eat!

LESSON REVIEW

Youngest students: Answer these questions:

1. What three things make up atoms?

2. What parts of an atom have electrical charge? What kind of charge does each have?

Older students: In your notebook, draw a hydrogen atom, a helium atom, and a carbon atom. Underneath each atom, write down the number of protons, neutrons, and electrons. Indicate the charge on each. Also, explain why the six electrons in a carbon atom cannot all fit into a circle that is close to the nucleus.

Oldest students: Do what the older students are doing. In addition, draw the ions that you made in your activity. Explain why they are ions and why one is negative but the other is positive.

Lesson 10: More about Atoms and Ions

In the previous lesson, you learned that atoms are made of protons, neutrons, and electrons. The protons have positive electrical charge, while the electrons have negative electrical charge. Neutrons have no electrical charge. An atom has equal numbers of protons and electrons. A carbon atom, for example, has six protons and six electrons. That means it has equal numbers of positive and negative charges, which means that they all cancel each other out. So an atom has no electrical charge. However, if an electron is removed, there are more protons than electrons, and you now have a positive ion. If an extra electron is added, there are more electrons than protons, and you now have a negative ion.

I want you to see how this ends up working out in the real world. To do that, you will need to start the following experiment. Once you have it started, continue with the lesson, and we will see how the experiment turns out at the end.

Copper and Iron

What you will need:

- About 30 pennies (the duller and uglier, the better)
- A nail (It should not be a stainless steel or a galvanized nail.)
- White vinegar
- Salt
- A small glass, like a juice glass
- A ½-cup measuring cup
- A measuring teaspoon

What you should do:

1. Use the measuring cup to add ½ cup of vinegar to the small glass.
2. Use the measuring teaspoon to add 1 teaspoon of salt to the vinegar in the small glass.
3. Stir the vinegar and salt so that most (if not all) of the salt dissolves.
4. Put all 30 pennies in the vinegar/salt solution.
5. Stand the nail in the vinegar/salt solution so that part of the nail is in the solution and part stays above the solution. You should be able to lean it against the inside of the glass, and with all the pennies in the solution already, it should stay standing without you holding it the whole time.
6. Watch the pennies and the nail for a while. What do you see happening?
7. Let the experiment sit while you continue doing the lesson.

Now what does this experiment have to do with atoms and ions? Well, pennies are made of a bunch of **copper** atoms. Copper atoms have 29 protons and 29 electrons. Remember, an atom has to have equal numbers of protons and electrons. Otherwise, it would have electrical charge and would therefore be an ion. Most copper atoms in a penny have 34 neutrons, but some have 36 neutrons. This brings up a new, important point. If you change the number of protons in an atom, you change the kind of atom it is, but if you change the number of neutrons, you don't change the kind of atom it is. Copper atoms have 29 protons, but nickel atoms have 28 protons. Zinc atoms have 30 protons. So in the end, the number of protons really determines what kind of atom you have. If the number of neutrons varies, however, the kind of atom actually stays the same. As a result, even though some copper atoms have 34 neutrons in them, other copper atoms can have a different number of neutrons in them.

When you get a new penny, it is nice and shiny, because copper is nice and shiny. However, as the penny sits out in air, the copper atoms slowly combine with the oxygen atoms in the air, making a molecule called copper oxide. This molecule is not shiny at all, so over time, a nice, shiny penny will get dull, because the copper atoms on the surface of the penny slowly combine with oxygen in the air to make molecules of copper oxide.

The shiny parts of these pennies are from the copper atoms that make up the pennies. The dull parts are from copper oxide, which forms on the surface of pennies over time.

When you put the pennies in the vinegar, the vinegar started reacting with the copper oxide, releasing copper ions. Ions love to dissolve into water, so as the copper oxide turned into copper ions, the ions dissolved into the water, making a solution of copper ions.

You should have known that something was going on in the experiment, because you should have seen bubbles forming on the pennies. Bubbles mean there is a gas, so there was something going on that made a gas. Actually, several things were going on, but for the purpose of this lesson, just understand that copper ions were being formed. Those ions were positive, which means they had more protons than electrons. They dissolved into the vinegar and salt solution. In the process, the copper oxide was removed from the pennies.

Did you see bubbles forming on the nail? You should have. What does that tell you? It tells you that a gas was being formed on the nail as well. But nails aren't made out of copper. They are made out of iron. Iron atoms have 26 protons and 26 electrons. That's only three fewer protons and electrons than copper, but the atoms are completely different. They don't have the same color as copper, and as you will see in a bit, they most certainly don't behave like copper atoms. Most of the iron atoms in a nail have 28 neutrons, but some have 30, some have 31, and some have 32. Since a nail is made (mostly) of iron, and since iron atoms are different from copper atoms, you can probably guess that whatever happened to the nail in the experiment is different from what happened to the pennies, even though both had bubbles forming on them.

Take the pennies and nails out of the solution now. Look at them. The pennies should be shinier than they were before. That's because the copper oxide has been at least partly removed, revealing the shiny copper underneath. If you wash them off, let them soak in more vinegar and salt, wash them again, etc., you will eventually have really nice, shiny pennies. That's because eventually, all the copper oxide will be gone, leaving nothing but shiny copper atoms.

Now pull the nail out of the solution and look at it. Do you see a difference between the part of the nail that was above the solution and the part of the nail that was in the solution? If not, put the pennies back in the solution and let the nail soak for a few more minutes and check again. Eventually, you should see that the part of the nail that was soaking in the solution is colored like a penny, while the rest of the nail looks just like a nail. Why? Remember that there were copper ions in the solution. They came from the pennies. Well, the iron atoms in the nail love to give away their electrons. They saw that the copper ions in the solution were short on electrons, so they "kindly" gave some electrons to them.

Since the copper ions gained electrons, they ended up having the same number of electrons as protons again. That made them have no overall charge, which means they became atoms again. Those atoms came to rest on the surface of the nail, because that's where they got their new electrons. As a result, the nail started being covered in copper. However, think about what happened to the iron atoms. They gave their electrons to the copper ions floating in the solution. What happened then? The iron atoms became positively charged iron ions, because they gave away some of their electrons. As a result, the iron ions dissolved in the solution. So what really happened was that the iron atoms in the nail were being *replaced* by copper atoms. The iron atoms turned into ions and dissolved in the solution. At the same time, the copper ions turned into atoms, becoming solid copper right there on the nail.

In the experiment, then, you saw the effect of taking electrons away from copper. When you take electrons away from a copper atom, you make a positive copper ion. Ions love to dissolve in water, so the copper ions dissolved into the solution. Remember, however, that atoms and ions are in constant motion. So those copper ions eventually made their way over to the nail. The iron atoms in the nail "happily" gave up their electrons so that the copper ions could have them. As a result, the copper ions turned back into copper atoms. Since the iron atoms gave away their electrons, however, the iron atoms turned into iron ions, which ended up dissolving in the solution.

The lesson you need to draw from the experiment is that under the right conditions, atoms can exchange electrons. One atom can give away one or more of its electrons, making it a positive ion. Another atom might take those electrons. This is, in fact, how the salt you use on your food is formed. Sodium atoms each give one of their electrons to a chlorine atom. As a result, the sodium atom becomes a positive ion, and the chlorine atom becomes a negative ion, which is called the chloride ion. This makes sodium chloride, which is the scientific name for table salt.

LESSON REVIEW

Youngest students: Answer these questions:

1. What kind of atoms do you find in a penny?

2. If an atom loses an electron, does it become a positive ion or a negative ion?

Older students: Explain in your notebook what happened in your experiment. Explain why the pennies got shiny, and explain why the part of the nail that soaked in the solution looked like copper. Be sure to use the word "ion" in your explanation.

Oldest students: Do what the older students are doing. In addition, suppose you have negative ions in a solution of water. Write what you would need to do to make them come out of the solution as atoms.

NOTE TO THE TEACHER: The next lesson contains an experiment (on page 32) that needs to sit for two or more days. It would be best if you had the students start the experiment now so that its results will be ready when you read the lesson. You don't need to read the first part of the lesson. Just do the experiment down to the point where you are supposed to let it sit. However, if you need to be completely done with science for today, just realize that you need to plan your next few days so that you start the experiment at least two days before the lesson.

Lesson 11: Hippocrates (c. 460 BC – c. 370 BC) and the Beginning of Modern Medicine

We've spent a lot of time studying Democritus and atoms, but it is time to move on to someone who lived at roughly the same time as Democritus but was interested in medicine, not atoms. His name was **Hippocrates** (hih pah' kruh teez), and he lived on an island called Cos in Greece. Before I go any further, I want to point out that very little is known about what Hippocrates actually wrote and believed. Most of what we know about him comes from other people, so it's possible that some of what we think he believed actually represent the beliefs of someone else.

This drawing is of a sculpture of Hippocrates.

Taking that into account, most historians call Hippocrates the father of modern medicine, and with good reason. During the time that he lived, most people in Greece thought that sickness was caused by the gods. In case you don't know, most Greeks believed there were many, many gods, and these gods were easily offended by people. Also, some of them were just plain mischievous. As a result, most Greeks thought that people who were sick were either being punished by one of their many gods or were just unlucky enough to have a mischievous god make them sick for no apparent reason.

Hippocrates didn't believe that. He believed that sickness was caused by some sort of natural interference with a person's body, and if that interference could be dealt with, the person would get better. He rightly thought that the body had its own ways of dealing with sickness, and one of the best ways to treat it was to treat the patient gently and encourage lots of rest. He thought that if the patient would just rest so the body could use its energy to fight the sickness, the person would get better. Of course, we now know that for some sicknesses, that's exactly the right thing to do. When you have a cold, what are you supposed to do? You are supposed to drink lots of fluids and rest. That way, your body can fight off the cold. Believe it or not, the idea of rest being an important part of getting better is thought to have started with Hippocrates about 400 years before Christ was born!

Even though Hippocrates had a lot of good ideas about medicine, he is probably most famous for an oath that historians think he wrote. It is now called the **Hippocratic** (hip' ih krat' ik) **Oath**, and most physicians today still take a version of the oath before they start practicing medicine. It essentially says that a physician will work hard to help sick people, will not share a patient's personal information with anyone else, and will not give a deadly drug to a patient, even when the patient asks for it. One of the most famous parts of the Hippocratic Oath is that the physician will not do harm to a patient. If you have heard the phrase, "First do no harm" in reference to medicine, most historians think it was inspired by the Hippocratic Oath.

Now remember, Hippocrates thought that a person's body had the ability to fight off sickness, and we now know that is correct in most cases. Even though he recognized this, he also understood that there were ways that physicians could help the body in its fight. For example, Hippocrates understood that if a person was cut, the cut would eventually heal. However, he also understood that this healing would take place better and faster if certain guidelines were followed. For example, he stressed that a cut should be kept clean, and if the cut was covered by a bandage, the bandage should be kept clean. He didn't really understand *why* this helped a cut heal, but he knew that it did. He also thought that bandages should be sprinkled with wine. Why? Hopefully, you have already started the following experiment, which will show you why. If not, start it now.

Wine and Wounds

What you will need:
- Rubbing alcohol
- A small amount of raw ground beef
- Four small glasses, like juice glasses
- Two paper towels
- Two rubber bands that will stretch to fit around the glasses
- A blank sheet of white paper
- A pencil
- An out-of-the-way place to let the experiment sit for 2 days
- A ½-cup measuring cup
- Water

What you should do:
1. Put a small amount of ground beef into each of the glasses. It doesn't need to be a lot – just enough to cover the bottom of the glass is fine. Try to make the amount in each glass roughly the same.
2. Cover one of the four glasses with a paper towel.
3. Pull the paper towel down so that it covers the sides of the glass, and then stretch the rubber band around the glass so that it holds the paper towel to the glass.
4. Pour some rubbing alcohol onto the hamburger in one of the other glasses. Swirl it around for a bit, and then carefully pour as much of the alcohol as you can out of the glass without dumping the ground beef out as well.
5. Repeat step 4 for another one of the glasses.
6. Take one of the two glasses that you just poured alcohol into and cover it with a paper towel. Like you did in step 3, pull the paper towel down around the glass and then use the rubber band to hold it there.
7. You should now have four glasses. One is open and has ground beef in it. Another is open and has alcohol-soaked ground beef in it. One is covered with a paper towel that is held to the glass by a rubber band, and it has plain ground beef in it. The last one is also covered with a rubber-band-held paper towel, and it has alcohol-soaked ground beef in it.
8. Find an out-of-the-way place that your experiment can sit for at least two days.
9. Put the paper down on this out-of-the-way place and arrange the four glasses on it.
10. Use the pencil to write "plain ground beef" in front of the glass that has just ground beef in it.
11. Use the pencil to write "alcohol-soaked ground beef" in front of the glass that is open and has alcohol-soaked ground beef in it.
12. Use the pencil to write "covered ground beef" in front of the glass that is covered by the paper towel and has regular ground beef in it.
13. Use the pencil to write "alcohol-soaked, covered ground beef" in front of the glass that is covered with a paper towel and has alcohol-soaked ground beef in it.
14. Let the glasses sit in the out-of-the-way place for at least two days, checking them once a day to see if anything has happened. To check the covered ground beef, just raise the paper towel below the rubber band and look in through the sides of the glass. Don't uncover the glass.
15. When at least two days have passed, take off the paper towels.

16. Look at the ground beef in each glass. Do you notice any differences?
17. Add ½ cup of water to each glass and swirl the ground beef around.
18. Observe the water in each glass. Do you notice any differences?
19. Clean up your mess. Wash the ground beef down the drain with lots of soap. Wash your hands.

What did you see in the experiment? You should have seen the plain ground beef in the open glass was really ugly-looking, and it smelled really bad. The covered ground beef was probably not as bad, the alcohol-soaked ground beef wasn't nearly as bad, and the covered, alcohol-soaked ground beef was in the best shape. When you added water and swirled, the water in the open, plain ground beef glass was probably pretty disgusting. Once again, the water in the glass that had covered ground beef was probably better, the water in the glass with alcohol-soaked ground beef was better still, and the water in the glass with covered, alcohol-soaked ground beef was probably the cleanest.

What explains these differences? Well, there are a lot of very tiny organisms called **bacteria** (bak tihr' ee uh) that float around on the dust that is in the air. Many of them love to eat the remains of dead animals, so when the dust they were riding on landed on the ground beef, they started eating it. This allowed them to grow and multiply, and they infested the ground beef. Even the covered ground beef got dust on it before you covered it, so it got infested, too. However, since not as much dust fell on it, not as many bacteria landed on it, so it wasn't as badly infested. Alcohol tends to kill a lot of bacteria, so the alcohol-soaked ground beef was pretty clean, since few bacteria could live on it. Of course, the covered, alcohol-soaked beef was the cleanest, since it had both the protection of the cover and the protection of the alcohol.

Now even though wine doesn't contain the same alcohol that is in the rubbing alcohol you used, it does contain a different kind of alcohol, which kills bacteria. Hippocrates didn't know anything about bacteria, but he did understand that an alcohol-sprinkled bandage protects a cut better than a bandage that doesn't have alcohol sprinkled in it. So he reasoned that the alcohol was preventing something bad from happening, which is why he recommended it. We now know that alcohol kills bacteria, so a bandage soaked in alcohol will protect a wound against infection. However, alcohol on a cut is very painful, so we usually use a less painful way of protecting cuts from infection, like a mild salve (sav) that kills bacteria. Nevertheless, if such a salve is not around, you can still help prevent an infection by using alcohol. So you see that Hippocrates not only figured out that the body can heal itself if a sick person rests and is kept clean, he also figured out that alcohol can prevent the infection of wounds. That's not bad for someone who lived more than 2,000 years ago!

LESSON REVIEW

Youngest students: Answer these questions:

1. What is the Hippocratic Oath?

2. Why does rest help a sick person get better?

Older students: Write down in your notebook what the Hippocratic Oath is and who is thought to have written it. Also, explain why rest can help a sick person get better and why bandages sprinkled with alcohol can be good for healing cuts.

Oldest students: Do what the older students are doing. In addition, find the Hippocratic Oath on the internet or in a book. Read it, and write down your thoughts on what it means.

Lesson 12: Hippocrates and Blood

For his time, Hippocrates really had some great ideas about medicine. However, as you might have already guessed, he had some really bad ideas as well. For example, he believed that there were four important liquids in the body, which were called **humors**. In Hippocrates's mind, these liquids were not funny! Instead, they strongly affected a person's mood and health. The four humors were **blood**, **yellow bile**, **black bile**, and **phlegm** (flem). If a person had those four humors in perfect balance, the person was healthy and happy. However, if one of the humors was out of balance, the person would have problems with mood, health, or both. Now remember that Hippocrates thought the body could heal itself, given enough energy and time. The way he thought the body did that was to bring these four humors back into balance. Once again, however, he thought a physician could help the body by doing certain things to help bring the humors back into balance.

To Hippocrates, each humor had a specific characteristic. Blood was warm and moist and tended to make a person hopeful or courageous. Yellow bile was warm and dry and tended to make a person angry or bad tempered. Black bile was cold and dry and tended to make a person sad and sleepless. Phlegm was cold and moist and tended to make a person calm and unemotional. So if a person was really sad all the time, Hippocrates thought the person had too much black bile. His job, then, would be to help the body get rid of its excess of black bile.

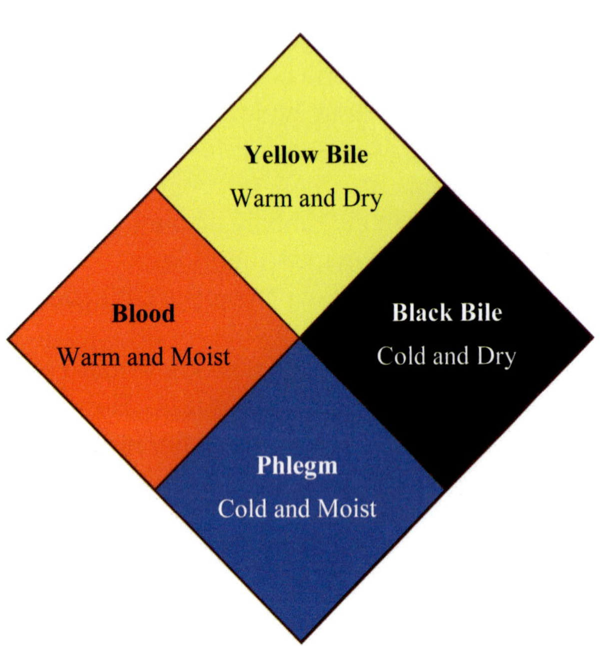

This diagram illustrates Hippocrates's view of the body's humors. Note how phlegm is the opposite of yellow bile.

This same reasoning applied to other problems, such as a fever. If a person has a fever, that person is hot and dry. What would that tell Hippocrates? It would tell him the person had too much yellow bile, which is the warm and dry humor. As a result, his job would be to help the body get rid of the excess of yellow bile. How would he do that? By trying to get the body to make more phlegm. After all, phlegm is cold and moist, which is the opposite of warm and dry. So encouraging the body to make phlegm would reduce the excess of yellow bile. As a result, he might tell the patient to take a lot of cold baths. That would add cold and moist exposure to the body, which should make more phlegm.

Modern medicine, of course, understands that this reasoning is incorrect. However, it is important to realize that even though it is incorrect, it allowed medicine to improve. This is one very important aspect of science: *Even wrong ideas can help science progress, because even wrong ideas can sometimes lead us to the right ideas*. For example, even though Hippocrates was wrong about their effects, he was right that the body does contain liquids.

Blood, for example, is one of the liquids that the body contains, and we now know it is critically important for health. So even though Hippocrates was wrong about what blood actually does in the body, his view of medicine encouraged physicians to study blood, which eventually led us to the right idea about what blood does for the body. So what does blood actually do for the body? Let's start with a simple experiment.

Blood and the Body

What you will need:

- Your wrist, if it has visible blue veins on it. If you can't easily see the blue veins on your wrist (see the photograph below), find someone whose wrist has visible blue veins.

What you should do:

1. Turn your wrist (or the other person's wrist) so that the palm of the hand is facing upwards.
2. Locate a nice, easy-to-see vein.
3. Use the middle finger on your other hand to press on the vein. Press hard, but not so hard that it causes you (or the other person) pain. Don't hold it for more than a minute, because holding it too long could damage the vein.
4. Use your index finger to press on the vein right below where your middle finger is, but then slide your finger toward your elbow along the vein. What happened to the nice, visible vein that was there?
5. Lift your middle finger up so it is no longer pressing on the vein. What happens now?
6. Repeat the experiment a couple of times to see the effect. You can do it on other people too.

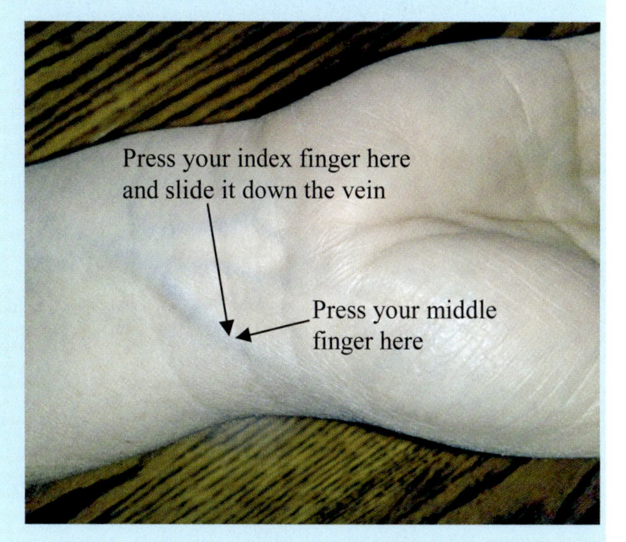

What did you see in the experiment? If all went well, you should have seen the vein "disappear." However, once you raised your middle finger, the vein should have "reappeared" again. Obviously, the vein didn't disappear and reappear. Instead, what the vein was carrying went away and then came back.

As you probably already know, the blood in your body is found in tubes called **blood vessels**. The blue veins you see in your wrist are one type of blood vessel. They look blue to you because of the blood that is in them. You can see the blood because the veins that carry it run fairly close to your skin, and your skin is a bit thin there. As you look down your arm, you can't see the blood anymore, because the vessels run a bit deeper, and your skin is a bit thicker.

Interestingly enough, while your veins appear blue, they are actually dark red, because the blood they are carrying is dark red. However, as the light is traveling through your skin, its color changes, so you end up seeing blue instead of dark red. This is an instance in which your eyes can deceive you. The veins in your wrist *appear* to be blue, when in fact, they are actually dark red.

When you pressed your middle finger on the vein, you closed it. Then, when you slid your index finger toward your elbow, you squeezed all the blood out of that part of the vein, moving it toward your elbow as well. You didn't see the blood there, however, because the veins were too deep and your skin was too thick to see them. That's why the blue "disappeared." Why didn't it reappear before you lifted your middle finger? Because your middle finger was closing the vein, keeping any new blood from flowing into the vein where the blood that you had just squeezed out had been. As a result, your vein seemed to "disappear," when in fact, what really happened was that the blood had been pushed out of that part of the vein and had not been replaced by new blood.

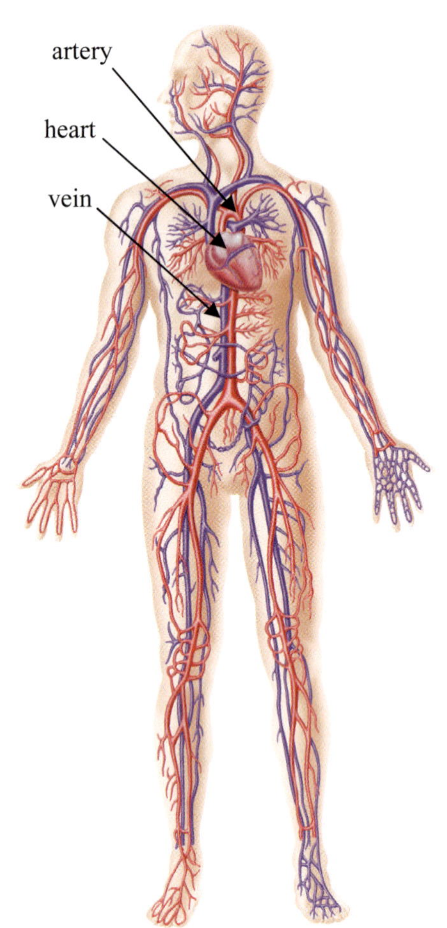

This is an illustration of the major arteries (red) and veins (blue) in the human body.

What does this tell you? It tells you that the blood in your veins flows only in one direction. You only closed one end of the vein. However, until you opened that end up, no new blood flowed in. So in that vein, the blood flows in only one direction: from your hand toward your elbow. This is actually very important, and it helps you understand what your blood does for you.

Your body has three types of blood vessels, but right now, you are going to learn about only two of them: **veins** and **arteries**. The blood is pumped through those blood vessels by your **heart**. Arteries carry blood *away from your heart*, while veins carry it back *towards your heart*. Why is this important? Because your blood's job is to carry oxygen and nutrients all over your body. When it is full of oxygen and nutrients, it needs to move away from your heart to the various parts of your body, so it can drop off these important chemicals where they are needed. However, once it has dropped off those chemicals, it needs to head back to the heart. That way, it can pick up new oxygen and nutrients along the way and once again be sent back out to the rest of your body.

So the blood is actually a transport system. It carries oxygen and nutrients throughout your body, dropping them off where they are needed. As the heart pumps your blood away from it so that the blood can drop off its important chemicals, the blood travels in arteries. Once it drops off its chemicals, it comes back to the heart by traveling in the body's veins. The drawing on the left illustrates this wonderful system of veins and arteries, which is called the **circulatory** (sur' kyoo luh tor' ee) **system**. The arteries are shown in red, and even though we know they are not really blue, the veins are in blue to make them easier to tell apart from the arteries.

LESSON REVIEW

Youngest students: Answer these questions:

1. What does blood do for the body?

2. What vessels carry blood away from the heart? What vessels carry it towards the heart?

Older students: In your notebook, write about a drop of blood traveling from the heart to the tip of the pinky finger and back. Explain why it is making the trip, what it does, and what kinds of blood vessels it travels in during the different parts of its trip.

Oldest students: Do what the older students are doing. In addition, notice that the arteries and veins close to the heart are very large, but they get smaller the farther they get from the heart. Write in your notebook why you think this is. Check your answer and correct it if you got it wrong.

Note to the parent/teacher: The next lesson (a challenge lesson) has an experiment that has to sit for five hours. It also requires that Jell-O or some other kind of gelatin be made and ready.

Lesson 13: Hippocrates and Bile, Part 1

If you read the introduction, you know that this is a "challenge lesson." You don't need to do it, but it is available for those who want to learn more about the fluids Hippocrates studied.

In this lesson, you are going to start to learn about bile and what it really does in your body. You need to start the experiment first, however, so it can sit for a while. Then you can do other things and return to science when it is finished.

Digestion

What you will need:

- Four small glasses, like juice glasses
- Jell-O or some other form of gelatin
- Vinegar
- A pineapple (It *cannot* be canned. It needs to be an actual pineapple.)
- A knife to cut the pineapple
- A spoon

What you should do:

1. If it hasn't already been made, make the Jell-O according to the instructions. It needs to be completely gelled and ready to eat before you can do the rest of the experiment.
2. Fill two of the glasses a third of the way full of vinegar.
3. Have an adult cut the pineapple into chunks that you can squeeze.
4. Squeeze the pineapple chunks so that the juice goes into the other two glasses. Pulp and seeds can go in as well; you aren't going to drink it. Squeeze the pineapple chunks until you have the other two glasses a third of the way full of pineapple juice.
5. Use the spoon to pull out a spoonful of Jell-O, and then put it into one of the glasses that contains vinegar. Try to keep it in one glob, but if it splits into a couple of globs, that's no big deal.
6. Do the same thing again, but this time put the spoonful of Jell-O into a glass that has pineapple juice in it. The glob should be about the same size as the one you used in step 5.
7. Use the spoon to pull out a spoonful of Jell-O, about the same size as the other two you got. This time, use your fingers to break the Jell-O into several smaller globs, and put those globs into the remaining glass of vinegar.
8. Do the same thing again, but this time put all the smaller globs into the remaining glass of pineapple juice. You should now have four glasses. One contains pineapple juice and a single glob of Jell-O. One contains pineapple juice and several small globs of Jell-O. One contains vinegar and a single glob of Jell-O. The last one contains vinegar and several globs of Jell-O.
9. Let the four glasses sit for about five hours. Every hour or so, come back and look at each glass to see if anything is happening. Also, when you are done looking, gently swirl each glass to mix the contents.

Now that you've let the experiment sit for about five hours, you are ready to start learning about **bile** and what it does for your body. First, you need to know that bile is a real fluid in your body. It is often yellow, but sometimes it is green. It is also called **gall**, and even though no one wants to taste it, every now and again you are forced to, because it is what comes up when you vomit on an empty stomach. When you are forced to taste it that way, it tastes incredibly bitter.

Notice that I said bile is yellow or green. I never said it was black. Hippocrates, however, thought there were two kinds of bile in the body: yellow bile and black bile. It turns out that bile is never black, so what Hippocrates thought was black bile was not bile at all. What was it? Most likely, it was dried blood. You see, Hippocrates determined the fluids in the body by observation. He knew blood existed in the body because he saw people bleeding. He knew yellow bile was in the body because when someone vomits on an empty stomach, you can see the yellow bile. Well, people who are really sick can vomit up dried blood as well, which is usually black. So even though we aren't sure, most people who study the history of medicine think that Hippocrates mistook dried blood for a second form of bile. So even though bile is a fluid in the body, it doesn't come in two forms. It only comes in one form, and it is typically greenish-yellow in color.

In order for your body to get energy from the food you eat, you must digest the food.

What does bile do? To learn about that, you first need to learn a bit about **digestion**. You probably already know that you eat food in order to get energy. There is chemical energy stored in the food, and your body converts that chemical energy into another form of energy that it can use to run, work, and do all the things it needs to do to keep you alive. Well, the first step in turning your food into energy is to digest it. When your body digests food, it breaks the food down. It doesn't just tear the food into tiny pieces, although that is a part of the digestion process. Instead, it breaks the food down into its individual molecules.

The digestion process is very interesting, and you will learn more about how it happens in your body later on. For right now, you just need to know the basics. To learn that, it is time to think about the results of your experiment. What happened? Most likely the Jell-O in the pineapple juice slowly "disappeared," and the color of the pineapple juice changed, becoming more like the color of the Jell-O. However, very little (if any) of the Jell-O in the vinegar was gone. What explains these results?

Jell-O starts out as a liquid solution. You dissolve the Jell-O mix into hot water, and then you let the water cool. As the water cools, the Jell-O sets and becomes a wobbly solid. This actually happens because the molecules of gelatin in the solution start to join together, making larger molecules. Those larger molecules eventually intertwine until they form the wobbly solid. To turn Jell-O back into a liquid, then, you need to break those molecules apart.

It turns out that pineapple juice contains a specific chemical called an **enzyme** (en' zime). That chemical attacks the molecules in gelatin and breaks them apart. This turns the gelatin back into a liquid. Vinegar doesn't have that enzyme. So even though the Jell-O sat for a long, long time in the vinegar, the vinegar wasn't able to break the molecules apart, so the Jell-O stayed in its wobbly solid form. Since the pineapple juice contained the right enzyme, it was able to break down the Jell-O molecules, turning it back into a liquid.

That's what digestion is all about. Your food has a lot of big molecules in it, but for your body to use the energy that is stored in those molecules, they need to be broken down. Your body makes all sorts of chemicals (including many enzymes) that break the food down into simple molecules that can then be sent throughout your body and used for many things, including making energy. Do you know what carries these molecules around your body? Your blood, of course! The molecules travel in the circulatory system so they can go anywhere they are needed in the body.

Now you might ask why I had you get a pineapple for the experiment. Why didn't I just have you get a can or jar of pineapple juice? It turns out that the enzyme needed to digest Jell-O is very sensitive to temperature, and when juices are canned or jarred, they are typically heated in order to kill bacteria that might have gotten in them. This makes them safer, but it destroys the enzyme that is needed to digest Jell-O. For the experiment to work, then, you need pineapple juice that hasn't been heated up, and usually, you can only get that from a pineapple.

So the vinegar wasn't able to digest the Jell-O, but the pineapple juice was. Did you notice, however, that the Jell-O you had broken into smaller globs got digested faster than the Jell-O that was in one large glob? That's because in order to attack the molecules in the Jell-O, the pineapple juice had to actually touch the Jell-O. It could only do that on the surface of the glob. Where there was only one big glob, only a certain amount of pineapple juice could touch the Jell-O. However, when you broke it up into smaller globs, there was a lot more access to the Jell-O, so more pineapple juice could touch it.

This should tell you something. Since your body also digests things by mixing them with a bunch of chemicals that have to touch the food to digest it, small chunks of food digest faster and better than large chunks of food. That's why you should always chew your food thoroughly before swallowing it. Fairly large chunks of food can fit down your throat, but they don't digest very easily. They tend to sit in your stomach for a long time. The smaller the chunks of food, the easier they are to digest.

Now remember, the reason I taught you about digestion is so that you can learn what bile does in your body. Obviously then, bile has something to do with digestion. You will learn exactly what it has to do with digestion in the next lesson.

LESSON REVIEW

Youngest students: Answer these questions:

1. When you eat food, what must happen to it before your body can use it?

2. Which digests faster, large chunks of food or small chunks of food?

Older students: In your notebook, write down a definition of digestion, and explain why your body digests food. Also, explain why small chunks of food digest faster than large chunks of food.

Oldest students: Do what the older students are doing. In addition, write down the following question and your answer to it: "If you heated up the pineapple juice before you put the Jell-O into it, would the experiment have turned out the same? Why or why not?" Check your answer and correct it if you were wrong.

Lesson 14: Hippocrates and Bile, Part 2

The red parts of this beef steak are meat. The white parts are fat.

Now that you have learned a bit about digestion, it is time to learn what bile actually does for your body. When you eat food, it must be digested. However, there are a lot of different chemicals in food, and some of them are more easily digested than others. Have you ever heard that a steak is harder to digest than vegetables? That's really true. One reason meat is harder to digest is that it contains a lot of **fat**. Look at the picture of raw beef on the left. The red part is the meat. Do you know what the meat is? It's actually the cow's muscle. When you eat meat from any animal, you are actually eating its muscles. The white strips running through the meat are strips of fat. Even though you might cut the fat off the meat that you eat, there is no way to get rid of it all. If you eat meat, you are eating fat as well, because there is always fat associated with muscle. Some meat contains less fat than other meat, and it is called "lean" meat. Nevertheless, even lean meat has fat in it.

Well, it turns out that fat can be pretty hard to digest. To see why, perform the following experiment.

Fat and Water

What you will need:

- Butter (Real butter is better than margarine, but margarine will work.)
- Water
- Dish soap
- Two tall glasses
- Two spoons
- A magnifying glass

What you should do:

1. Fill each glass halfway with water.
2. Put a small amount of butter on each spoon. It doesn't need to be a lot, but the amount on each spoon needs to be about the same.
3. Put one spoon in one glass of water and the other spoon into the other glass of water.
4. Add several squirts of dish soap to one of the glasses.
5. Stir the spoons in each glass vigorously for about the same amount of time.
6. Set the glasses on a counter or table and watch them as the water slowly stops spinning. What differences exist between the two glasses?
7. Hold each glass up to a sunny window or a bright light, and look at the water in each of them with the magnifying glass. **Don't look at the sun or a bright light directly with the magnifying glass!** What differences do you see between the glasses?
8. Dump the contents of both glasses down the drain.
9. Thoroughly wash the glasses and the spoons with lots of soap.

What happened in the experiment? Let's start with the glass that had no soap in it. You stirred and stirred, but not a lot happened to the butter. It might have come off the spoon, and it might have split up a little, but if it did, large pieces of butter just floated to the top of the water, right? If you have done anything with butter and water before, these results were probably not surprising. Butter and water just don't mix, do they? When you looked at the water with a magnifying glass, you might have seen a few small bits of butter in the water, but there wasn't much, was there?

What happened with the glass that had dish soap in it? First, the butter probably came off the spoon much more easily. In addition, while a large amount of butter was probably floating on the surface (along with a lot of bubbles), you should have seen more bits of butter floating in the water. When you looked at the water with the magnifying glass, you should have seen even more bits of butter floating around.

What explains the difference? Well, even though butter and water don't mix, butter does mix a bit with soap. In addition, soap mixes well with water. When you added soap to the water, the molecules in the soap mixed a bit with the butter, and a lot with the water. As a result, it pulled bits of butter into the water. So the soap acted as a sort of "go between." It allowed the butter to mix a bit better with the water. It still didn't mix incredibly well with the water, but it mixed better with soap and water than just plain water.

When you eat fat, the same basic thing happens. Butter is actually made of fat, and fat doesn't mix well with water. So when you eat fat, it doesn't mix well with the water that holds the chemicals your body uses to digest things. Without some help, the fat would float around in big chunks, just like the butter did in water. This would make it hard for the digestive chemicals to get to the fat, and as a result, a lot of the fat you ate would not get digested well.

The bile in your body fixes that. It acts like the soap did in your experiment. It is attracted to both fat and water. As a result, it acts as a "go between." It breaks up the fat into tiny bits, allowing those tiny bits to mix better with the enzymes that are needed to digest the fat. Just as the Jell-O digested faster in last lesson's experiment when it was broken into bits, the fat you eat digests much faster because the bile breaks it up into tiny bits so that it can more easily mix with your body's digestive chemicals.

In the end, then, bile doesn't play anything close to the role that Hippocrates thought it played in the body. Nevertheless, it is an important fluid in the body. If your body didn't have it, the fats that you eat would not digest very well. As a result, your body couldn't use them for the chemical energy that is stored there. This is important, because of all the chemicals you eat in your food, fats contain the most energy. If you couldn't digest them well, you would have to eat a lot more food in order to have the energy you need to live.

It turns out that there is something you probably drink regularly that is very similar to a mixture of water, bile, and fat. It obviously doesn't taste like that, or you wouldn't drink it! Nevertheless, it is a mixture of water, some other chemicals, fat, and a "go between" that keeps the fat in the solution so that it doesn't rise to the top. Can you think of what I am talking about? Believe it or not, I am talking about milk!

This is a picture of milk that has been magnified. The little circles you see are the tiny globs of fat that are floating in the milk.

Yes, a glass of milk is a mixture of many things dissolved in water. However, not everything in milk is dissolved. For example, unless you drink skim milk, the milk contains fat. Like butter, fat does not dissolve in water. So in order for it to mix well with the milk, a "go between" is included so that the fat gets broken up into tiny globs that float around in the milk. That "go between" performs the same duty in milk that bile does in your body. It splits the fat up into tiny pieces so that it can more easily mix with the water. We call this **homogenization** (huh moj' uh nih zay' shun).

Have you ever eaten really spicy food? If you have, you might have experienced your tongue getting really "hot." The temperature of your tongue didn't really rise, but it felt like it did. If you didn't like how your tongue felt, you probably tried to "cool it off" by drinking a lot of water. Did that work? Not really. Why? Because the spices in the food are what causes your tongue to feel that way, and the spices (like fat) don't mix well with water. As a result, the water did very little to remove the spices.

The next time you eat spicy food, try washing it down with milk. You will find that your tongue doesn't feel nearly as hot. That's because of all the fat floating around in the milk. The spices don't like to dissolve in water, but they love to dissolve in fat. So as the little globs of fat that are in milk run over your tongue, they pick up the spices, taking them down as you swallow. As a result, your tongue doesn't feel nearly as bad.

The main point here is that some chemicals (like fat) don't mix well with water. However, if you add a "go between," like bile, they will mix a bit better with water. When it comes to digesting fat, this is important, because it allows the digestive chemicals to get at the fat much more easily, which allows the fat to dissolve more quickly.

LESSON REVIEW

Youngest students: Answer these questions:

1. When you eat meat, what do you also eat along with it?

2. What does bile do in the body?

Older students: In your notebook, make a drawing of what fat would look like in a glass of water. Then make a drawing of what it would look like if you added bile. Use your experiment observations as a guide. Explain in your notebook how bile helps your body digest fat more quickly.

Oldest students: Do what the older students are doing. In addition, write the following question and answer it in your notebook: "If you drink skim milk after eating spicy food, will your tongue feel any better than if you drank water instead?" Check your answer and correct it if you were wrong.

Lesson 15: Hippocrates and Phlegm

Remember that Hippocrates believed the body contained four humors: blood, yellow bile, black bile, and phlegm. You have already learned that blood is definitely a fluid in your body, but unlike what Hippocrates thought, it is actually a transport system that carries oxygen and nutrients to different parts of your body. You also learned that bile is a real fluid in the body, but there is only one kind. Unlike what Hippocrates thought, it aids in the digestion of fat. You probably won't be surprised to know that phlegm is a real fluid in the body as well, but it doesn't do what Hippocrates thought.

Have you ever coughed so much that something greenish-yellow came out of your mouth? It looked pretty gross, didn't it? That's phlegm. Since Hippocrates treated patients who coughed a lot, he saw a lot of phlegm. As a result, he decided it must be an important fluid in the body. It turns out that it is really important. Do the following experiment to see why:

Phlegm's Job

What you will need:
- A white paper plate
- Elmer's glue (or any other white, all-purpose glue in a bottle)
- Pepper in a pepper shaker
- A sink

What you should do:
1. Set the paper plate on a counter or table.
2. Use the glue bottle to draw a design on the paper plate with glue. Don't use a lot of glue – just enough to make a design. I drew a smiley face, but you can draw whatever you want. Notice that it's pretty hard to see the design, because the glue is white, and the paper plate is white too.
3. Use the pepper shaker to sprinkle pepper over the entire paper plate. Use a lot of pepper.
4. Once the paper plate is covered in pepper, take it over to the sink and tilt it so that the pepper falls into the sink. Blow on the plate gently to get rid of more pepper.
5. Now look at what you have. You should be able to see your design much more clearly now.
6. Clean up your mess.

What happened in the experiment? The glue held onto the pepper, while the rest of the paper plate didn't. As a result, once you tilted the paper plate and blew on it, the pepper that landed on your design didn't move, but the rest of the pepper fell into the sink. In the end, your design appeared much more clearly, because it was highlighted with pepper! Believe it or not, that's essentially the purpose of phlegm in your body. It doesn't highlight designs, but it does hold on to stuff!

Before I explain further, let's use modern terms instead of the terms Hippocrates used. Today, the term "phlegm" refers to the gross stuff that you cough up when you are sick. It's a specific example of something called **mucus** (myoo' kus), which can be found in many different parts of the body. For example, it is found in your nose as well. When you have a runny nose, it's because your nose is producing too much mucus. When it comes out of your nose, however, we don't generally call it mucus or phlegm – we call it **snot**.

Since phlegm is just a specific kind of mucus (the kind you cough up), the real question is, "What does mucus do for your body?" As I said before, it does essentially what the glue did in your experiment – it holds on to things. What kinds of things does it hold onto? To answer that question, you need to know where a lot of your body's mucus can be found.

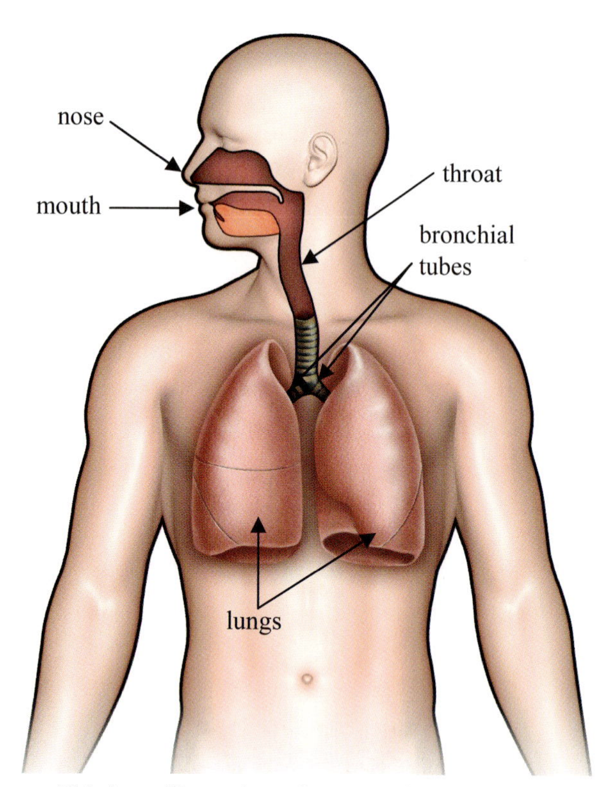

This is an illustration of your respiratory system.

In order to live, you need to breathe, right? That's because your body needs oxygen, and the air around you contains oxygen. In order to get that oxygen, you take air into your body, your body processes it, and then you push out of your body what remains. We call that breathing, and it is taken care of by your **respiratory** (res' puh ruh tor' ee) **system**. As you can see from the drawing on the left, your respiratory system is made up of your **mouth**, **nose**, **throat**, **lungs**, and some tubes called **bronchial** (brahn' kee uhl) **tubes**. Air can be taken into your respiratory system through your nose or your mouth. It is carried through the bronchial tubes to your lungs, where the oxygen in the air is added to the blood in your circulatory system.

Mucus lines your nose, throat, and bronchial tubes. Why? Because there are a lot of nasty things in the air that you breathe. Dust and other particles float in the air, and if they got down into your lungs, lots of bad things could happen. Remember, for example, the experiment you did in Lesson 11. Bacteria on dust particles in the air were responsible for the awful changes in the meat. So your body has to stop those particles before they reach your lungs. Guess what stops them? The mucus. Dust and other particles get stuck on the mucus, much like the pepper got stuck on the glue in your experiment. That way, they don't make it into the lungs.

But wait a minute. If the mucus keeps trapping dust and other particles, it gets dirtier and dirtier, right? Once it gets really dirty, it won't be very sticky any more, and it won't be able to do its job. What does your body do to take care of that? It gets rid of the old mucus and makes new mucus. You see, your respiratory system isn't only lined with mucus. It is also lined with tiny "hairs" that beat back and forth. Scientists call them **cilia** (sil' ee uh). Those tiny hairs push the mucus up the bronchial tubes or down the nose until it reaches your throat. Then, you swallow it. Your body makes new mucus to replace the mucus you swallowed, so that you always have mucus lining your respiratory system.

Now it might sound strange to you that you swallow your own mucus, but it is true. It probably sounds odd because you never taste any mucus. However, think about where the mucus is going. It is going up the bronchial tubes to your throat or down the nose to your throat. As a result, it misses the part of your body that supplies taste. Do you know what that is? It's the tongue. The tongue has taste buds, and since your mucus never hits your tongue, you don't taste it.

Not only does the mucus trap dust and other particles, it also traps nasty things that are trying to infect your body. If you have a cold, what happens? You get a runny nose and you cough a lot, don't you? Your nose gets runny because it is producing a lot of mucus to trap the virus that is causing the cold. It runs out of your nose so the virus is taken out of your body. In the same way, you cough up that mucus, and when it leaves your body, it is called phlegm. Once again, your body is producing excess mucus to fight the virus that is trying to take over your body, and as a result, you tend to cough it out. So the mucus in your respiratory system is a means by which your body protects itself from dust, particles, and things that want to infect you and make you sick.

Before I end this lesson on mucus, there is one more thing I want to point out. When you are outside in the cold, your nose often runs. Since your nose also runs when you have a cold, a lot of people think that you get a cold by being cold. That's not true. You get a cold by getting a virus, which is something that wants to infect your body. You can get a virus when you are warm or cold, so the temperature has very little to do with catching a cold.

Why does your nose run when you are outside in the cold? It is a side effect of something the nose does. You see, one of the most important things your body has to do to the air that it takes in is heat it up. Your body is very warm: about 37° Celsius (99° F). Your lungs work best if the air that reaches them is also at that temperature. As it travels through your nose, then, your nose tries to warm it up. In order to do that, your nose has to stay warm.

Well, when it is cold outside, the inside of your nose starts to get cold. To fix that, your body sends it more blood. Blood is warm, so more blood to your nose means the inside of your nose will get warm as well. As more blood gets sent to your nose, your nose makes more mucus. Also, there is a lot of water vapor in the air you are breathing. The colder the air, the more likely that water vapor is to condense and form liquid water. That makes your nose's mucus thinner, which means your nose runs more easily. So even though your nose runs when you are outside in the cold, it is for a completely different reason than why your nose runs when you are sick. The two things are not related.

This is actually an important point about understanding the world around you. Often, you will see situations that are very similar, and it is natural to think that the causes are fairly similar as well. Since you get a runny nose when you are outside in the cold, and since your nose runs when you have a cold, it kind of makes sense to think that you can get a cold from being cold. Even though it kind of makes sense, it isn't true. A good scientist will not assume two things that are similar are the result of similar causes. Instead, the scientist will try to investigate both things to see what causes each of them. They might be the result of similar causes, but they might not. Until you learn enough about each, you shouldn't assume anything!

LESSON REVIEW

Youngest students: Answer these questions:

1. What does mucus do for the body?

2. Why don't you taste the mucus that you swallow regularly?

Older students: In your notebook, explain what mucus does for your body. Explain how your body gets rid of old mucus and replaces it with new mucus. In addition, explain why you cough up phlegm and your nose runs when you are sick.

Oldest students: Do what the older students are doing. In addition, people who work in very dirty air, such as those who mine coal in dirty coal mines, cough up a lot of phlegm. Explain why this happens. Check your answer and correct it if it is wrong.

Science Before Christ, Part 2

This is a statue of Aristotle that was fashioned for Aristotle University in Thessaloniki, Greece.

Lessons 16-30: Science Before Christ, Part 2

Lesson 16: Plato (c. 424 BC – c. 348 BC)

About 30 years after Democritus and Hippocrates were born, a man named **Plato** (play' toh) was born in Athens. When he was still a young man, he heard the teachings of a man named Socrates (sah' kruh teez) and decided to devote his life to them. Unfortunately for Socrates, many of those teachings went against the religious and political views of the day, and as a result, he was condemned to death by the government. This upset Plato, and he left Athens for many years, traveling to Italy and perhaps even parts of Africa.

This statue of Plato is in Athens, Greece.

Most of what we know about Socrates comes from Plato's writings. He was obviously devoted to Socrates and wished to preserve the man's teachings. Interestingly enough, however, many of Socrates's teachings (and Plato's writings) were opposed to science. Plato thought that the world which we see, feel, and sense is not the *real* world. Instead, it is a vague shadow of the real world, which is more perfect than anything we see, feel, and sense. Those who try to understand things by studying the world with their senses, then, will never really learn anything useful. In order to learn useful things, you need to neglect your senses and use only your mind.

Obviously, this is pretty much exactly opposite of how we do science today. In order to learn about the world, we study it with our senses. We observe things, experiment on them, and try to reason out how things work based on the information we collect with our senses. Plato and his teacher Socrates didn't think that was a reasonable way to learn anything. Why, then, am I talking about Plato in this science book? Because even though most of his ideas contradict science as we know it today, he actually did two things that ended up advancing the science that he continually argued against!

The first thing he did was emphasize that there was something more to mathematics than just usefulness. Do you remember the great feat that Thales accomplished with math? He used math to measure the height of the pyramids. In other words, Thales used math as a tool that replaced a really big measuring stick. Lots of people in Plato's day used math as a tool. However, Plato thought there was something more to mathematics than that. He thought that mathematics held basic truths about the world. In his mind, mathematics was not invented by people. Instead, it was *discovered* by them. In other words, mathematical truths have always existed. What we call "mathematics" is simply our way of finding out what those truths are.

This might sound like a strange idea to you, but it is actually very powerful, and very important for science. If people invented math, there is no reason to believe that it could describe the world around us. However, if mathematics is a part of the world around us, and we are just discovering it, then there is every reason to think that it is an important part of the world and can be used to describe the world. In order to get you thinking about this a bit more, I want you to do a math activity. I want you to start with *any number* you choose. I will have you do math on that number, and regardless of the number you choose, the final answer will be 5. Do you believe I can do that? See for yourself:

The Answer Is 5

What you will need:

- Your notebook (a blank sheet of paper if you are doing the review exercises for the youngest students)
- A pencil
- If you can't do multiplication and division well, use a calculator to help you

What you should do:

1. Start a page in your notebook that is titled, "A Mathematical Exercise."
2. Below the title write, "Starting with any number, the answer is 5."
3. Choose any number, as long as it is not 0. It can be small if you want the math to be easy, or it can be large if you want to test how well this works. Write that number down.
4. Multiply that number by itself. If you chose "2," for example, do 2x2 = 4.
5. Add the number you chose to the result of step 4.
6. Divide the result of step 5 by the number you chose.
7. Add 24 to the result of step 6.
8. Subtract the number you chose from the result of step 7.
9. Divide by 5. What is the result? I told you!
10. If you want, try this again with any other number. Your parent/teacher has an explanation for why this works.

If you don't want to know why this works or can't understand the explanation your parent/teacher has, don't worry. To understand the explanation fully, you need to understand a bit of algebra. The point is that I used a mathematical truth to make you turn any number you chose into the number 5. Now here's the question to consider: Is that mathematical truth a real truth or something that was just invented? If it was just invented, how can it work on *any* number, even ones I haven't tested? In the end, Plato would say that the mathematical truth is a real truth that existed long before anyone discovered the trick.

Why is this important to science? Because science deals with mathematics quite a lot, and it often uses mathematics to describe the world. As philosophers like Hippocrates were discovering truths about nature (truths that Plato didn't think were real), Plato encouraged philosophers to discover mathematical truths. Without the mathematical truths that were discovered by people who followed Plato's teachings, modern scientists would not have been able to use math to describe the physical world. As a result, all sorts of science, like physics and a lot of chemistry, would never have existed.

Now to you and me, Plato's insistence that mathematical truths are real and not invented should not be very surprising. After all, we know that God created the universe. As a result, we know that it follows the laws that He laid down. Many of those laws are mathematical in nature, so God created mathematics as He formulated the laws that govern nature. Just as scientists discover those laws, mathematicians simply discover the rules of mathematics.

You have to understand, however, that there are some scientists and mathematicians who do not believe God created the world. As a result, they have no idea why mathematics does such a great job of describing the physical world. These mathematicians call this "The Question." They have no reason to believe that mathematics should describe the real world, but it does. As Dr. Barry Mazur, a mathematician at Harvard University says, "One thing is—I believe—incontestable: if you engage in mathematics long enough, you bump into The Question, and it won't just go away." ["Mathematical Platonism and its Opposites," Barry Mazur, http://www.math.harvard.edu/~mazur/papers/plato4.pdf, retrieved 12/14/11].

I told you that Plato did two things that helped advance science. Proclaiming that mathematical truths should be investigated is one of them. What was the other? Well, he eventually returned to the city where he was born (Athens) and founded the Academy, which was the first school of higher learning in the world. It attracted many smart people whose main goal was to simply learn more about life and the world around them. One of those people, named **Aristotle** (air' ih stot' uhl), became Plato's student. While he disagreed with much of what Plato said, he recognized that Plato could teach him quite a bit, so he gladly studied under him. You will learn about Aristotle and his contributions to science in the next few lessons. If Plato had not founded the Academy, Aristotle's contributions to science might never have happened. For that alone, Plato deserves to be mentioned in a science book like this one.

This is how Italian artist Raffaello Sanzio imagined the Academy that Plato founded. Plato and Aristotle are the two men at the center. Plato is in orange robes, and Aristotle is in blue robes.

LESSON REVIEW

Youngest students: Answer these questions:

1. What did Plato think you could learn by studying the physical world?

2. Did Plato think that mathematics was discovered or invented?

Older students: Underneath the mathematical exercise you wrote in your notebook, write that there is a mathematical truth that tells us the answer will always be 5. Explain what Plato thought about whether this mathematical truth was invented or discovered. Explain why this idea works very well for those who believe the world was created.

Oldest students: Do what the older students are doing. In addition, see if you can understand the explanation of the mathematics exercise you did. If not, don't worry about it. If you do understand, however, write the explanation in your own words.

Lesson 17: Aristotle (384 BC – 322 BC)

This is a statue of Aristotle that is found in the town where he was born: Stageira, Greece.

I told you in the previous lesson that one of the great things Plato did for science was to form the Academy in Athens and attract Aristotle, who became Plato's student. This was an important step for science, because Aristotle is generally considered one of the most important philosophers in all of history. He not only made incredible contributions to the progress of science, but he also wrote important works on poetry, theater, music, logic, language, government, and ethics. He truly is a towering figure in history. Even though he was born in 384 BC and died in 322 BC, many of his ideas influenced science well up until the 1600s. That's almost 2,000 years of influence!

Unlike Plato, Aristotle believed you could learn a lot by observing the world around you, so he made lots and lots and lots of observations. For example, he watched how things moved, and he tried to understand why they moved the way that they did. He came up with a rather interesting idea, which we now know is completely wrong. Nevertheless, it is worth learning about. Aristotle believed there were five basic elements in nature: **earth**, **air**, **fire**, **water**, and **ether** (ee' thur).

According to Aristotle, everything in nature was made up of a combination of these elements, and more importantly, each element had its "natural" place. The earth element belonged near the center of the earth, the water element belonged on the surface of the earth, the air element belonged above the earth, the fire element belonged with the moon, and the ether element belonged with the planets and the stars. The elements would strive to get back to where they belong, and that's what explained a lot of motion.

If you dropped a rock, for example, Aristotle would say that it belongs at the center of the earth, because rock is mostly composed of the earth element. To get to where it belongs, it would fall towards the center of the earth. The surface of the earth would stop it, of course, but it would at least get as close as possible to where it belongs. Bubbles rose in water, according to Aristotle, because bubbles contain gases, which are made with the air element. Air belongs above water, so when gases are trapped in water, they rise. Flames rise in a fire because they are made of the fire element, and they are trying to get to where they belong – all the way up to the moon! You can see how convincing Aristotle's reasoning was.

Of course, we now know that this isn't true at all. You probably already know why bubbles rise in water. A given volume of air weighs a lot less than the same volume of water, so the air floats in water. It floats so well that it rises and escapes the water. A rock will sink when placed in water, not because it is trying to get to the center of the earth, but because it weighs a lot more than an equal volume of water. As a result, it sinks. This same reasoning applies to things in air. A helium balloon floats in air because it is lighter than an equal volume of air. An air-filled balloon sinks because it is heavier than an equal volume of air that is not in a balloon. "But wait a minute," Aristotle might say, "the balloon has some earth element in it, so that's why it falls when it has air in it." However, we can make a gas fall even when it is not in a balloon, as long as we use the correct gas!

A Sinking Gas

What you will need:

- A 2-liter bottle, like the kind soda comes in
- Vinegar
- Baking soda
- A funnel
- A measuring teaspoon
- A measuring cup
- A candle
- Something you can use to light the candle
- An adult to help you

What you should do:

1. Use the funnel to put one cup of vinegar into the bottle.
2. Add one teaspoon of baking soda to the bottle. You don't have to get it all in there. Just do the best job you can. Observe what happens as you add the baking soda.
3. Have an adult help you light the candle.
4. Slowly tilt the bottle so that the bottle's opening is near the flame but not touching it (see the picture on the right). Tilt the bottle as much as you can without letting any of the liquid pour out. What happens?
5. Repeat the experiment if you like.
6. Clean up your mess and put everything away.

What happened in the experiment? If everything went well, the flame should have gone out, despite the fact that liquid never came out of the bottle. Why did the flame go out? Well, when you added baking soda to the vinegar, you should have seen lots of bubbles. What do bubbles tell you? They tell you that a gas is being formed. In this case, the gas was carbon dioxide. Well, it turns out that a group of carbon dioxide molecules is heavier than an equal volume of the oxygen and nitrogen that make up most of the air around you. What does that tell you? Carbon dioxide sinks in air.

Once you added baking soda to the vinegar, the carbon dioxide that was produced filled up the bottle. So when you tilted the bottle, you were tilting a bunch of carbon dioxide gas. Since it is heavier than an equal volume of air, the carbon dioxide didn't rise out of the bottle. Instead, it sank. So what you were really doing, then, was "pouring" carbon dioxide gas out of the bottle and onto the flame.

Why did that put out the flame? Because fire needs oxygen in order to burn. Fire is a chemical reaction that causes oxygen to combine with whatever is burning. In this case, the candle wax is burning, so in order for the candle to stay lit, it needs wax and oxygen. When the carbon dioxide poured out of the bottle, it pushed all the air out of the way. That, of course, means there was no oxygen for the candle, so the flame went out. Now Aristotle would never have expected the result you got, because he would have said that the carbon dioxide was already where it should be (in the air). As a result, it shouldn't have moved down to the flame. It should have stayed where it was.

In this photograph, a firefighter is holding a carbon dioxide fire extinguisher. "CO_2" is the chemical symbol for carbon dioxide.

Believe it or not, your experiment actually demonstrates one way we put out fires today. There are many different kinds of fires, and they require different ways of putting them out. However, one kind of fire extinguisher is called a **carbon dioxide fire extinguisher**, and it just has lots and lots of carbon dioxide in it. When you point the nozzle and squeeze the trigger, it shoots out carbon dioxide. If you point the nozzle at the fire, the carbon dioxide will push the air out of the way, which gets rid of the oxygen around the fire. This puts out the fire, just like the carbon dioxide put out the flame in your experiment. Now I have to tell you that this kind of fire extinguisher doesn't work on all fires, because there are other things to consider. However, it is a good general-purpose fire extinguisher.

LESSON REVIEW

Youngest students: Answer these questions:

1. Did Aristotle agree with Plato about studying the world around us?

2. Name at least two of the elements that Aristotle thought made up everything around you.

Older students: Write down the five elements that Aristotle thought existed in nature and where each belonged. Then explain how Aristotle used this to describe motion. Finally, explain why your experiment shows that this idea is not correct.

Oldest students: Do what the older students are doing. In addition, see if your home has a fire extinguisher. Read the label to see what is in the fire extinguisher. Is it carbon dioxide? See what kinds of fire the extinguisher can be used on, according to the label.

Lesson 18: Aristotle and Things that Fall

I've told you that things float in water because they are lighter than an equal volume of water and that they sink because they are heavier than an equal volume of water. In addition, you learned in the previous lesson that things rise in air because they are lighter than an equal volume of air, and they fall in air because they are heavier than an equal volume of air. However, that doesn't answer a more basic question: What makes things fall or sink to begin with? Do you see the difference in the question? Knowing that something weighs more than an equal volume of air tells you it will fall, but it doesn't tell you what is actually making the object fall or rise in the first place.

Aristotle thought he had the answer to that. He thought that things fell to earth because they were mostly made of earth and *wanted* to get back to where they belonged. As your experiment demonstrated, the idea that things rise or fall based on "getting back to where they belong" is wrong. However, I want to stick with that idea for another lesson, because it leads to a conclusion that was held for about 2,000 years after Aristotle, so it is worth noting. It also teaches us a couple of things about the nature of science.

Aristotle thought that heavy objects fell faster than lighter objects. His reasoning was quite simple: light objects had less of the earth element in them, and heavy objects had more of the earth element in them. As a result, heavy objects wanted to get back to where they belong more strongly than lighter objects, and therefore they fell more quickly. Now when you first read about Aristotle's belief that heavy objects fall faster than lighter objects, it makes sense. After all, if you drop a rock and a feather, the rock will land on the ground much sooner than the feather. However, if you subject Aristotle's belief to experiment, you find that it just doesn't work.

How Quickly Do Things Fall?

What you will need:

- A heavy rock (You need to be able to lift it easily.)
- A medium-weight rock (It needs to be noticeably lighter than the one above.)
- A very light rock (It needs to be a rock, but it needs to be noticeably lighter than the one above.)
- A small board or tray that you can set your rocks on and push them off the edge easily
- A ruler or sturdy stick
- A stepstool, ladder, or something that is safe to stand on so that you can reach high up
- A person to help you

What you should do:

1. Have your helper hold the board up as high as he can. The board should be out in front of your helper, not directly over his head.
2. Stand on the stepstool or ladder so that you can easily put things on the board that your helper is holding up.
3. Put all three rocks on the board. Position them so that they are all as near to the edge of the board as possible without them falling off.
4. Use the ruler to push all three rocks off the edge of the board at the same time. Note when each hits the ground.
5. Repeat steps 3-4 three more times, but when you position the rocks, position them in a different order. For example,

NOTE: It is best to do this experiment outside, as it involves falling rocks.

if the heaviest rock was closest to you the first time, arrange the rocks so that the heaviest one is farthest from you the next time.

6. Clean up your mess and put everything away.

In the first five steps of the experiment, you repeated the same thing four times. In science, we call each of these repeats a **trial**. You might think of a trial as something that happens in court, but that's not the way a scientist thinks of a trial. A scientist often repeats the same experiment many times, and each time she does it, she calls it a trial. Why would you repeat the same experiment over and over again? Because experiments can have errors in them, and one way to see if there is any error in an experiment is to repeat it over and over again, looking for different results. If there are different results in the same experiment, we say that the experiment contained **experimental error**.

If an experiment has experimental error, does that mean you can't learn anything from it? Surprisingly, the answer is, "No!" Most experiments have experimental error in them, so if we couldn't learn anything from those kinds of experiments, scientists wouldn't be able to learn much at all. The key is to *understand what causes the experimental error*. If you can do that, you can think through how the error would affect your experiment and then take it into account.

For example, in your experiment, you might have found that in one trial, the heaviest rock hit the ground first. In another trial, it might have been the lightest rock. In other trials, all rocks might have hit the ground at the same time. If the results were different between trials, there is obviously some kind of error going on, right? In the case of your experiment, what could the error be? Well, you were using the ruler or stick to try to push all three rocks off the board at the same time. However, that's a hard thing to do. Most likely, one of the rocks got pushed off the board slightly earlier than another. Since it started falling first, it probably hit the ground first. However, since you had four trials, most likely, in at least one of the trials, they all started falling at about the same time. As a result, they all hit the ground at about the same time. In the other trials, the rock that started falling first probably changed, because you changed the order of the rocks. As a result, in the other trials, the rock that hit the ground first changed.

So what does this tell you? If, as Aristotle thought, heavier objects fall more quickly, the heavier rock should have been the one always hitting the ground first. If you did every trial without error, however, you should have seen that all the rocks hit the ground at the same time. If you had so much error in your experiment that each trial had one rock hitting the ground first, most likely, a different rock hit the ground first each time. One trial might have had the heaviest rock hitting the ground first, while another trial might have had the medium-weight rock hitting the ground first. The point is that even if you had error in your experiment, you should have seen that there was no real pattern between the weight of the rocks and which one hit the ground first. That tells you Aristotle was wrong. The weight of the rock does not affect how quickly it falls.

Most likely, you know what really causes a rock to fall. It's a force called **gravity**. You'll learn more about gravity when you start studying the time period when scientists proposed the concept of gravity. For right now, just understand that things don't fall for the reason that Aristotle thought. It's not that they want to get back to where they belong. Instead, gravity attracts them to the earth. The most important thing you need to understand about gravity right now is that, unlike Aristotle thought, heavier things do not necessarily fall more quickly than light things. Your experiment should have shown you that.

But wait a minute. You know that *sometimes*, heavier things do fall more quickly than light things. A rock will fall more quickly than a feather. Why? Most likely, you already know the answer to this. You probably already know that there is something called **air resistance**. In order for something to move through the air (falling things must move through the air), it has to move the molecules that make up the air. Those molecules don't really have any reason to move, so the thing that is traveling through the air must move them out of the way. They resist that movement, which tends to slow an object down. That's air resistance.

A feather falls slowly because of air resistance.

It turns out that because of shape and many other factors, some things are good at moving air molecules out of the way, and other things aren't so good at it. As a result, some things can fall more quickly than others, but not really because of their weight. It's because of how easily they can shove air molecules out of the way. Most rocks have the right shape and the right arrangement of molecules that they can shove air molecules out of the way very easily. As a result, they fall at the same speed, regardless of their weight. A feather can't shove air molecules out of the way very easily, so it cannot fall nearly as quickly as a rock.

There are two important lessons here. First, experiments can have error in them. That doesn't necessarily make them useless, however. If you can figure out the errors, usually by doing several trials of the same experiment and looking for different results between the trials, you can take those into account and still learn something good from the experiment. Second, what "makes sense" in science isn't always correct. When you first read that Aristotle thought that heavy objects fell more quickly than light objects, it probably made sense to you. However, a quick experiment hopefully showed you that it was not correct. Scientists can't conclude things just because they make sense. Scientists must test their conclusions with experiments to make sure they are correct.

LESSON REVIEW

Youngest students: Answer these questions:

1. Why do scientists often repeat the same experiment many times?

2. Why does a feather fall more slowly than a rock?

Older students: Write down how Aristotle thought the weight of an object affects how quickly it falls. Explain how your experiment shows that this is not true. Also, explain why scientists often repeat the same experiment several times.

Oldest students: Do what the older students are doing. In addition, think about this: It is actually possible to remove almost all the air from a container. Suppose you could remove almost all the air from a tall container. Suppose you could also drop a rock and a feather at the same time from the top of that container. In your notebook, explain the situation and then write whether or not the rock would hit the bottom of the container first. Explain your reasoning. Check to see if you are right, and correct your answer if you are not.

Lesson 19: Aristotle and the Basic Rule of Optics

Aristotle was wrong about a lot of things, but he was also right about a few things. In this lesson, you will learn about one of the things he got right – the basic rule of optics. His teacher, Plato, thought that people see because light comes out of their eyes, interacts with light from the sun or a flame, and forms the image. Aristotle didn't think that was right. Based on his observations, he thought that light reflected off something, entered our eyes, and then our eyes did something to make us see the object. He supported this belief (which you should know is correct) by observing light as it traveled through small openings, such as the holes in a sieve (a cooking utensil with lots of tiny holes in it). He saw that images of large objects were formed by the light, despite the fact that the holes were tiny. You can see this yourself by building a really neat device called a **camera obscura** (ob skur' uh).

A Camera Obscura

What you will need:

- An empty soup can with one open end
- A hammer
- A reasonably thin nail, like a finishing nail
- Wax paper
- Scissors
- Tape
- A dark towel or blanket
- An adult to help you
- A sunny window (If it's not sunny, you can use a dark room and a lamp that has a shape which makes it easy to recognize the top and bottom of the lamp.)

What you should do:

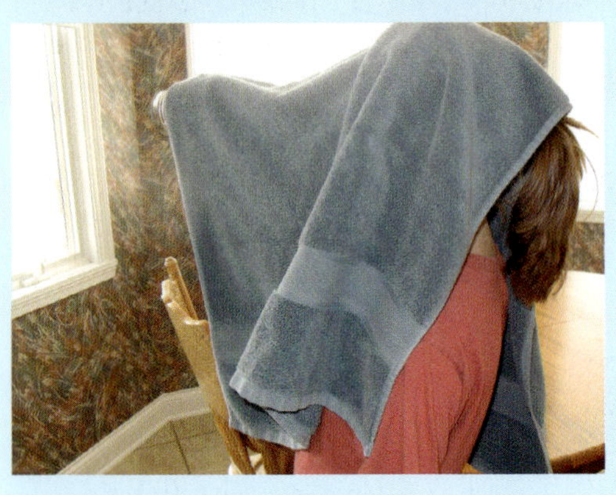

1. If the soup can is dirty, clean it out and dry it completely.
2. Have an adult help you hammer the nail through the center of the closed end of the soup can (see the top left picture).
3. Cut a square of wax paper much larger than the opening of the soup can.
4. Lay the square of wax paper over the opening of the soup can and use the tape to fasten it to the can. The wax paper should be stretched nice and flat over the opening of the can, and it should be held tightly in place by the tape (see the top left picture).
5. Go to a window that has an interesting scene outside. You want to be looking at something like trees or a house – something that has an easy-to-recognize top and bottom.
6. Hold the can in both hands so that the closed end with the hole is away from you and the open end covered in wax paper is in front of your eyes. You should hold it with your arms pretty much stretched out.
7. Have the adult help cover you in the towel or blanket so that your head and arms are covered but the end of the soup can sticks out (see the bottom picture on the left).
8. Point the soup can at the window, and look at the waxed paper. You should see an image of what's outside the window on the wax paper. If you don't see an image, try to reposition the towel or

blanket so that your face is in as much darkness as possible. The image on the wax paper will be a bit faint, so too much light will overwhelm it. However, if the towel or blanket darkens things enough, the image should be very clear. As mentioned above, if it is not sunny, you could instead go into a dark room with a lamp. Make sure the lamp is the only source of light in the room, and point the soup can at the lamp. You should see an image of the lamp on the wax paper.

9. Does anything look odd about the image, besides the fact that it might be a bit faint?
10. Look at other things with your camera obscura. You won't be able to see everything, because whatever you are looking at needs to be reasonably bright. Nevertheless, you should be able to see a lot!
11. Clean up your mess and put everything away.

What did you see on the wax paper? You should have seen an upside-down image of whatever you were looking at. Why was the image upside down? Because of the way we see things. Remember, in order to see something, light must reflect off of it and hit your eyes. Well, when the sun (or a light) shines on something, light reflects off of it in all directions. However, the hole in your camera obscura limited what light you could see. You couldn't see just any light reflecting off the object. You could only see light that reflected off the object and passed through the hole. This is very important. Look at the drawing below to see what I mean:

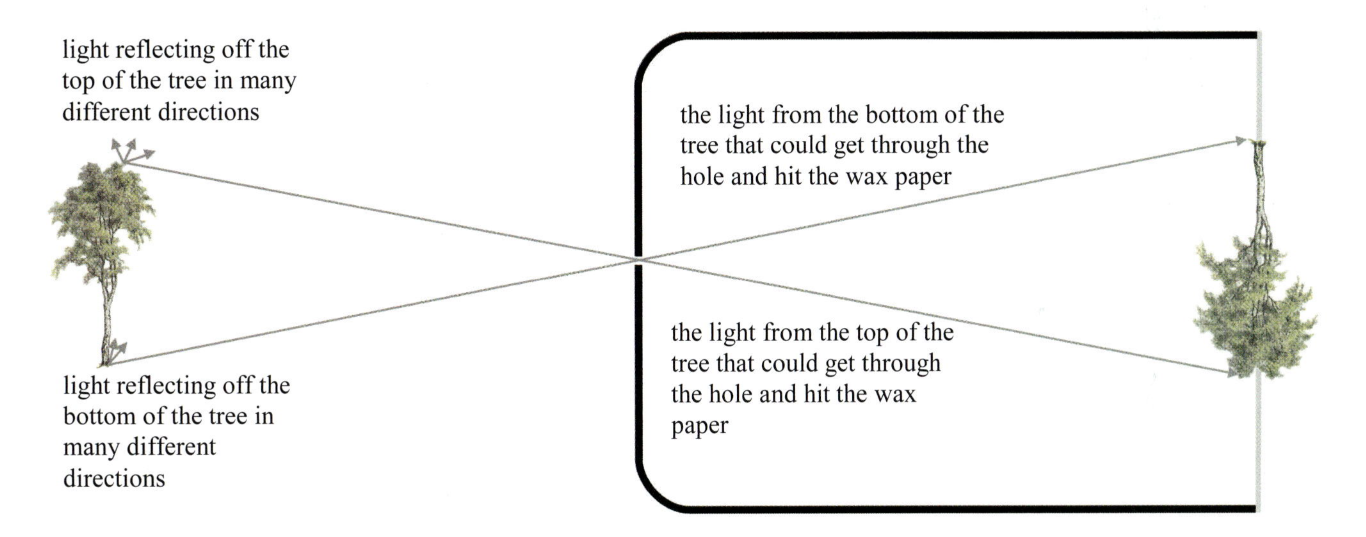

Notice that even though light is reflecting off the top of the tree in many directions, only one beam of light can pass through the hole and hit the wax paper. It hits the wax paper near the bottom. In the same way, even though light is reflecting in many directions off the bottom of the tree, the only light beam that can pass through the hole and hit the wax paper ends up hitting the wax paper near the top. As a result, light from the bottom of the tree shows up near the top of the wax paper, and light from the top of the tree shows up near the bottom of the wax paper. In other words, the tree appears upside down on the wax paper.

Now, Aristotle didn't make a camera obscura, but because he observed light passing through tiny holes, he saw the same result – the images were upside down. He didn't reason through it the way I just did, but he decided that the only way his this could possibly work was if he was looking at light that was bouncing off what he saw, and therefore he reasoned that we see an object because light reflects off the object and enters our eyes. This is considered by many historians to be the beginning of the science we call **optics** (op' tiks), which studies how light is made, how it travels, and how it can be guided by various devices.

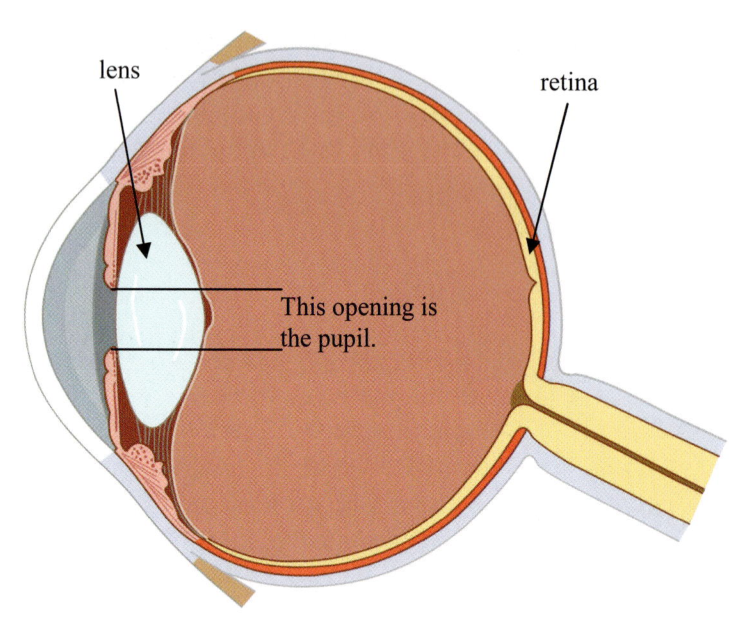

This is a simplified drawing of a human eye.

Now I want to take this moment to point out something that might be a bit surprising to you. Each of your eyes is, essentially, a camera obscura. Your eye has a small opening, like the hole you put in your soup can. It is called your eye's **pupil**. Light has to travel through that opening before it can hit the **retina** (ret' nuh), which is the part of your eye that detects light and sends information about it to your brain. Light also passes through a lens, but that just bends the light a bit in order to focus it better. In the end, then, for you to see something with your eyes, light must travel through a small hole and hit the retina at the back. Guess what that means? It means that, like the camera obscura, the images that hit your retina are upside-down!

But wait a minute. You don't see the world upside down, do you? Why is that? It's because when you are a tiny baby, you start touching things and figuring out which side is up and which side is down. Because of this, your brain starts understanding the images better, and as you develop, it figures out that it must flip all the images it is getting from your eyes. It does this automatically, without you even thinking about it. If your brain didn't do that, you would see the world upside down, because the way light has to travel into your eye is very, very similar to the way it traveled into your camera obscura!

LESSON REVIEW

Youngest students: Answer these questions:

1. What is the opening in your eye called?

2. Since our eyes work a lot like the camera obscura you made, why don't we see the world upside down?

Older students: Make a drawing like the one on page 57. Instead of a camera obscura, however, draw an eye, like the one on this page. Label the pupil, lens, and retina. Use it to show how images that appear on our retinas are upside down. Explain why you don't see the world upside down.

Oldest students: Do what the older students are doing. In addition, suppose you did your experiment on a very bright, sunny day and used your camera obscura to look outside. How important would the towel or blanket be? If you did your experiment on a cloudy day, how important would the towel or blanket be? Explain your reasoning in your notebook. Check your answer and correct it if it is wrong.

Lesson 20: Aristotle and Classification

As I already told you, Aristotle studied and wrote about many different things. Even though you have spent three lessons learning about Aristotle, you haven't yet learned about his greatest contribution to science. Aristotle developed the first reasonable **biological classification system**. You probably already know that scientists tend to group living organisms based on their similarities. This is called **classification**, and anyone who studies living organisms in depth needs to know how they are classified by modern scientists. While some philosophers before Aristotle attempted to classify living things, they didn't carefully examine the living things they were classifying. As a result, their classification systems didn't work very well.

Since Aristotle believed that you can learn by studying the world around you, he studied living things, especially animals, in great detail. As a result, he came up with a reasonable classification system that worked really well. Before I explain his system to you, I want to see what kind of classification system you can come up with by doing the following activity.

Classifying Animals

What you will need:

- The pictures shown here
- Your notebook (a blank sheet of paper if you are doing the review exercises for the youngest students) and a pen or pencil

What you should do:

1. Look at the following pictures of different animals:

2. Think about the things you know about each of these animals. What do they have in common? What are their differences? If you don't know anything about some or all of these animals, just look at their pictures and try to answer those two questions.

3. Split the animals into two groups. Animals in each group should have a lot of things in common, and they should have little in common with the animals of the other group.
4. Name each of the two groups.
5. In your notebook, write "My Classification of Animals," and underneath that, write your two groups and what animals you put in each group.

Did you find it hard to take those six animals and split them into only two groups? You might have, because you didn't have the actual animals to study in detail. How did you end up splitting them? You might have split them according to those that can fly and those that can't. If so, the bird and wasp were in the "can fly" group, while the others were in the "can't fly" group. You might have split them up according to where they lived. If that's the case, the crayfish and starfish were in the "live in water" group, while the others were in the "live on land" group.

Aristotle studied the actual animals, and he studied them in detail. In fact, he studied more than 500 different animals! He studied how they looked, and he cut many dead ones open to see what they looked like on the inside. Do you know what that's called? It's called **dissection** (dy sek' shun), and it's something scientists do to this day in order to learn more about living things. It might sound kind of gross, and you might even think it is cruel. However, we wouldn't know much about plants and animals if scientists hadn't dissected them.

In Genesis 1:28, God gave Adam and Eve dominion over nature. This means we must care for nature, and part of caring for nature is learning about it. Do you have a pet? If so, the pet probably brings you a lot of joy. Suppose your pet gets sick. What do you do? You take it to a veterinarian, don't you? What does the veterinarian do? He examines your pet and determines how to treat your pet to make it better. If it weren't for scientists dissecting animals and learning how their insides work, the veterinarian couldn't do that. You and the veterinarian would not be able to care for your pet if scientists hadn't dissected animals. Believe it or not, even a lot of medicines and treatments for people were developed based on knowledge gained by dissecting animals. Dissection is not a bad thing. It is a part of what we must do in order to have proper dominion over nature.

Since Aristotle made very detailed observations of animals, he came up with a classification system that was probably a lot different from yours. He split animals into two groups: those with blood and those without blood. Today, we know that most animals have blood. However, when Aristotle talked about blood, he talked about the kind of blood that people had – a thick, red fluid. All the animals pictured in the experiment above have blood, but only three of them have blood that is very similar to a person's blood: the horse, the bird, and the lizard. Aristotle classified them as animals with blood. The others (the wasp, crayfish, and starfish) have blood that doesn't look much like a person's blood, so Aristotle called them animals without blood.

Now even though Aristotle wasn't right about the wasp, crayfish, and starfish having no blood, he ended up splitting those animals into the same two groups that modern scientists split them into. Can you see what those groups are? The horse, bird, and lizard all have backbones. As you probably already know, modern scientists call animals with backbones **vertebrates** (vur' tuh braytz). The wasp, crayfish, and starfish do not have backbones, and modern scientists call them **invertebrates**. So even though he described the groups incorrectly, he ended up splitting animals into the same two groups that modern scientists recognize. This shows you how good Aristotle was at observing and studying animals!

Once Aristotle grouped the animals into two main groups, he then put each of them into subgroups. For example, he understood that fish and whales belonged to the same basic group of animals, the vertebrates. However, he studied these two kinds of animals well enough to know that there was something fundamentally different about them. For example, he studied fish well enough to know that they lay eggs, and their young hatch from those eggs. However, he had also traveled with sailors and had studied whales well enough to know that they didn't lay eggs. Instead, they gave birth to live young, just like cats, dogs, and people do. As a result, even though he put them in the same basic group (animals with blood), he put them into two different subgroups.

Even among the whales, he recognized two different kinds. He saw that some whales had teeth, while other whales had what looked to be feathers or bristles in their mouth. He used that to split them into two different groups, as do modern scientists. The ones with teeth are called **toothed whales**, and they use their teeth to cut and tear the flesh of the animals they eat. The others are called **baleen** (bay' leen) **whales**. They eat by opening their mouths and taking in a bunch of water. They then close their mouths and push the water out through those "hairs," which trap any tiny animals that were in the water. They then swallow those trapped animals. Isn't it amazing how well Aristotle understood animals, even though he lived more than 2,000 years ago?

The white whale (left) is a toothed whale, but the gray whale (right) is a baleen whale. Notice how its mouth looks like it has "feathers" or "hairs" in it.

LESSON REVIEW

Youngest students: Answer these questions:

1. What do we call it when scientists put living things into different groups?

2. What two basic groups of animals do modern scientists recognize?

Older students: Right after the classification system you made in your notebook, write how Aristotle split those same animals into two groups. Explain what was right about his two groups and what was wrong about them. Explain what modern scientists call these two groups of animals.

Oldest students: Do what the older students are doing. In addition, have you read about the famous whale called Moby Dick? If not, find out some information about him. Indicate in your notebook whether he was a toothed whale or a baleen whale.

Lesson 21: Aristotle and the Heavens

Aristotle wrote about the stars and planets as well. Interestingly enough, however, he didn't spend much time observing them. I guess you can only observe so much in one lifetime, and Aristotle was so busy studying light, animals, motion, etc., that he didn't get around to observing the planets and stars and how they changed in the night sky. Nevertheless, he obviously read what others said about these things, and that allowed him to make conjectures about them.

Not surprisingly, Aristotle got a lot of things wrong when it came to the planets and the stars. Nevertheless, his views about them are important, because they affected the views of many other scientists for about 2,000 years. To get an idea of his view, do the following activity.

A Model of the Universe

What you will need:

- A square or rectangle of cardboard (It needs to be larger than a paper plate.)
- Four paper plates
- A nail
- Scissors
- Crayons, colored pencils, or markers
- Some modeling clay, like Play-Doh
- An adult to help you

What you should do:

1. Draw some stars around the outer edge of one of the paper plates.
2. Cut another one of the paper plates down so that it is still a circle, but smaller than a paper plate.
3. Draw a sun at the very edge of the circle you just cut out (see the picture on the bottom left to see what you are trying to do – the sun is orange).
4. Cut another paper plate down so that it is still a circle, but smaller than the circle you cut in step 2.
5. Draw a small planet at the very edge of the circle you just cut out.
6. Cut the last paper plate down so that it is still a circle, but smaller than the circle you cut in step 4.
7. Draw the moon at the very edge of the circle you just cut out.
8. Have an adult help you push the nail through the center of the cardboard so the head is in contact with the cardboard.
9. Set the cardboard square down so the pointed end of the nail is sticking straight up, as shown in the picture on the top left.
10. Have an adult help you push the paper plate with stars on it down on the nail so the nail sticks through the center of the plate and the plate rests against the cardboard.

11. Have an adult help you push the circle with the sun down on the nail so the nail sticks through the center of the circle. Push it down on the nail, but don't push it so far down that it touches the paper plate that is already there. There should be some air between the plate and the circle.
12. Repeat step 11 with the circle that has the planet on it. Once again, there should be air between it and the circle below it.
13. Repeat step 12 with the circle that has the moon on it.
14. Make a small ball of clay and stick it on the end of the nail. This will represent the earth. In the end, your project should look like the picture on the bottom left of the previous page.
15. Spin the paper plate that has the stars on it. None of the other circles should move. Think about the ball of clay as the earth. What would a person on earth see as the paper plate spins? The person would see the stars changing position in the sky.
16. Spin the other circles one by one. Once again, think about what a person on earth would see as each circle spins.
17. Think about how fast each circle needs to spin in order to get back to where it started. For example, think about the circle that has the sun on it. On earth, we see the sun rise once per day, right? So the circle that has the sun would have to spin fast enough so that the sun would get back to where it started in exactly one day. However, think about the moon. We see its position in the sky change, but not nearly as quickly as the sun. In fact, it takes about a month for us to see the moon return to the position it started at. So think about how slowly the circle with the moon would have to spin compared to how quickly the circle with the sun would have to spin.
18. Keep your project, because you will use it in the next lesson. Clean up the rest of your mess, however.

The model you just built is very much like how Aristotle thought the stars, planets, and earth were arranged. The ball of clay you put on the nail represents the earth. Aristotle thought that the earth was so important that it must be at the center of everything. We call this a **geocentric** (jee' oh sen' trik) view, because "geo" usually refers to the earth. So a geocentric view puts the earth at the center.

Aristotle thought that the earth was surrounded by many, many hollow **spheres**. Do you know what a sphere is? It is a perfectly round object, like a ball. The sun, the moon, and each planet were fastened to a sphere, and each of those spheres spun at a different speed. So the sun moved in the sky, according to Aristotle, because the sphere it was fastened to spun around the earth. That sphere spun at a different speed than the sphere that held the moon, however. So even though the moon moved in the sky because it was attached to a sphere, it moved in the sky at a different speed because its sphere moved at a different speed. Each planet had its own sphere that

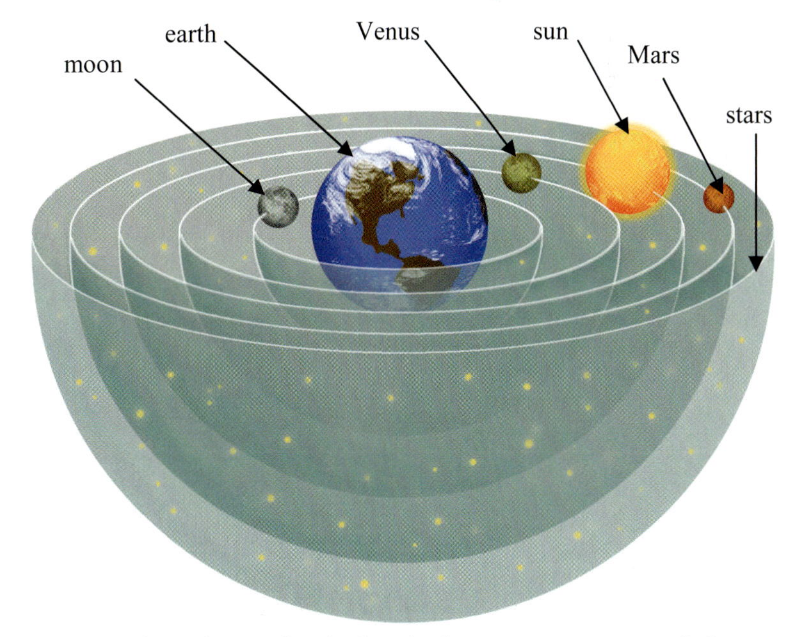

This drawing shows how Aristotle thought the sun, stars, moon, and planets were arranged. The drawing shows only half of each sphere to make the planets easier to see.

moved at its own speed, so that's why the planets moved in the night sky at different speeds. Aristotle knew that the planets moved differently from the stars in the night sky, so he thought all the stars were fastened to the very largest sphere at the very edge of the universe.

Do you see how your model shows this? Each circle that you drew represents one of the spheres. Since each circle could spin on its own, the sun could move at a different speed from the moon, which could move at a different speed from the planet, which could move at a different speed from the stars. As you played with the model, you should have seen that based on where Aristotle thought the sun was, its sphere had to rotate really quickly compared to the moon's sphere. After all, the moon takes almost a month to get back to where it starts, but the sun does its entire trip in a day. In Aristotle's view the spheres moved at incredibly different speeds.

Have you heard the Christian hymn, "This Is My Father's World?" It is one of my favorites. It starts out this way:

This is my Father's world,
And to my listening ears,
All nature sings, and round me rings
The music of the spheres.

Do you know what the line, "The music of the spheres" means? It refers to a very ancient concept that is best understood through Aristotle's view of the universe. Many ancient philosophers, like Pythagoras, thought that the universe was full of many spheres that each turned at a different speed. According to them, this turning produced beautiful music. This music wasn't necessarily music that you hear with your ears. Instead, it was a "symphony" of nature working beautifully.

Even though we now know that Aristotle was wrong about the stars, moon, sun, and planets being fastened to turning spheres, the concept of nature working so beautifully that it produces a sort of harmonius music is still around today. We also know that Aristotle was wrong about the order in which the stars, moon, sun and planets were arranged, and you'll learn about that in the next lesson.

LESSON REVIEW

Youngest students: Answer these questions:

1. What do we mean when we say the word "geocentric"?

2. Why did the spheres in Aristotle's view of the universe need to spin?

Older students: Draw a picture of Aristotle's view of the universe in your notebook, but make it a bit more detailed than your model. Include all the planets that were known to him at the time. In his view, the moon was closest to the earth, followed by Mercury, Venus, the sun, Mars, Jupiter, Saturn, and the stars. Explain why it is called a geocentric view.

Oldest students: Do what the older students are doing. In addition, answer the following question in your notebook: Suppose two different planets fastened to different spheres took exactly the same amount of time to return to their original position. In Aristotle's universe, would their spheres move at the same speed? You might want to play with your model more if you don't know

Lesson 22: Aristarchus (c. 310 BC – c. 230 BC)

Almost everyone who studies science learns about Aristotle. That's because his ideas were very important in their time and affected how scientists looked at the world for more than 2,000 years after his death. However, almost no one has heard of **Aristarchus** (air' ih star' kus), a scientist who was born about 12 years after Aristotle died. He did a lot of his scientific work between 280 BC and 240 BC, but unfortunately, most of it has been lost. However, we do know a little about him. Other people who lived around the same time (and after) have written about his work, and one of his books has survived to the present day.

Interestingly enough, even though he is not nearly as well-known as Aristotle, he got one thing right that Aristotle got completely wrong. As you learned in the previous lesson, Aristotle believed in a geocentric universe. He thought the earth was at the center of everything. Aristarchus spent a lot of time observing the sun, moon, planets, and stars, and he disagreed with Aristotle. He believed that the sun was at the center of everything. We call this a **heliocentric** (hee' lee oh sen' trik) view. The Greek god of the sun was called "Helios," so the term "helio" usually refers to the sun. This tells us that the word "heliocentric" means "sun-centered." Since we now know that Aristarchus was closer to the truth than Aristotle was, I want you to change your model so that it demonstrates what Aristarchus thought.

A Better Model of the Universe

What you will need:

- The model you made in the previous lesson
- A couple of paper plates
- Scissors
- Crayons, colored pencils, or markers
- An adult to help you

What you should do:

1. Pull the ball of clay off the end of the nail.
2. Remove the first three circles (the ones with the moon, a planet, and the sun drawn on it)
3. Use the scissors to cut out a circle on a paper plate that is as large as the one that has the sun drawn on it.
4. Draw the earth on the edge of this new circle.
5. Have an adult help you push the circle with the earth drawn on it down on the nail so the nail sticks through the center of the circle. Push it down on the nail, but don't push it so far down that it touches the paper plate that has the stars drawn on it.
6. Put the other two circles (the one with the planet and the one with the moon) back just like they were before.
7. Replace the ball of clay so that it is stuck to the end of the nail again.
8. The ball of clay now represents the sun, and the earth is now on the third circle.
9. Look at your model. This is closer to correct than Aristotle's model, but there are several things wrong with it. Can you tell your parent/teacher what those things are? Try. Don't worry if you get it wrong. Just look at the model, think about what you already know about the sun, planet, moon, and stars, and see if you can figure out what is wrong about your model so far. Your parent/teacher can look at the "helps and hints" book to see if you are right.
10. Clean up your mess, but keep your model (and the paper plate that has the sun on it), because you will use it again in Lesson 35.

As I told you in the activity, there are a lot of things still wrong with your model, but it is at least more correct than it was before. Rather than completely correcting your model (which would actually be too hard), I just want to talk about the corrections Aristarchus would probably make to it. First, Aristarchus realized that if the sun really was at the center of everything, it would not move in the sky the way we see it unless the earth rotated. You probably already know that the sun appears to move in the sky not because the sun is actually moving, but because the earth spins like a top. This changes what part of the earth faces the sun, which turns night into day. We call this **rotation**, and we know that it takes 24 hours for the earth to make one complete rotation. That's why a full cycle of night and day is 24 hours long. Your model is incorrect because the earth is drawn on the circle and can't rotate so that night will turn into day.

There is one other incorrect thing that Aristarchus would probably point out. We are not completely sure he knew this, since we don't have any of his writings on this subject. We are just going by what others wrote about him. However, most likely he realized that the moon could not travel around the sun the way it is shown in your model. Instead, it actually travels around the earth. So while the planets travel around the sun, the moon actually travels around the earth. In the end, then, his view of the universe looked more like the drawing you see below.

Now it's very important to understand that while this view is closer to correct than Aristotle's view, we know that it is still not right. For example, there are more planets than the ones drawn here. In addition, the planets aren't drawn correctly in terms of their size or their actual distance from the sun. However, at least their order is correct. In other words, we know that the planet Mercury is closest to the sun, Venus is the next closest, earth is the next closest, etc.

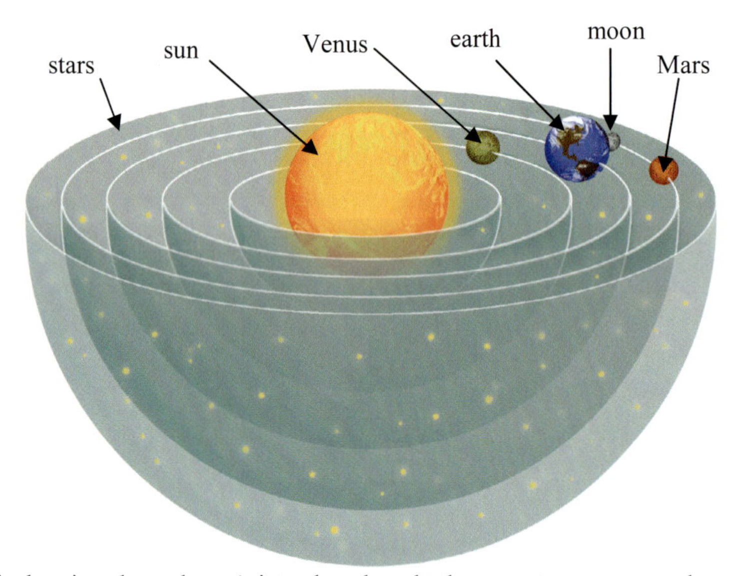

This drawing shows how Aristarchus thought the sun, stars, moon, and planets were arranged. It doesn't include all the planets that Aristarchus knew about, however. Also, it is still far from correct. It is just closer to the truth than Aristotle's view.

Another thing that is wrong with this picture is that the stars don't orbit the sun. Instead, their position in the night sky changes because the earth travels around the sun. The stars do move, but they don't move around the sun. Instead, most of them move away from the earth and away from each other. Finally, the spheres you see in the picture don't really exist.

Even though there are a lot of things wrong with Aristarchus's view, it is still closer to the actual arrangement of the sun, moon, earth, planets, and stars than Aristotle's view. Unfortunately, this more correct view was dismissed until 1543 when a natural philosopher named Copernicus argued strongly in favor of it. Why was it dismissed for so long? There were probably several reasons. However, one important reason was that Aristotle was considered to be a much better natural philosopher than Aristarchus. As a result, people were more willing to believe Aristotle's ideas. This is one of the problems you run into when you believe something just because someone who everyone admires believes it. Even really good, really smart scientists can be wrong. You should not believe in something just because someone else who everyone looks up to believes in it. You should believe something because there is a lot of good evidence to support it.

Now before I end this lesson, I want to give you the correct view of how the sun, planets, and moon are arranged. However, I want to make sure you understand the right terminology. The **universe** refers to everything God created. The sun, moon, planets, stars, and all other things we see in the night sky are part of the universe. However, the stars are very far away from the earth, and they don't travel around the sun like the earth and the planets do. So when scientists talk about how things are arranged, they typically talk about our **solar system**, which is the sun, everything that travels around the sun, and everything that travels around the planets (like the moon). A mostly correct view of the solar system is given in the drawing below:

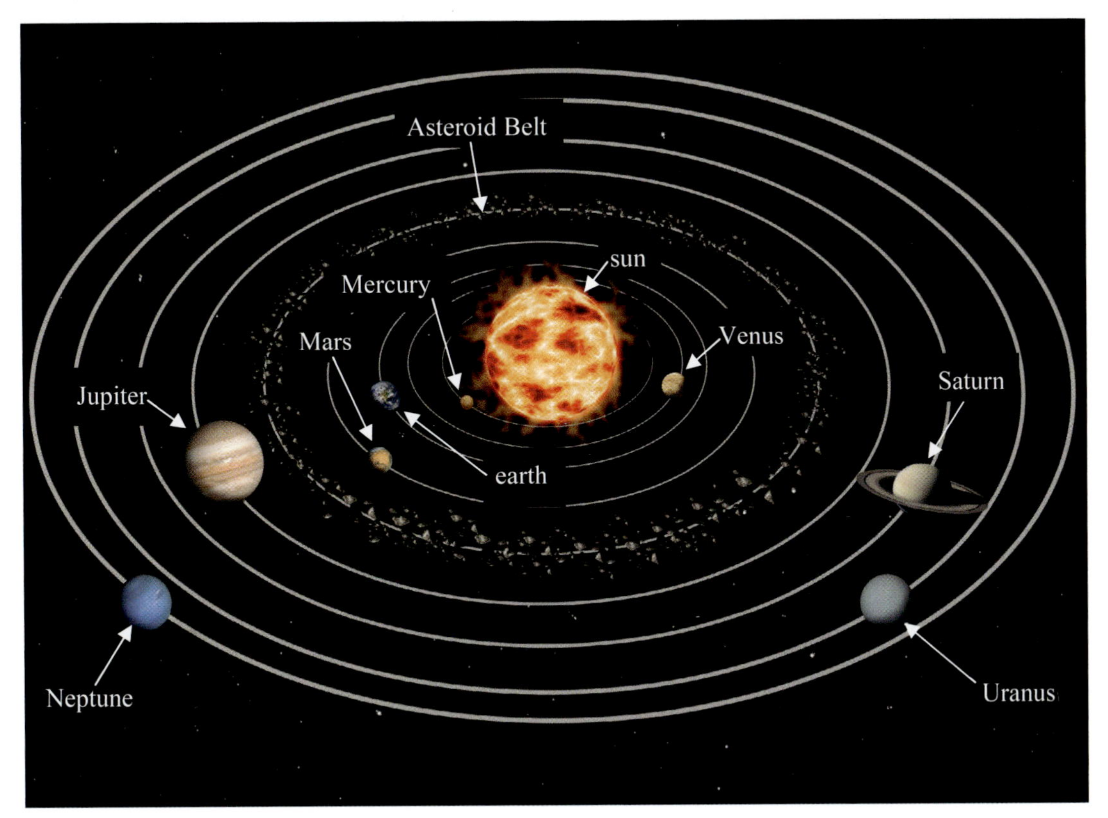

As you learn more about science, you will learn more about our solar system.

LESSON REVIEW

Youngest students: Answer these questions:

1. What do we mean when we say the word "heliocentric"?

2. Which is correct: a geocentric view or a heliocentric view?

Older and Oldest students: In your notebook, draw the universe as Aristarchus saw it. He thought the order of the planets known in his time was Mercury, Venus, earth, Mars, Jupiter, and Saturn. If you look at the drawing above, you can see that he was right about that! Explain why this is a heliocentric view of the universe. Also, note what is still wrong with it.

Lesson 23: Euclid (c. 300 BC) and the Law of Reflection

This is how 15th-century artist Justus van Gent imagined Euclid working on his geometry.

There are some things we don't know very well when it comes to ancient history, because a lot of information has been lost. You might be surprised to know that we don't really know when the person I want to discuss next actually lived. Based on what we do know, we think **Euclid** (yoo' klid) lived at about the same time as Aristarchus. However, all we know for sure is that he was alive around 300 BC. Even though we don't know exactly when he lived, we know that he was an incredibly influential mathematician. When you get to high school, you will learn a lot about geometry, and Euclid is considered the father of geometry. He didn't discover it, but he wrote a book that ended up being the main textbook for teaching geometry for about 2,000 years!

Remember, of course, that back then, science and math were really considered the same thing. So although Euclid is best known for his work in mathematics, he was also important in science, especially in the field of optics. Now, you should remember that Aristotle did some work in optics as well. Euclid could have learned something from Aristotle, because although Euclid did get a lot of things right, he still believed Plato's idea that we see because light comes from our eyes. Aristotle showed that Plato's idea was wrong. He figured out that we see things because light bounces off them and hits our eyes.

Even though Euclid was wrong about how we see things, he was right about a lot of other issues. For example, he gave a detailed explanation of the fact that when things are close to us, they appear larger than when they are far away from us. As a result, a small object can seem to be bigger than a large object if the small object is close to us and the large object is far away. However, what I want to concentrate on is that Euclid was the first to describe the **Law of Reflection**. To understand this law, however, you first have to understand what an **angle** is and how it is measured.

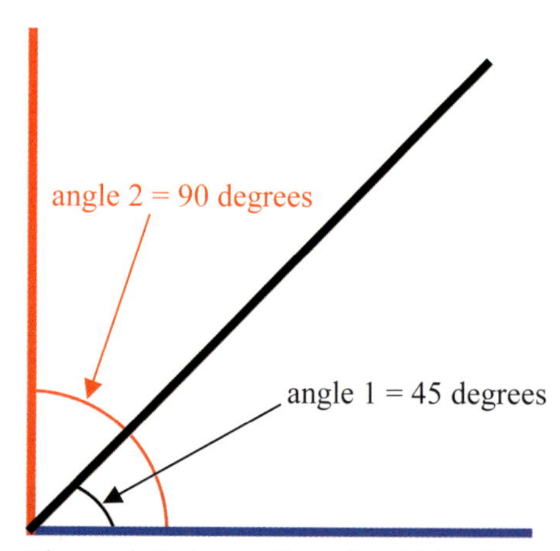

The angle between the red and blue lines is 90 degrees, while the angle between the black and blue lines is 45 degrees.

Look at the drawing on the right. The blue line is horizontal, and the red line is vertical. The black line is drawn so that it is directly in between the other two lines. Now notice that there is a space between the blue line and the black line. I have marked it with a black curve and labeled it "angle 1." That space is called an angle. Whenever two lines are tilted compared to each other, they will form an angle at the point where they touch. There is also an angle between the blue line and the red line. I have marked that with a red curve and labeled it "angle 2." Notice it is a lot larger than angle 1.

Since one angle is larger than the other, there must be a way to measure them. There is. We measure angles in **degrees**. This is not the same as the degrees we use in temperature. In geometry, a full circle is 360 degrees. One-fourth of a circle, then, is 90 degrees. Look at the red curve. What does it look like? It looks like one-fourth of a circle, doesn't it? So angle 2 is 90 degrees, because it represents one fourth of a circle. Angle 1 is only half as large as angle 2, so it is 45 degrees.

Now if you don't understand the degrees part, don't worry. The key is to understand that when two lines that are tilted compared to one another touch, they form an angle. The more tilted they are compared to each other, the larger the angle. What does this have to do with the Law of Reflection? Perform the following experiment to find out.

The Law of Reflection

What you will need:

- A flashlight
- A flat, hand-held mirror
- A thick piece of cardboard, like the side from a cardboard box
- A white piece of paper
- A black piece of construction paper
- Scissors
- Tape
- A dim room where you can work on the floor against a wall

What you should do:

1. Use the scissors to cut a circle out of the construction paper. The circle needs to be slightly larger than the face of the flashlight.
2. Cut a small slot in the bottom of the circle.
3. Tape the circle to the face of the flashlight so the only light that comes out of the flashlight will come out of the slot, as shown in the picture on the top right.

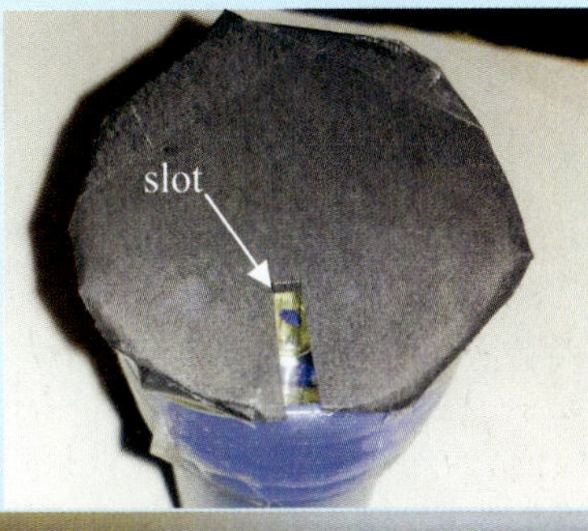

4. Put the mirror on the floor against the wall.
5. Slide the cardboard so it pushes the mirror up against the wall. The edge of the mirror should fall below the top of the cardboard.
6. Put the white paper on top of the cardboard so it also pushes against the mirror.
7. Turn the flashlight on and lay it on the paper so a beam of light can be seen on the paper.
8. Adjust the flashlight so the beam of light hits the mirror and reflects off the mirror, making another beam of light that comes off the mirror. Your experiment should look like the picture on the bottom right.

9. Look at the angle between the beam of light coming from the flashlight and the mirror. Now look at the angle between the mirror and the beam of light reflecting off the mirror. How do they compare?
10. Change the aim of the flashlight so that the angle between the beam of light coming from the flashlight and the mirror is different from what it was before. Now look at the beam of light reflecting off the mirror. How has the angle there changed? Once again, how do the two angles compare?

11. Explore more angles, constantly comparing the angle between the line of light coming from the flashlight and the mirror to the angle between the mirror and the line of light reflecting off the mirror.
12. Clean up your mess and put everything away.

What did you see in the experiment? You should have seen that the angle between the mirror and the light coming from the flashlight was always the same as the angle between the mirror and the light reflecting off the mirror. That's the **Law of Reflection**, which Euclid discovered.

When light reflects from a smooth surface, the angle of reflection will be equal to the angle at which the light hit the surface.

Now while you studied reflection using an experiment, Euclid actually studied it mathematically. That's what allowed him to make a precise conclusion. After all, you didn't actually measure the angles in your experiment. You could have, but that would have required you to learn how to use a protractor. Euclid demonstrated mathematically that the Law of Reflection is always true, and the angles are always equal.

LESSON REVIEW

Youngest students: Answer these questions:

1. What does the Law of Reflection tell us?

2. If a beam of light hits a mirror at an angle of 35 degrees, what will be the angle between the mirror and the reflected beam of light?

Older students: In your notebook, draw your experiment, including the mirror and the beams of light. Use curves to identify the angles that you were looking at in the experiment, and explain what the Law of Reflection says about them.

Oldest students: If you look at paper through a microscope, you will see that it is bumpy and rough. Would you expect the Law of Reflection to work if you reflected light off white paper? Why or why not?

Lesson 24: Archimedes (c. 287 BC – c. 212 BC)

This is part of a painting by Domenico Fetti. It is how he imagines Archimedes deep in thought.

The next scientist I want to discuss lived at the same time as Aristarchus. He was born a bit later (in about 287 BC) and died a bit later (in about 212 BC), but he did a lot of his work while Aristarchus did his. His name was **Archimedes** (ar' kuh me' deez), and like Aristotle, he was one of the greatest natural philosophers who lived before the birth of Christ. Like all natural philosophers of his day, he worked with math. He made several interesting contributions to our understanding of math, but in the next few lessons, I want to concentrate on what he taught us about science.

Probably the thing that Archimedes is most famous for is telling us about how liquids affect objects that are put in them. To see what Archimedes discovered, perform the following experiment.

The Buoyant Force

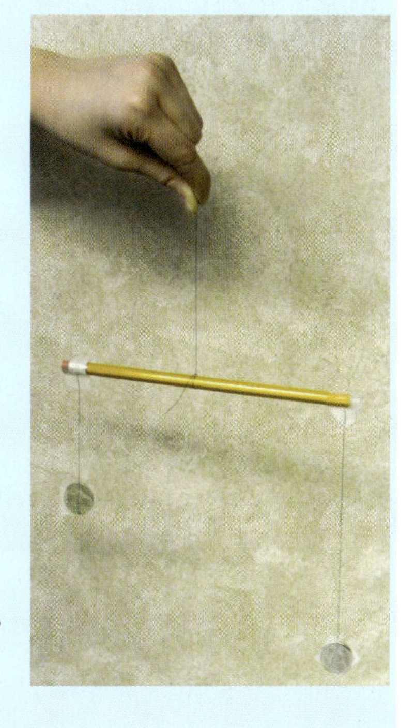

What you will need:

- Thread
- A pencil
- Two nickels (Two of any coin will work.)
- Tape
- Someone to help you
- A sink that can be filled with water

What you should do:

1. Plug the sink and fill it with water. As long as you keep checking the sink so that it doesn't overflow, you can do most of the other steps while the sink is filling.
2. Cut three lengths of thread. Two of them should be about 18 centimeters (7 inches) long. The third should be about 10 centimeters (4 inches) long.
3. Tie one of the 18-centimeter-long threads to the middle of the pencil.
4. Tape a nickel onto the end of the remaining two threads.
5. Tape the 10-centimeter-long thread to one end of the pencil so that its nickel dangles at the bottom of the thread.
6. Tape the 18-centimeter-long thread to the other side of the pencil so that its nickel dangles at the bottom of the thread.

7. Your contraption should look something like the photograph on the previous page.
8. Have your helper hold the thread that is attached to the middle of the pencil (like the hand is holding it in the picture on the previous page) while you hold the pencil to keep it from tilting wildly.
9. Adjust the position of the thread so that you can let go of the pencil, and the pencil stays balanced. It doesn't have to exactly balance; it can be tilted a bit to one side or the other. The key is that it should be able to hang steadily when it is held only by the middle thread.
10. When you find the position where the pencil balances or almost balances, tape the middle thread in place so it doesn't move. You should now have a contraption that looks just like the picture, and it should stay stable when held by just the middle thread.
11. Carry your contraption over to the sink and hold it over the water.
12. Slowly lower the contraption so that the nickel on the longer thread sinks into the water. What happens to the pencil?
13. Continue to lower the contraption until both nickels are fully underwater, but neither of them is touching the bottom or edges of the sink. What happens to the pencil?
14. Slowly raise the contraption again so that the first nickel comes out of the water, but the second nickel is still underwater. What happens to the pencil?
15. Pull the contraption up so that both nickels are out of the water. What happens to the pencil?
16. Use your finger to touch the bottom of the nickel on the longer thread. Gently push up on the nickel. What happens to the pencil?
17. Drain the sink, put everything away, and clean up your mess.

What happened in your experiment? When the nickel that was on the longer thread went underwater, the pencil tilted towards the other nickel, didn't it? When both nickels were underwater, the pencil went back to the way it was before. Then, as you raised it out, it once again tilted towards the nickel that came out of the water first. When both nickels came out of the water, it once again returned to the way it was before.

How do we explain this? Well, think about what happened in the last part of the experiment, when you pushed up on the nickel with your finger. What happened to the pencil? The same thing as when that nickel went underwater. It tilted towards the other nickel. What does this tell us? It tells us that the water was acting just like your finger. It was *pushing up* on the nickel, causing the pencil to tilt toward the other nickel. Once both nickels were underwater, the water was pushing up on both of them. That caused the pencil to return to the way it was before.

So when an object is in water (or any liquid), the water actually pushes up on that object. We call this the **buoyant** (boy' uhnt) **force** that water exerts, and Archimedes was the first one to figure it out. His observations of objects in water (and even himself while he was in the bathtub) helped him to realize that water pushes up on anything that is in it, and that push is now called the water's buoyant force. Archimedes did more than that, however. He actually determined how much force the water pushes with. He figured out that the water pushes up with a force equal to the weight of the water that the object had to move in order to sink. This is now called **Archimedes's principle**.

When something goes underwater, it has to push the water molecules out of the way so that it can go where the water was. The more scientific way to say this is that the object **displaces** the water. The weight of the water the object displaces is equal to the force with which the water pushes up. But how much water does an object displace? Well, the object has to take up a certain amount of volume, which is determined by the size and shape of the object. The bigger the size, the more volume it has. To go underwater, then, it has to displace an equal volume of water. When the nickels sunk in the water in your experiment, then, they each displaced a nickel-sized amount of water. Whatever a nickel-sized amount of water weighs, that's how much the water pushed up on each nickel.

Do you remember why things sink or float? It all depends on how much they weigh compared to how much an equal volume of water weighs. If an object weighs more than what an equal volume of water weighs, the object sinks. If it weighs less, the object floats. You can see why this is the case when you think of Archimedes's principle. If water can push up on an object with more force than what the object weighs, then the water will "hold up" the object, and the object will float. If the object weighs more than the force with which water pushes up on it, then the water won't be able to hold up the object, and it will sink.

But Archimedes's principle tells us more than just that. Have you ever been playing with something in a pool and noticed that it is easier to lift when it is underwater than when it is above water? That's because of Archimedes's principle. When something is underwater, the water is pushing up on it with a buoyant force equal to the weight of the water displaced. In other words, it is helping to lift the object. This makes it easier for you to lift the object. Once you pull it out of the water, it will be much harder to lift, because the buoyant force is no longer helping you lift it. To give you an idea of how big an effect this can be, consider the largest fish on the planet – the whale shark. Out of water, a whale shark weighs about 20 tons! However, it displaces an almost equal weight of water in the ocean. As a result, even though it is a very heavy creature, you could probably lift it in the water, if it didn't swim away from you!

This is a photo of a whale shark and a scuba diver. You get some idea of how huge the whale shark is by comparing it to the diver.

LESSON REVIEW

Youngest students: Answer these questions:

1. What is Archimedes's principle?

2. How much water does an object displace when it goes underwater?

Older students: In your notebook, draw a picture of your experiment contraption before you sank one of the nickels in the water. Draw another picture of your experiment contraption when only one of the nickels was underwater. Use Archimedes's principle to explain why the pencil tilted like it did.

Oldest students: Do what the older students are doing. Also, write the following question and your answer in your notebook: "A 150-pound object is put underwater. It displaces 145 pounds of water in order to sink. If you try to pick it up while it is underwater, how much will it feel like it weighs?" Check your answer and correct it if you are wrong.

Lesson 25: Archimedes and the Lever

In the previous lesson, you balanced two nickels on a pencil that was hung from a thread. The thread was attached pretty much to the center of the pencil, and if you played with where the thread was tied for a while, you could get the pencil to hang straight, because each nickel balanced the other out. This idea was well known in the ancient world, and it was used to make a **balance**, which could weigh things very accurately. An example of a balance is shown in the picture on the right. Look at how it is made. The bar rests on a tiny point called the **fulcrum** (ful' krum). The fulcrum doesn't move, but the bar can tip one way or the other, depending on the weights that are hung on its ends. If the weights are equal, the bar is horizontal, as shown in the picture. If the bar tilts, it will tilt toward the side that has the heavier weight.

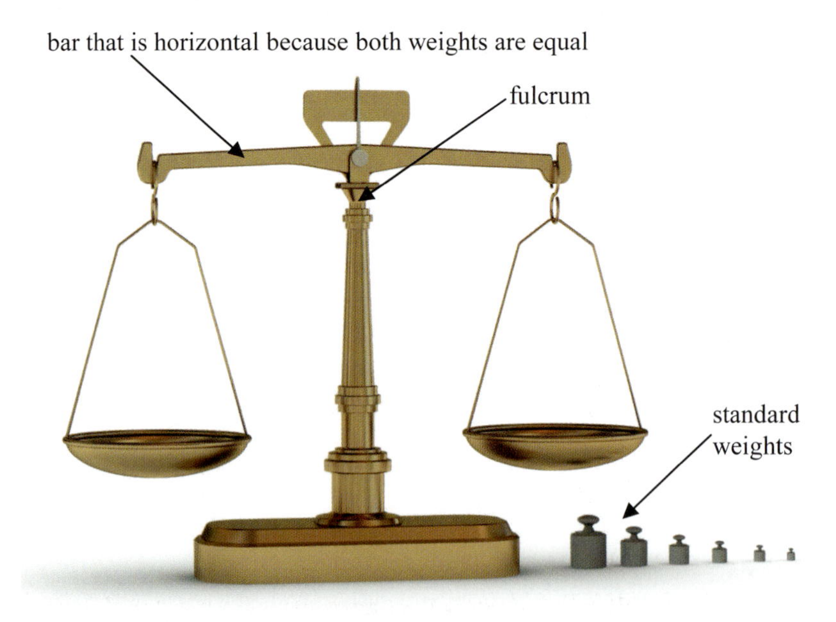

A balance weighs things using the idea that the bar from which the weights are being hung will be perfectly horizontal when the weights are the same.

The little gray things on the right side of the picture are **standard weights**. This means their weight is known. So if you have something you need to weigh, you put it on one side of the balance. The balance will tilt to that side, because there is nothing on the other side of the balance. However, as you start adding standard weights to the other side of the balance, the bar will start to hang more horizontally. When the bar is perfectly horizontal, you know that the weights on each side are equal. If you add up all the standard weights you had to use to make the bar perfectly horizontal, you now know the weight of whatever you put on the other side.

This was well known long before Archimedes; however, Archimedes did lots of work trying to understand how you could get the bar to rest horizontally even if the weights were not the same. He found that you could do this easily as long as you could move where the fulcrum was on the bar. He found that as long as you moved the fulcrum towards the heavy object, you could get the bar to balance, as shown in the illustration below:

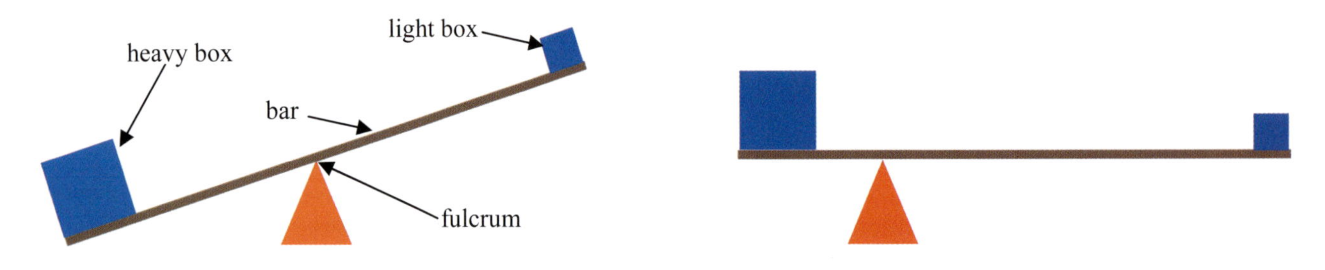

Even though the box on the left is much heavier, you can get the bar to rest horizontally as long as you move the fulcrum closer to the heavy box.

In the end, then, Archimedes determined that the position of the fulcrum is absolutely critical in a situation like this. If you want to make a balance, the fulcrum needs to be right in the middle of the bar. However, if you wanted to balance unequal weights, you could just move the fulcrum so that it was nearer to the heavier of the two weights. You probably knew this already, but you might not realize what this really means. Think about it. When you have a balance, whatever heavier weight you have moves down, tilting the balance's bar. There are two ways you can fix that. You can either add more weight to the other side so that it lifts the heavier weight, or you can move the fulcrum. If you move the fulcrum, *a lighter weight ends up moving a heavier weight.* Why is this important? Perform the following experiment to see.

A Lever

What you will need:

- A shovel, rake, or other tool that has a long, strong, wooden handle
- Some old hardcover books
- A heavy couch or other piece of furniture that has a small gap between its bottom and the floor
- A ruler
- Someone to help you

What you should do:

1. Grab the bottom of the couch on one side and try to lift it up a little. *Don't hurt yourself!* Just try to tilt the side of the couch you are grabbing up a bit. Can you do it? If you can do it, how hard is it to do?
2. Pile the books so that they are right next to where you grabbed the couch and are just a bit higher than the gap between the couch's bottom and the floor.

3. Put the top of the shovel's handle under the couch where you grabbed it before, but between the couch and the pile of books, as shown in the picture on the left.
4. Put one of your feet on top of the books to hold them in place.
5. Grab the shovel's handle near the other end and push down, remembering to hold the books in place with your foot. What happens to the couch? How easy is it to tilt the end of the couch using the shovel?
6. Try using just one finger to push on the shovel. Can you lift the couch using only one finger on the shovel? If not, can you lift it with two fingers on the shovel?
7. Have your helper measure how far the end of the couch has been lifted off the floor.
8. Have your helper measure the distance between the floor and the hand you are using to push on the shovel.
9. Slowly reduce the force you are pushing the shovel with so that the couch comes to rest on the floor again.
10. Once the couch has come to rest on the floor, have your helper measure the distance between the floor and the hand you are using to push on the shovel.
11. Put everything away.

What did you find out in your experiment? The couch was much easier to lift with the shovel, wasn't it? While you could push on the end of the shovel with one or two fingers and lift the couch, there was no way you could lift the couch by grabbing it with just one or two fingers. Why? Think about what you did with the shovel. The books were a fulcrum. The shovel handle was a bar. Since the fulcrum was put very near the couch (which was heavy), you could use a small weight (the force you used to push down on the shovel) to lift a large weight!

What you actually made in the experiment is called a **lever**. The kind of lever you made in the experiment is used to make things much easier to lift or move. Because the fulcrum is placed very close to the heavy weight, a small force on the opposite end of the lever will lift a large weight. Now that's really great, but the ease with which you can lift things with a lever comes at a cost. How far did you lift the couch? How far did you have to push the shovel? You can determine that by subtracting the two measurements of the distance between the floor and your hand. You should see that you had to push the shovel much farther than the couch lifted up. That's the "cost" of using the lever. You can lift the couch with a smaller force, but you have to apply that force over a longer distance.

Archimedes realized that a lever could be used to lift incredibly heavy objects, provided there was a place to stand, a large enough bar, and the fulcrum could be placed close enough to the object.

Archimedes figured all this out more than 200 years before Christ! Using his studies of a balance and applying some mathematical reasoning, he realized that as long as you could get a bar that was long enough and place the fulcrum close enough, you could use a tiny, tiny force to lift an incredibly heavy object. In fact, he is reported to have said, "Give me a place to stand on, and I will move the earth." Even though he was overstating his case, he made an important point. A lever can be used to lift incredibly heavy objects, as long as the fulcrum is positioned properly!

LESSON REVIEW

Youngest students: Answer these questions:

1. What two things do you need to make a lever?

2. If you want to lift something heavy with a lever, where should the fulcrum go?

Older students: In your notebook, draw a picture of a lever. Label the fulcrum and the bar. Also, indicate where the fulcrum needs to go in order to lift something heavy. In addition, explain the relationship between the distance you need to push the lever and the distance the heavy object moves.

Oldest students: Do what the older students are doing. Also, suppose you were to measure the energy that a person uses lifting a heavy object with his hands and lifting it with a lever. Would the energy be the same in both cases, or would it be different? Explain what you think in your notebook. Check your answer and correct it if it is wrong.

Lesson 26: Archimedes and Pulleys

Note that this lesson and the following one are challenge lessons. If you aren't doing challenge lessons, you should skip ahead to Lesson 28.

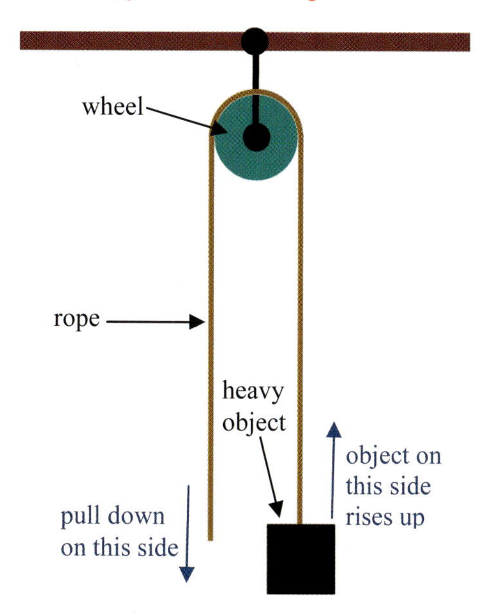

This simple pulley changes the direction of the force you use, allowing you to pull down in order to lift the object up.

Archimedes must have spent a lot of time trying to think about how to move heavy objects. Not only did he explain how a lever works, he also invented a system that allowed sailors to lift heavy objects high in the air. To understand this system, you first have to understand what a **pulley** is. Look at the picture on the left. A wheel is attached to a ceiling so that it can spin. A rope is thrown over the wheel, and one end of the rope is attached to something heavy. When you pull down on the other end of the rope, the heavy object rises. This is what's called a **simple pulley**. All it does is change the direction of the pull that is used. A person pulls *down* on the rope, but the object moves *up*.

Now a simple pulley doesn't really make it much easier to raise the object. Some people find it easier to pull down on a rope than to grab onto an object and lift it up. However, when you determine the amount of force you have to use, it is the same whether you lift the object up directly or you use a simple pulley. This was all known before Archimedes. What Archimedes figured out was how to actually use less force to lift the object. To understand what he figured out, let's do an experiment.

A Block and Tackle

What you will need:

- Thread
- At least six quarters and at least two pennies
- Super glue
- Several hardcover books
- A plastic pen (It should be made of slick plastic.)
- Tape
- A ruler
- An adult to help you

What you should do:

1. Have the adult help you use the super glue to glue the penny to the center of one quarter.
2. Have the adult help you use the super glue to glue another quarter on the other side of the penny. In the end, you should have a penny sandwiched in the middle of two quarters, as shown in the picture on the right.
3. Make two stacks of books that are each roughly 18 centimeters (7 inches) high. Stack them close enough so that you can lay the pen across them but there is still a good-sized gap in between.
4. Lay the pen across the stacks of books.

5. Put another book on top of each stack so that the pen is held in place. If you are having a hard time visualizing what you are doing, look at the pictures on this page.
6. Cut a length of thread that is about 25 centimeters (10 inches).
7. Tape the "penny sandwich" you made to one end of the thread.

8. Tape two quarters and one penny together. It doesn't matter how they are arranged. They just need to be securely taped together.
9. Tape the two quarters and one penny to the other end of the thread.
10. Drape the thread over the pen so that one set of coins hangs on one side of the pen, and the other hangs on the other side of the pen, as shown in the picture on the right.
11. Notice that none of the coins move up or down, because each side of the thread weighs the same. That means each side is being pulled down with equal force, so nothing happens.
12. Take the thread off the pen.
13. Remove two quarters from the side that has three coins taped together. Now you have the "penny sandwich" on one side and just a penny on the other. Can you guess what will happen when you drape the thread over the pen again?
14. Drape the thread over the pen again and see what happens. In this case, the thread and pen are acting like a pulley. The pen doesn't spin like the wheel of a pulley, so it isn't as good as a real pulley. However, it works pretty much like a pulley. The "penny sandwich" was heavier than the penny, so its weight pulled down on one side of the pen, making the penny on the other side of the pen rise. This happened because, unlike in step 11, the two sides didn't have equal weight. As a result, the heavier weight (the "penny sandwich") lifted the lighter weight (the penny by itself).

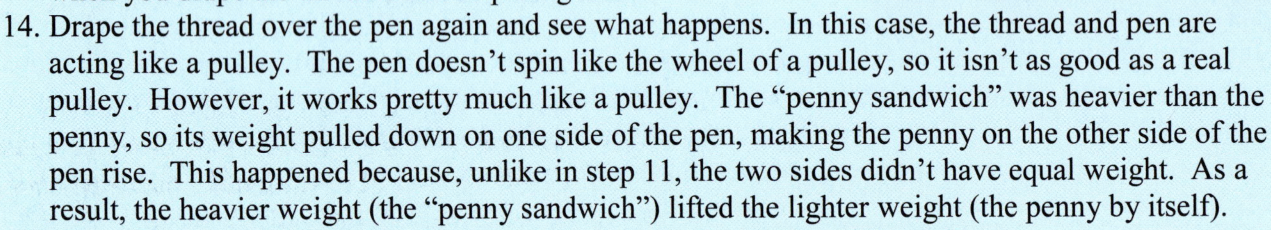

15. Now we are going to change the system a bit. Remove the thread from the pen.
16. Cut another length of thread that is about 43 centimeters (17 inches) long.
17. Tie one end of the thread directly to the pen.
18. Tape two quarters and one penny together and then tape them all to the other end of the thread.
19. Drape the end of the thread over the pen, and hold the two quarters and penny so there is a loop of thread hanging down from the pen.
20. Have the adult put the "penny sandwich" in the loop of thread so that the thread is between the two quarters and rests on the penny. See the photo on the right.
21. Raise or lower the two quarters and penny you are holding. The "penny sandwich" should spin freely, rising or falling, depending on how you are moving your hand.
22. Release the two quarters and penny. What happens?
23. Remove the penny that is taped to the two quarters and set the system up so it is like you had it before, but now you are holding only two quarters.
24. Release the two quarters. What happens?
25. Clean up your mess and put everything away. If you want to separate the coins in the "penny sandwich," soak it in nail polish remover overnight.

What did you see in the experiment? If things worked well, when you used the pen like a simple pulley, the coins did not move when there was equal weight on both sides of the pen. Since both sides of the pen had equal forces pulling on them, nothing happened. However, what happened when you set the "penny sandwich" in the thread so that it rolled as the thread moved? Suddenly, it rose, even though the weight pulling down on the other end was the same as before. Even when you removed the penny, the "penny sandwich" should still have risen, despite the fact that the weight pulling down was *lighter* than the penny sandwich.

This block and tackle has three pulleys, which makes the force with which you pull three times greater.

The system that allowed the "penny sandwich" to move is called a **block and tackle** system. It contains one simple pulley (the pen) and one or more movable pulleys (the "penny sandwich"). When you add a movable pulley to a simple pulley, the moveable pulley makes it easier to lift the load. If you added more pulleys to the system, the load would be even easier to lift. Look at the picture on the left, for example. There are three pulleys in the picture. Because of the way they are arranged, you have three ropes pulling up on the pulley at the bottom of the picture. As a result, any force used to pull the rope on the left is magnified three times. That's why a block and tackle makes it easy to lift heavy things. For each added pulley, there is an added segment of rope that pulls up with the force that is being used. A five-pulley block and tackle system, then, will make the force being used five times as strong! Archimedes figured this out, and as a result, he made the lives of many sailors (and eventually other laborers) much easier.

LESSON REVIEW

Youngest students: Answer these questions:

1. What does a simple pulley do?

2. What does a block and tackle do?

Older students: In your notebook, draw a picture of a simple pulley, and then next to it, draw a block and tackle with three pulleys, like the one pictured above. Explain what the simple pulley does, and then explain what the block and tackle does. Explain the significance of the number of pulleys.

Oldest students: Do what the older students are doing. Also, suppose you were to measure how much rope you pull when you use a block and tackle. If you compared it to the distance that the object rises, would you expect it to be longer, shorter, or the same? Explain your answer. Check your answer and correct it if it is wrong.

Lesson 27: Archimedes and Pi

As I told you in another lesson, mathematics and science are very closely related to one another. Indeed, many ancient Greek philosophers such as Archimedes really didn't see much difference between them. As a result, Archimedes made a lot of contributions to our current understanding of mathematics. I want to talk about one of those contributions, because it is very important to many calculations that we do in science today.

Think about circles. They seem to be everywhere. Cars, trucks, bicycles, and most other kinds of vehicles run on circles that we call wheels. Coins are essentially circles of metal. Compact discs (CDs) are circles of plastic, and even the plates you eat food from are circles. One thing you can immediately say is that circles come in all sizes. Coins are small circles of metal, while the wheels on an 18-wheel truck are huge, and there are all sorts of circles in between.

How do you measure the size of a circle? Well, there are actually two ways. You can measure the length of a line that goes from one side of the circle to the other, passing through the very center of the circle. That's called the circle's **diameter** (dye 'am uh tur). The other way you could measure a circle is to measure the distance all the way around the edge of the circle. That's called the circle's **circumference** (sur kum' fuh rents).

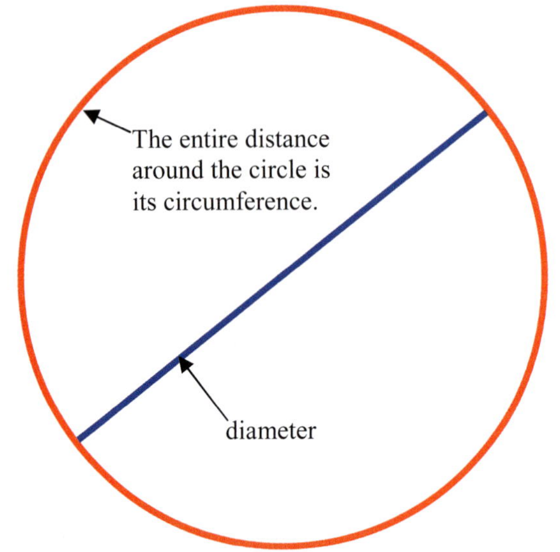

If you measure the distance around the red circle, you have its circumference. If you measure the distance of a line that goes from one side of the circle to the other while passing through the center of the circle (the blue line), you have its diameter.

Now if you think about it, these two things have to be related to one another. If I draw a circle with a big diameter, it must have a big circumference. If I draw a circle with a small diameter, it must have a small circumference. But how, exactly, are they related? That's what Archimedes figured out! To give you an idea of what I am talking about, do the following experiment.

A Circle's Diameter and Circumference

What you will need:

- A wheel (It can be from anything that rolls - a bicycle, wagon, wheelchair, etc. It works best if you can remove it from what is attached to so that there isn't a significant bump in the center.)
- String
- Tape
- Scissors
- Someone to help you

What you should do:

1. Stretch the string so that it goes from one end of the wheel to the other, passing over the very center of the wheel. Remember, the wheel was attached to the bicycle, wagon, wheelchair, etc. at the center, so the string needs to pass over where the wheel was attached to its vehicle.

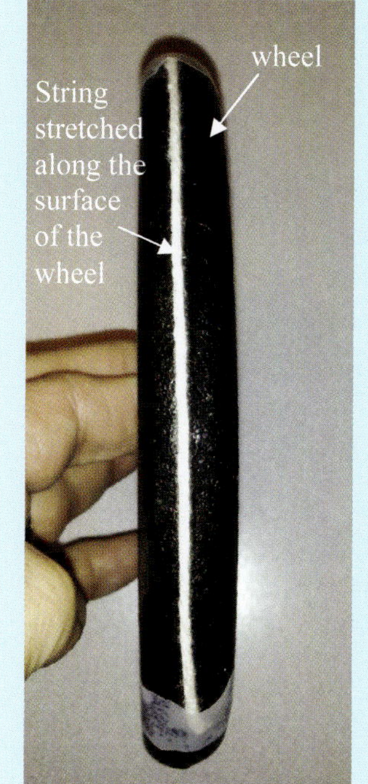

2. While you are holding the string, have your helper cut the string so that it is just long enough to stretch from one end of the wheel to the other while passing over the center. You now have a string that is as long as the diameter of the circle.
3. Use that string to measure out and cut two more strings of the same length. You should now have three strings, each of which is as long as the diameter of the wheel.
4. Tape one end of the string to the surface of the wheel. The surface of the wheel is the part that touches the ground.
5. Pull the string so that it follows the surface of the wheel and is tight. Tape the other end of the string down so it is firmly attached to the wheel. In the end, the string should be stretched over the surface of the wheel, as is shown by the white string taped to the surface of the black wheel in the picture on the left.
6. Hold the end of the second string so that it touches one taped-down end of the string that is already on the wheel.
7. Tape the end of the second string down so that it is just a continuation of the first string that you taped down.
8. Once again, pull the string so that it follows the surface of the wheel and is tight. Tape the other end of the string down so it is firmly attached to the wheel. You now have two strings taped to the surface of the wheel.
9. Repeat steps 7 and 8 with the third string. You now have all three strings taped to the surface of the wheel so that they are running around the circumference of the wheel.
10. Are the three strings able to run around the entire surface of the wheel? They aren't, are they? Instead, there is still a little bit of the wheel's surface that is not covered with string. If you used a fourth string of the same length, it would be way too much string, wouldn't it? So three strings were not quite enough to go all the way around the surface of the wheel, but four strings would go a lot farther than once around the surface of the wheel.
11. Clean up your mess and put everything away.

In your experiment, you tried to determine the relationship between a circle's diameter and its circumference. The strings were each the length of the wheel's diameter, but three of them were not quite able to make it all the way around the circumference of the wheel. If you repeated this experiment with other wheels, you would always find the same thing: three diameter-length strings would not make it all the way around the wheel, and four diameter-length strings would make it all the way around with a lot of string to spare.

What does this tell us, then? It tells us that the distance around a circle (its circumference) is slightly more than three times the length of its diameter but less than four times the length of the diameter. That tells us something about the relationship between a circle's diameter and its circumference, but it's not very exact. Can we be any more exact than that? Yes we can, and Archimedes was the first to demonstrate how to do that. He didn't use string and tape, however. In fact, he didn't use an experiment to figure out how many diameters it would take to go all the way around a circle. He used *only mathematics*.

The math he used is pretty complicated, so I won't go into it. Once you have taken high school geometry, however, you will be able to understand what he did, so you can always find out his reasoning then. What his mathematics told him was that the number of diameters it would take to go all the way around a circle is somewhere between 3.143 and 3.141. It turns out that the actual value is about 3.142! Not bad for someone who was doing math about 200 years before Christ was born!

If you don't know already, the relationship between a circle's diameter and its circumference is given a special name. It is called **pi**, and it is represented by the Greek letter "π." It tells you exactly how many times a circle's diameter will fit around its circumference. So had you been able to cut a string that was 0.142 times as long as the other three strings, you would have been able to cover that last part of the wheel's surface that didn't have any string on it.

Who cares? Why is this important? Well, the value of π turns up all over the place in science. Lots of scientific formulas use π, because there are lots of circles in nature. For example, suppose you want to know the distance from the surface of the earth to its center. Since we can't drill through the earth, how would you learn it? Well, you can travel along the earth's equator. If you measure that distance, you find that it is 40,075 kilometers (24,901 miles). If you divide that number by π, you get the diameter of the earth, which is 12,757 kilometers (7,926 miles). Divide it by 2, and you will get the distance to the center of the earth: 6,379 kilometers (3,963 miles). Now it turns out that because of things like mountains, valleys, and other effects, the distance from the surface of the earth to the center is different at different places, but 6,379 kilometers (3,963 miles) is a pretty good average.

Since π is found in many scientific formulas, it is good to know what it means and what its value is. I hope the experiment you performed helps you to visualize exactly what this important number really means.

LESSON REVIEW

Youngest students: Answer these questions:

1. What do we call the length of a line that goes from one end of a circle to the other while passing through its center?

2. What does the number called pi tell you?

Older students: Draw a circle in your notebook. Now draw and label the diameter of that circle. Below your drawing, indicate what the circumference of the circle is. Finally, explain what the number pi is and indicate its value.

Oldest students: Do what the older students are doing. In addition, pi is called an **irrational number**. That's not because it's hard to reason with pi. Instead, it's because the number actually has no end of digits. Do some research and find out the value of pi with at least six digits in it.

Lesson 28: Eratosthenes (c. 276 BC – c. 195 BC)

Archimedes died in about 212 BC. The next important natural philosopher, **Eratosthenes** (air' uh tas' thuh neez'), was born a bit later than Archimedes, so he managed to live until 192 BC. This means that while Archimedes was a philosopher in the third century BC, Eratosthenes did some of his work in the second century BC. If you find it odd that a date like 212 BC is called "third century BC.," just remember that a century is 100 years. So the third century BC starts in 300 BC and counts down to 201 BC. The second century BC starts in 200 BC and counts down to 101 BC, and the first century BC starts with 100 BC and counts down to 1 BC.

You just finished a lesson on circles, so you understand what a circle is. However, think about a baseball or tennis ball. Would you call them circles? Not really. A circle is flat – it's round, but like a wheel, it's flat. A baseball is not flat. It's round, but it's a ball. In other words, it's a sphere. Now a sphere is like a circle. It has a circumference (the distance around the sphere) and a diameter (the length of a line drawn from one end of the sphere to the other and passing through its center), but it is not exactly like a circle, because it is not flat.

It turns out that the earth is much like a baseball or tennis ball. It is mostly spherical. Thus, the earth is really just a big ball. Have you ever heard that scientists used to believe the earth was flat? That's really not true. I am sure some really bad scientists throughout history thought the earth was flat, but most scientists have not. In fact, even in the third century BC, scientists knew that the earth is a sphere. In fact, the most important thing Eratosthenes is known for is measuring the circumference of the earth, and to do that, he had to know that the earth is a sphere. Perform the following experiment to see how he did it.

Shadows on a Sphere

What you will need:

- A baseball or tennis ball (It can be some other ball, as long as it is about the same size as a tennis ball and has a thick outer coating into which pins can be stuck without harming the ball.)
- Two straight pins
- A flashlight
- A room that can be made dark when necessary

What you should do:

1. Set the ball on a stable surface in the room. It needs to stay still, so if it is rolling around, use some small books or other objects to hold it in place.

2. Push a pin into the ball so that it sticks straight up from the ball.
3. Push the other pin into the ball close to where the first pin is. In the end, your experiment should look something like the picture on the left. Notice that the two pins are tilted compared to one another. That's because the ball is curved.
4. Make the room dark and turn on the flashlight.
5. Hold the flashlight as high as you can above the ball. The light should shine down so that both pins cast a shadow. If you can't hold it high enough for both pins to cast a shadow, find someone who is a bit taller to help you.

6. Move the flashlight around a bit to see how the pins' shadows change.
7. Now you want to move the flashlight so that it is straight above the pin on the left. Do that by moving the flashlight around and playing with its tilt until the pin on the left casts no real shadow. The head of the pin might cast a shadow, but if it does, it should be a tiny circle centered on the pin itself. When the light is shining straight down on the ball but there is basically no shadow coming from the pin on the left, the flashlight is shining down directly above the pin.
8. Look at the shadow cast by the other pin. Get a good idea of how long it is.
9. Turn on the lights in the room.
10. Move the pin on the right so it is stuck in the ball again, but it is *farther away* from the pin on the left. Notice that the pins are tilted compared to one another, and the tilt is greater than it was before.
11. Darken the room and turn on the flashlight.
12. Once again, adjust the flashlight so there is basically no shadow coming from the pin on the left. That means the flashlight is shining directly above the pin on the left.
13. Look at the shadow cast by the pin on the right. How long is it compared to the shadow it cast in step 8?
14. Repeat steps 9-13 twice more, each time moving the pin on the right farther from the pin on the left so that they become even more tilted compared to each other. How does the length of the shadow change as the tilt between the pins increases?
15. Put everything away.

What did you find out in the experiment? You should have seen that the farther the two pins were from each other, the longer the shadow of the pin on the right became. Why is that? Well, think about the fact that the pins were stuck into a ball, which is a sphere. Since the sphere is curved, the pins were tilted compared to each other. This means there was an angle between the two pins. Each pin stuck out straight, but the curve of the ball caused one to be at an angle compared to the other. Well, you kept making sure that the flashlight was directly above the pin on the left. That meant there was an angle between the flashlight and the pin on the right. The farther the pin on the right was from the pin on the left, the larger the angle became. As a result, the longer its shadow became.

So what does this tell you? Think about a tall pole sticking straight up out of the ground. When the sun is directly above it, the pole will not cast a shadow, just like the pin on the left did not cast a shadow. However, another pole sticking straight up out of the earth would cast a shadow (just like the pin on the right did), provided it was far from the first pole. More importantly, the larger the angle between the two poles, the longer the shadow of the second pole would be. This is because the farther the second pole was from the first pole, the more tilted the second pole would be compared to the first pole. This tells you something very important: the length of the shadow cast by the second pole is a measure of the angle between the two poles!

Eratosthenes realized this, and he decided to use it to measure the distance around the entire earth. He heard that there was a city called Syene (sy ee' nee) in Egypt in which the sun was directly overhead at noon on a certain day each year. He traveled to Syene to make sure this was correct. On that date at noon, he looked down a deep well and saw that its sides cast no shadow onto the water at the bottom of the well. That told him the sun was directly above the city at noon on that date. If there had been a pole sticking straight out of the ground, it would not have cast a shadow.

The next year, he traveled to a city called Alexandria. At exactly noon on the same date, he measured the length of the shadow of a pole that was sticking straight up out of the ground. That's all he needed to determine the circumference of the earth! How in the world could he get the distance around the earth just by measuring the length of the pole's shadow? Remember that the length of the shadow is a measure of the angle between a pole sticking out of the ground at Syene and the pole that cast the shadow. So by measuring the length of the pole's shadow, he measured the angle of the tilt between Syene and Alexandria.

How did this help him measure the distance around the earth? He realized that since the earth is a sphere, when you start walking in a straight line, your tilt changes. Because of this, the angle between you when you started and you as you walk will increase. However, even though you walk in a straight line, you will end up back where you started, because you are walking on a sphere. Geometry told him that if you did that, your angle would change by 360 degrees. Well, based on the length of the shadow in Alexandria, he calculated that the angle between Syene and Alexandria was just over 7 degrees, which is 1/50 of 360. That told him he had walked one-fiftieth (1/50) of the way around the earth.

Well, he knew the distance between Syene and Alexandria. So he reasoned that if he multiplied that by 50, he should get the distance you would have to travel to go all the way around the earth. That, of course, is the earth's circumference. So Eratosthenes was the first to measure the earth's circumference. How did he do? He did really well. His measurement was 39,690 kilometers, or 24,662 miles. Of course, Eratosthenes didn't use kilometers or miles to measure distance. He used something called "stadia." However, his measurement in stadia works out to those measurements in kilometers or miles. Modern science tells us that the earth's circumference is 40,008 kilometers or 24,860 miles. So even though Eratosthenes made this measurement about 200 years before Christ was born, he ended up being very, very close to the value given to us by modern science! That's very impressive.

Now if you didn't understand exactly how the tilt of a pole in Alexandria compared to Syene tells you the distance around the earth, don't worry about it. The important thing to realize is that Eratosthenes understood that you didn't have to walk all the way around the earth, measuring the distance as you went, to determine the circumference of the earth. Instead, all you had to do was walk across part of the earth, know that distance, and know the tilt of where you ended compared to where you started. After that, a little math would tell you the distance around the earth, assuming that the earth is a sphere. That's the power of math and science together. By using them both, you can measure things that would be nearly impossible to measure directly!

LESSON REVIEW

Youngest students: Answer these questions:

1. What is the proper scientific and mathematical term for a ball?

2. What is the circumference of a sphere?

Older and Oldest students: In your notebook, write a short story explaining how Eratosthenes measured the circumference of the earth. Be as creative as you want, perhaps including what you think Eratosthenes might have been thinking as he did his work or what other people thought of what Eratosthenes was doing.

Lesson 29: Hipparchus (c. 190 BC – c. 120 BC) and the Moon

The last scientist I want to discuss in this section is named **Hipparchus** (hih par' kus). We don't know exactly when he was born or when he died, but later scientists mention his work as occurring between 162 and 127 BC. He was devoted to a study of the sun, moon, and stars, as shown in the drawing on the left. Like most ancient scientists, he also studied a lot of math. In fact, he developed a field of math called "trigonometry," which is an advanced way of analyzing angles. In high school, you will find that there are some science subjects you cannot learn unless you first learn trigonometry. As a result, even his mathematical studies had a strong impact on science today.

This is a 19th century engraving of how one artist imagined Hipparchus studying the moon and stars.

Of course, what I want to concentrate on are his scientific studies. Let's start with his observations about the moon. Have you ever noticed that the moon's size in the sky seems to change? When the moon is high in the sky, it looks very small. However, when it is closer to the ground, it looks very large. Why is that? We know that when things are closer to us, they look like they are larger than when they are far away from us. Because of this, some people say that when the moon looks larger, it is closer to the earth, and when it looks smaller, it is farther away from the earth. However, that's not true.

Do you remember Aristotle? Do you remember that he believed the moon was attached to a sphere that rotated around the earth? If that were really true, then the distance between the earth and the moon couldn't change. So he decided that the change in size of the moon had nothing to do with it being closer to or farther away from the earth. Instead, he thought that the air around the earth magnified the moon when it appeared to be close to the ground. It turns out that's wrong too, but my point is that even as far back as Aristotle, scientists didn't think the change in the moon's size had anything to do with the distance between the earth and the moon.

You might already know that the change in the moon's size depending on where it is in the sky is actually an illusion. Hipparchus was able to demonstrate that. How? Perform the following experiment to see.

Blocking Balls

What you will need:

- A dime
- A ruler
- Two balls of different size (like a baseball and a basketball)
- Two people to help you (one should be an adult)

What you should do:

1. Have your helper who is not an adult stand several feet away from you and hold the smaller ball in front of his or her face.
2. Hold the dime with the thumb and forefinger of one hand.
3. Close one eye.
4. With the eye that is still open, look at the ball your helper is holding.
5. Hold the dime in front of your eye and move it back and forth. You should see that when the dime is very close to your eye, it completely blocks the ball from your sight. When the dime is farther from your eye, you can see part of the ball.
6. Move the dime so that it just barely blocks the ball from your sight. In other words, hold the dime so that if you were to move it any farther from your eye, you would see part of the ball.
7. Continue to hold the dime right where it just barely blocks the ball, and have an adult use the ruler to **carefully** measure how far the dime is from your face. The adult can **gently** rest the beginning of the ruler against your face right below the eye and then use it to measure how far the dime is away from your face.
8. If your helper moved around at all, have him or her get back to the original position so you are the same distance away from each other as you were in step 1.
9. Have your helper hold the larger ball in front of his or her face.
10. Repeat steps 2 through 8 so that you measure how far the dime must be from your eye to just barely block the larger ball.
11. Compare the two measurements.
12. Put everything away.

What did you see in the experiment? You should have noticed that the dime needed to be closer to your eye to just barely block the larger ball. It needed to be farther away from your eye to block the smaller ball. If you spent more time and were a bit more careful, you would find that to block a ball that was twice as large, the dime would have to be half as far from your eye. So…let's suppose the dime had to be 4 inches away from your eye to block the small ball. It would have to be two inches from your eye to block a ball that was twice as large. In other words, the distance the dime has to be from your eye to block the ball depends on how big the ball is.

What you built in your experiment is a very simple version of a device Hipparchus built. It allowed him to see how the size of the moon, sun, and stars changed in the night sky. When he used his device on the moon, he found something very interesting – the moon's size doesn't change very much when it is high in the night sky or near to the ground.

You can use what you made to see exactly what Hipparchus saw. When you see the full moon low in the sky one night, it will appear large to you. However, repeat the experiment you just did, using the dime to block the moon. Write down how far the dime was from your eye in order to just barely block the moon. On another night when the moon is high in the sky, repeat the experiment. You will see very little difference in how far the dime has to be from your eye in order to just barely block the full moon.

So in the end, Hipparchus showed that the big change in size of the moon is really just an optical illusion. As you probably already know, it's because of how your brain tries to judge the size of an object. When the moon is low in the sky, there are trees, houses, and hills near it. Your brain uses them to help it judge the size of the moon. When the moon is high in the sky, there are no trees, hills, or houses near the moon. As a result, your brain judges the moon to be smaller. A device like you used in your experiment isn't affected by how your brain judges the size of things, so it shows you that the actual size of the moon doesn't change much when it changes height in the sky.

Now notice I said that the actual size of the moon in the sky "doesn't change much." It actually does change a bit, because the moon doesn't orbit the earth in a perfect circle. The actual orbit of the moon is a slight oval, so sometimes the moon is closer to the earth, and sometimes it is a bit farther away. This does have a small effect on the size of the moon in the night sky.

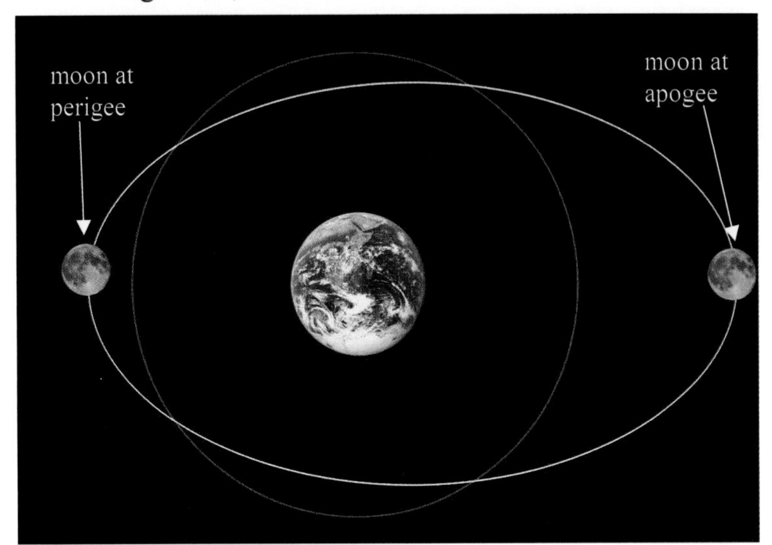

If the moon's orbit around the earth were a circle, it would follow the red path in this drawing. However, the moon's orbit is an oval, meaning that the distance between the moon and earth does change. The moon's orbit is not nearly as oval as what is shown in the picture. The drawing is exaggerated to make the difference easy to see.

When the moon is closest to the earth, scientists say it is at **perigee** (pehr' uh jee). When it is farthest from the earth, scientists say it is at **apogee** (ap' uh jee). If you have a hard time remembering which is which, remember that both "apogee" and "away" start with the letter "a." This should remind you that **a**pogee is when the moon is farthest **a**way from the earth.

Now even though the distance between the earth and moon changes, the change is small. At apogee, the moon is about 406,500 kilometers (252,500 miles) from the earth, and at perigee it is about 356,500 kilometers (221,500 miles) from the earth. This makes the moon a bit larger in the sky at perigee, but not a lot. Even though the change is small, Hipparchus was able to measure it using his device. In the end, then, he showed that the actual change in the moon's appearance in the night sky was not nearly as large as it seems to the human eye, but it still does exist. Hipparchus didn't realize that this was because of the moon's oval orbit, however. That knowledge didn't come until much later.

LESSON REVIEW

Youngest students: Answer these questions:

1. Does the moon orbit the earth in a perfect circle?

2. Does the moon's size in the sky really change as much as it looks like it does?

Older students: In your notebook, explain how your device was a way of measuring the size of a distant object. Then explain how Hipparchus was able to use his device to show that while the moon's size in the night sky does change a bit, it doesn't change nearly as much as it appears to change when you look at it with just your eyes.

Oldest students: Do what the older students are doing. In addition, make a drawing like the one above, showing that the moon's orbit around the earth is not a perfect circle. Point out the moon's apogee and perigee, and define those words below the drawing.

Lesson 30: Hipparchus and the Sun and Stars

As I told you in the previous lesson, Hipparchus was devoted to the study of the sun, moon, and stars. He used his device to see if the sun's size in the sky changes throughout the year, but he couldn't detect any change. That was consistent with the idea most scientists held at the time: that the sun was attached to a sphere that rotated around the earth. If that were true, the distance between the sun and earth would never change, so the size of the sun in the sky would never change.

We now know that the size of the sun does change a bit in the sky, but the change is incredibly small – a lot smaller than the change of the moon's size in the sky. Once again, this is because the earth doesn't orbit the sun in a circle. Instead, like the moon's orbit around the earth, the earth's orbit around the sun is an oval. As a result, sometimes the earth is a bit closer to the sun, and sometimes it is a bit farther away from the sun. However, the effect this has on the size of the sun in the sky is very, very small. It cannot be seen with your eyes. It takes a sophisticated instrument to notice it.

The size of the sun in the sky does change a bit, but you cannot see the change without very precise instruments.

You might think that the distance between the earth and the sun is what causes the seasons to change. However, that is not correct. Summer doesn't occur because the earth is closer to the sun and therefore warmer. In the same way, winter doesn't occur because the earth is farther from the sun and therefore cooler. It took scientists a long time to understand what causes the seasons to change, and you will eventually learn about that. I just wanted to make sure you didn't get the wrong idea in your head before you learned the correct explanation later on.

Speaking of the seasons, one thing everyone notices is that there is more daylight in summer and less daylight in winter. During the winter, the daylight hours can be very short, depending on where you live. During the summer, however, the daylight hours can be very long, depending on where you live. Hipparchus studied this quite a bit. He looked specifically at an event called the **summer solstice** (sohl' stis) – the day in which the sun reaches its highest noontime point in the sky. This happens on the longest day of the year and marks the beginning of summer. He made measurements of the summer solstice himself, and he examined the measurements of others.

From those measurements, he tried to determine the precise length of a year. He defined a year as going from summer solstice to summer solstice, and by that definition, a year was exactly 365 days, 5 hours, and 55 minutes long. Interestingly enough, that is very close to today's modern value. According to our best instruments, the time between summer solstices is 365 days, 5 hours, 48 minutes, and 45 seconds long. So more than 100 years before Christ, Hipparchus was able to measure the precise length of a year to within about 6 minutes of its correct value! That should give you an idea of what a careful scientist he was.

There is also something called the **winter solstice**, which is the day on which the sun reaches its lowest noontime point in the sky. Not surprisingly, this happens on the shortest day of the year and marks the beginning of winter. From the summer solstice to the winter solstice, then, the number of daylight hours decreases, so the days get shorter. From the winter solstice to the summer solstice, the number of daylight hours increases, so the days get longer. For those who live in the Northern Hemisphere (like those in North America, Europe, and the former Soviet Union), the summer solstice happens between June 20th and 22nd, and the winter solstice happens between December 21st and 23rd, depending on the year. For those living in the Southern Hemisphere (like Australia, New Zealand, and South America), the dates are reversed. So the summer solstice happens between December 21st and 23rd, and the winter solstice happens between June 20th and 22nd, depending on the year.

The shapes on the globe in this second-century statue represent constellations. Some astronomers think it was made based on the star catalog compiled by Hipparchus.

Hipparchus not only studied the sun; he also studied the stars. You probably know that the sun is actually a star – a very special star that provides the earth with just the right amount of energy to support life. However, scientists didn't know that back in his day, so when Hipparchus produced a listing of 850 stars, the sun was not on the list. Even with this oversight, however, the list that Hipparchus put together (usually called his **star catalog**) was very impressive. Like most star catalogs, the one he put together discussed the **constellations** (kahn' stuh lay' shunz), shapes that people imagined groups of stars make in the night sky. While his actual star catalog has been lost to history, many astronomers think that a statue from the second century AD called the Farnese Atlas (shown in the picture on the left) was made based on the catalog's contents.

One very important thing that Hipparchus listed in his star catalog was the **magnitude** of each star. Have you noticed that some stars in the night sky are brighter than other stars? Hipparchus wanted to record each star's brightness, so he came up with a measurement he called magnitude. He said that the brightest stars had a magnitude of 1, while the dimmest stars had a magnitude of 6. Stars with brightness in between these were listed with magnitudes of 2 through 5.

Today, astronomers still use magnitude to describe a star's brightness, even though they use a lot more than six numbers. Nevertheless, they still use the lowest numbers to indicate the brightest stars and the highest numbers to indicate the dimmest stars. In addition, they make a distinction. They say the **apparent magnitude** is the brightness of the star as it appears in the night sky. However, a star's brightness in the night sky depends on two things: how bright the star actually is, and how far it is away from the earth. The sun, for example, is not the brightest star in the universe. It is the brightest star in the sky, however, because it is so much closer to the earth than any other star. To get rid of the effect of distance, astronomers also define the **absolute magnitude**, which is the *actual* brightness of a star. Hipparchus, of course, listed the apparent magnitude of the stars in his catalog.

How did Hipparchus determine the apparent magnitude of the stars in his catalog? We don't know for sure, but you can make your own device that measures a star's magnitude.

Star Light, Star Bright

What you will need:

- A sheet of cardboard that is about 20 cm (8 inches) by 20 cm (8 inches)
- Some Ziploc bags
- Scissors
- Tape
- An adult to help you

What you should do:

1. Have the adult help you cut four rectangles out of the cardboard, as shown in the drawing on the right.
2. Have the adult help you cut out seven slightly larger rectangles of plastic from the Ziploc bags. Even though you are cutting them out of a bag, each rectangle should be just one strip of plastic.
3. Leave the top rectangle alone, but on the next lower rectangle, tape one strip of plastic so it completely covers the rectangle. The plastic should be tight so that you can see through it without any difficulty.
4. On the next rectangle down, tape one strip of plastic as you did in step 3, but then tape another strip of plastic over that. When you look through this rectangle, then, you are looking through two layers of plastic.
5. Tape four layers of plastic over the bottom rectangle so that when you look through it, you are looking through four layers of plastic.
6. When there is a relatively cloudless night, go outside and look at the stars. Find a section of sky with several stars in it, and look at them through the rectangle that has no plastic on it.
7. Look at the same section of the sky through the rectangle that has one layer of plastic over it. Did some of the stars disappear?
8. Repeat step 7 with the other two rectangles. You should see that the dimmer stars disappear when you are looking through more plastic. In the end, then, this is a way of measuring the magnitude of a star. The brightest stars (the ones with the lowest magnitude) can be seen through the bottom rectangle. The dimmest stars, on the other hand, can only be seen through the rectangle with no plastic on it. So you could call the stars you saw through the bottom rectangle magnitude 1 stars, while those you saw only through the top rectangle magnitude 4 stars.

LESSON REVIEW

Youngest students: Answer these questions:

1. What do we call the day on which the sun reaches its highest noontime point in the sky?

2. If one star has a magnitude of 2 and another has a magnitude of 4, which appears brighter?

Older students: In your notebook, define the winter solstice and the summer solstice. In addition, define the magnitude of a star, and make the distinction between apparent magnitude and absolute magnitude.

Oldest students: Do what the older students are doing. In addition, there are specific names for when the earth is closest to the sun and when it is farthest from the sun. Do some research to find those names, and discuss how they are similar to apogee and perigee, which you learned about previously.

Science Soon After Christ

This statue of Galen is in the Anatomical Theater at the medical school of the University of Bologna in Italy. The theater was designed for anatomy lessons back in 1637 but is not used for that purpose anymore.

Lessons 31-45: Science Soon After Christ

Lesson 31: BC and AD

So far, the natural philosophers I have been discussing were born and did their work before the time of Christ. I have not discussed *all* of the natural philosophers that existed before Christ, of course, but I do think I have hit the most important ones. Now we have reached the point where I will start talking about the science that went on after Christ was born. Before I do that, however, I want you to learn something strange about the way we mark history.

When I have given dates so far, I have put "BC" after them. Most people know this means "Before Christ." When you see a date like 150 BC, then, you know that it was 150 years before Christ. A date like 100 BC is 100 years before Christ. Which date is more recent? 100 BC. Why? Because the BC dates *count down* to the birth of Christ, so the smaller the BC year, the more recent it is. Now I will start putting "AD" in front of dates, which is an abbreviation for *Anno Domini*, a Latin phrase that means "Year of our Lord." For these dates, the larger numbers represent more recent dates, because the AD dates *count up* from the birth of Christ. So the date AD 150 is more recent than the date AD 100. Please also be aware that often, "AD" is left out, so when I say I was born in 1963, it is understood that this means AD 1963.

You probably knew most of that already. However, there's probably something you don't know about BC and AD, so I want you to do this activity.

BC and AD

What you will need:

- Your notebook (a blank sheet of paper if you are doing the review exercises for the youngest students)
- A pencil

What you should do:

1. Start a page in your notebook that is titled, "A Short Calendar."
2. A few inches below the title of the page, draw a timeline that looks like the drawing below. Each vertical dash represents the beginning of a year. If you go to the right on the timeline, you are going to more recent dates. If you go to the left, you are going to dates that are farther in the past.
3. Draw a nativity scene on the middle dash. That represents when Christ was born.
4. Label the dashes that represent 1 BC, 2 BC, and 3 BC.
5. Label the dashes that represent AD 1, AD 2, and AD 3.
6. Under the year you have labeled AD 1, write how old Christ would have been on his birthday that year.
7. Repeat step 6 for AD 2 and AD 3.
8. Now write "Fredicus the Silly" above the line you labeled 3 BC. This represents the birthday of someone I just made up. You can also draw a picture of Fredicus there, if you like.
9. Under each line (including the line that represents Christ's birth), write down how old Fredicus would be on his birthday that year.
10. Keep your notebook handy, because you will be writing in it soon.

How did your timeline turn out? It probably looked like the timeline drawn below. The year 1 BC starts on the dash that appears to the left of Christ's birth, and the year AD 1 starts on the dash that appears to the right of Christ's birth. Because of that, Christ would have been one year old on his birthday in AD 1, two years old on his birthday on AD 2, and three years old on his birthday in AD 3. Fredicus the Silly would have been one year old in 2 BC, three years old when Christ was born, and six years old in AD 3. That all makes sense, right? It does, but it turns out this is *not* the way our calendar works! You and I are very comfortable with the idea of zero. As a result, we think of Christ being born in year zero. That would make the year before Christ was born 1 BC and the year after Christ was born AD 1. However, that's not how the years were counted back when the calendar we use today was put together!

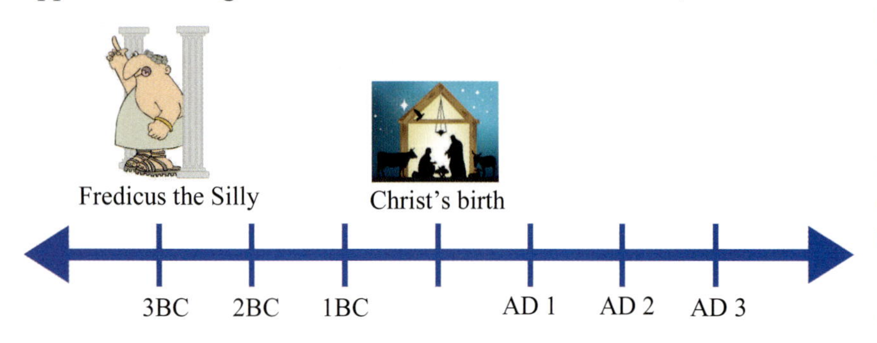

This timeline makes sense but *is not* the way our calendar works!

The complete history of how the calendar developed is incredibly interesting, but it goes beyond what I am trying to explain here. I will start the story about 1,500 years ago, when a monk named **Dionysius** (dye' uh nye' see us) **Exiguus** (ek sig' yoo us) was trying to calculate the date of Easter for the years AD 532 through AD 626. How do you determine the date for Easter? You look it up on the calendar, right? Well, someone had to make that calendar. How did that person determine the date for Easter? Because our calendar is different from the Jewish calendar, and because the Jewish calendar is what determines Easter, it actually takes a lot of math to figure out when Easter falls on the calendar we use. Back then, Dionysius Exiguus had to do that math by hand in order to determine the date of Easter for each year in that time span.

At that time, the calendar used by most Europeans marked everything as being before or after a Roman emperor named Diocletian (dye' uh klee' shun), who ruled Rome three hundred years previously. Dionysius didn't like the fact that the calendar used an emperor who had persecuted Christians as a reference, so he decided to change this. He calculated the year he thought Christ was born and used that as a reference. Other Christians like the idea, so pretty soon, everything that happened after Christ's birth was labeled "AD" for *Anno Domini* (Year of our Lord), and everything that happened before Christ's birth was labeled "AC" for *Ante Christum*, a Latin phrase that means "Before Christ." The "AC" was later changed to "BC."

Now here's the strange part. During the time of Dionysius, the mathematical concept of zero was not popular in Europe. That might sound really odd to you, but it is true. Back then, numbers were for counting, and if you didn't have something to count, there was no reason to use numbers. Have you ever studied Roman numbers? If not, you will someday. One thing to notice about Roman numbers is that there is no number for zero. If you want to represent a number like 130, you use a "C," which means "100," and three "X's," each of which means "10." Thus, in Roman numbers, 130 is written as CXXX. You can search and search, but there is no Roman number that means zero.

So when Dionysius decided to mark the calendar using the birth of Christ, he didn't use a zero. He said that the year before Christ was born was 1 BC, and the year that Christ was born was AD 1. There was no year zero. The timeline you drew probably had 1 BC, then the year that Christ was born, and then AD 1, much like the timeline drawn on the previous page. In that kind of calendar, the year

Christ was born would be year zero. But that's not how the timeline that Dionysius set up worked. Instead, there was no year zero. The year 1 BC was followed directly by the year AD 1.

What's the big deal? Well, it makes the way we count years a bit odd when we get around the time Christ was born. Look at the timeline below, which is the way our calendar really works. In this timeline, Christ was born in AD 1. As a result, He didn't turn one year old until AD 2. That means He was only one year old in the second year of our Lord! In the same way, the mythical Fredicus the Silly was born in 3 BC, but he was three years old in AD 1. That doesn't make a lot of sense, but it's the way our calendar works, all because Dionysius did not include a year zero in the calendar!

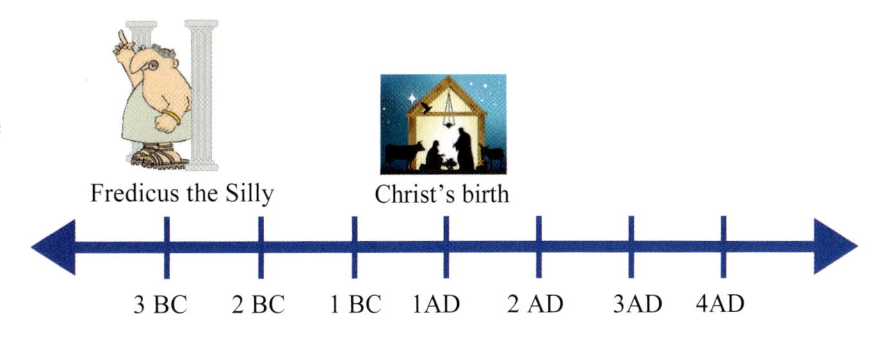

This timeline doesn't make sense but is the way our calendar works.

Now to make matters worse, there are two other things I need to point out about the calendar. First, Dionysius Exiguus made a mistake in his calculations. He got the year of Christ's birth wrong. We don't know exactly when Christ was born, but historians agree that it was somewhere between 7 BC and 2 BC. Most historians think it was in 4 BC. Of course, the mistake wasn't caught for a long time, so even though we know the calendar as Dionysius worked it out is wrong, we go ahead and use it anyway. As a result, we are stuck with the odd situation that Christ was probably born four years before Christ, at least according to our calendar!

The other thing I need to tell you is that there are those who wish to change the calendar to get rid of Christ's name. Those people don't use BC or AD. Instead, they call the years we mark with AD as being part of the "Common Era," and the years before then as being "Before the Common Era." So they use "CE" instead of "AD" and "BCE" instead of "BC." Obviously, I won't be doing that, because I don't want to take Christ out of the calendar. However, I wanted you to be aware of it, because you will probably run across it in museums or other books.

LESSON REVIEW

Youngest students: Answer these questions:

1. What does AD stand for (in English)?

2. In our calendar, what year comes right after 1 BC?

Older students: If you drew your timeline the way it is drawn on page 94 (or any way other than how it is drawn on page 95), make a new drawing under the old one showing how the calendar really works. Explain in your own words why it works like that, and give the Latin phrase for "AD."

Oldest students: Do what the older students are doing. In addition, write down this question and your answer in your notebook: "Suppose Christ was born in 4 BC, as most historians suggest. Luke 3:23 says that He was about thirty years old when He was baptized. Assume he was exactly 30 years old and that his birthday had already passed that year. In what year AD was He baptized?" Check your answer and correct it if it is wrong.

Lesson 32: Pedanius Dioscorides (c. AD 40 – c. AD 90)

NOTE: Today's experiment has a step where you must wait for an hour. Plan your day so that other things can be done during that time.

In the ancient world, diseases, hurts, and their treatments were not well understood. If you remember, Hippocrates discussed how to treat people who were ill, and sometimes, his treatments did work. However, if they did work, it was not because the explanation Hippocrates gave was correct. As time went on, Hippocrates became the main resource for physicians, and since many of his ideas were wrong, many of the treatments that physicians came up with were not very effective. Others were actually harmful! Remember, for example, that Hippocrates thought that most illnesses were caused by an imbalance of fluids in the body. If a patient was thought to have an illness related to too much blood, a physician would actually draw blood out of the patient, either by cutting the patient or using blood-sucking animals called leeches. Obviously, taking blood out of a sick person generally does more harm than good.

This is what French artist André Thévet (1516 – 1590) imagined Dioscorides looked like while working on his famous book.

Things started to change a bit for the better thanks to a physician named **Pedanius** (puh day' nee us) **Dioscorides** (dee uh skor' uh deez), who lived from about AD 40 to AD 90. He was a surgeon with the Roman army (Nero was emperor of Rome at the time), and that allowed him to examine all kinds of medical problems, from injuries to new diseases. As he traveled, he studied plants and animals and how they were used by local physicians. In addition, he read the accepted authorities on medicine, such as Hippocrates. However, one of the most important things that he did was to confirm everything that he wrote. He did not accept the word of any living physician or the word of any authority, even Hippocrates. Instead, if he learned that a specific medicine might treat a specific problem, he tested it first. After all, people (even authorities) can be wrong, and he didn't want any incorrect information in his book!

After 20 years of traveling, studying, and testing, he wrote one of the most important books in the history of medicine. It was called *De Materia Medica*, which means "On Medicinal Substances," and it covered *more than a thousand* natural things that could be used by physicians to treat their patients. Some of these things, like the seeds of the poppy plant, could ease a person's pain. Others, like the juice from a plant he called "Panax Heraklios," could treat a person's cough (and other ailments). Note that it wasn't the entire poppy plant that eased a person's pain, or the entire Panax Heraklios that helped a person's cough. That's an important point, and it is worth exploring.

Chemicals and Plants

What you will need:

- Red grapes (You cannot use green grapes. You need the grapes that have a red peel and a whitish, fleshy interior. They are the most common grape [other than the green grapes] sold in the produce section of the supermarket.)
- Clear vinegar

- Clear ammonia (sold with the cleaners in the supermarket)
- Six small glasses, like juice glasses (The glass needs to be clear, not colored.)
- Plastic wrap
- A sink with a water tap and a drain

What you should do:
1. Peel the skin off two grapes, putting the skin in one of the glasses and the inside of the fruit (usually called the "flesh" of the fruit) in another glass. You can pull the skin off in little bits if that's what you need to do, and it doesn't have to be perfect. If you get some flesh with the skin, that's no problem. The most important thing is to get all the skin off the grapes and into one of the glasses, while the other glass contains *only* the flesh of two grapes.
2. Repeat step 1 with two more grapes, putting the peels in another empty glass and the flesh in a different empty glass. You should now have two glasses that contain grape peels and two glasses that contain grape flesh.
3. Add clear vinegar to one of the glasses that has peels in it. Add enough so that the glass is about one-third full. Make sure all of the peel pieces are in the vinegar.
4. Add clear vinegar to one of the glasses that has grape flesh in it. Once again, the glass should be about one-third full.
5. Add clear ammonia to the other glass that has peels in it. Add enough so that the glass is about one-third full. Make sure all of the peel pieces are in the ammonia. The ammonia will smell terrible, so don't breathe too deeply, and keep things far away from your face.
6. Add clear ammonia to the other glass that has grape flesh in it. Add enough so that the glass is about one-third full. Once again, don't breathe too deeply, and keep things far away from your face.
7. Cover the glasses with plastic wrap. You can cover them individually or all together – it really doesn't matter. This is just to reduce the smell while you wait.
8. Wait for about an hour. Do something else while you are waiting.
9. After about an hour, remove the plastic wrap (keeping your face far from the glasses) and look at each glass.
10. Carefully pour the liquid from one of the glasses that had peels in it into an empty glass. You want to pour it so that only liquid (no peels) ends up in the new glass.
11. Repeat step 10 for the other glass that had peels in it. You should now have two glasses, each of which contains the liquid from the glasses that had peels in them.
12. Note the colors of the two liquids.
13. The darker liquid is ammonia. Slowly add some clear vinegar to the ammonia until you notice a change in color.
14. Slowly add some clear ammonia to the other liquid, which is vinegar. Once again, you should eventually see a color change.
15. Clean up your mess.

What did you see in your experiment? First, nothing interesting should have happened in the glasses that contained the flesh of the grapes. The liquid in them should have been clear when you added it, and it should have been mostly clear an hour later. However, the glasses with peels in them should have been a completely different story. The liquid in the glass with ammonia and peels should have been dark green or blue, while the liquid in the glass with vinegar and peels should have been pink. These colors should have been easy to see once you poured the liquid into two new glasses, leaving the peels behind.

What explains this effect? Well, there are different chemicals in the grape peels than in the grape flesh. As a result, the grape peels had a reaction to the ammonia and vinegar that the grape flesh did not have. This is important, especially when it comes to using plants as medicine. Since different parts of a plant have different chemicals in them, it's not enough for a physician to say, "Use this plant to help reduce your pain." It is quite possible that some parts of the plant contain chemicals that will make the pain worse! Other parts of the plant will contain chemicals that will neither help nor harm you. Only certain parts of the plant will contain the chemicals that will help you, so those parts need to be specified.

This was something Dioscorides was careful to make clear. Sometimes, it was the root of a plant that helped a patient. Sometimes, the leaves had to be dried and tea had to be made from them in order to help the patient. Sometimes, the seeds had to be pulled from the plant and used. Sometimes, the plant had to be squeezed to get out oil or some other liquid, because that's what held the chemical that would help the patient. In the end, Dioscorides not only told physicians *what* plants (or sometimes animals) to use in order to make medicine, he also told them how to prepare the plant so that only the proper part was used. For these (and other) reasons, Dioscorides's book was considered the most important book on medicines for the next 1,500 years!

Now before I end this lesson, I must tell you why the colors changed again in steps 13 and 14 of your experiment. The specific chemicals in the grape peels that caused the color change are sensitive to two broad groups of chemicals called **acids** and **bases**. You will learn more about acids and bases as you study more science. For right now, just realize that there are many chemicals that act as acids and many that act as bases. Vinegar, for example, is an acid, while ammonia is a base. In the presence of a base, the chemicals in the grape peels turn dark colors, and in the presence of an acid, they turn light shades of red.

In your experiment, then, the vinegar turned the chemical in the grape peels pink because it is an acid. However, when you added a lot of ammonia to the vinegar, the ammonia became more chemically important than the vinegar, so the solution became a base. As a result, the color changed. In the same way, the ammonia was originally green or blue because it is a base. As you added vinegar to it, however, the acid properties of the vinegar became more important, and the chemicals in the grape peels turned pink. We call such chemicals **indicators**, because they indicate whether a solution is mostly acid or mostly base.

LESSON REVIEW

Youngest students: Answer these questions:

1. When using a plant for medicine, why is it important to use only a specific part of the plant?

2. Why did Dioscorides test everything he used instead of accepting the word of someone else?

Older students: In your notebook, write a summary of what you did in your experiment, and explain why only the glasses that had peels in them produced an interesting effect. Explain how that relates to using plants in medicine.

Oldest students: Do what the older students are doing. In addition, write down the following question and your answer to it: "Based on the color of the grapes, would you say that grapes contain acid or base?" Check your answer and correct it if it is wrong.

Lesson 33: Hero of Alexandria (c. AD 10 – c. AD 70) and Siphons

The next scientist I want to discuss might be out of order. We aren't sure. Some think he lived around 150 BC. Others think he lived around AD 250. However, most historians think he lived from about AD 10 to about AD 70, so those are the dates I will use. One reason it is a bit hard to pin down when he lived is that his name (Hero) was fairly common back then, so there are a lot of references to men named **Hero of Alexandria**.

The particular Hero who I want to discuss was a teacher, and he wrote at least seven books. One of them was on mathematics, one was on how to measure the length of something, one was about optics, and the rest were about various machines. For example, one of the books discussed war machines, while another discussed ways in which builders could lift heavy objects. However, the book I want to concentrate on was called *Pneumatica*, and it was about how you could use air pressure, steam pressure, or water pressure to do work.

After the introduction, his book starts out with a discussion of a **siphon**. It is thought to be the first book that gives the proper explanation for why a siphon works. Perform the following experiment to understand what a bent siphon is and how it works.

Siphoning Water

What you will need:

- Four bendable straws (You could use flexible plastic tubing instead, as long as it's okay to ruin th e tubing in the course of the experiment.)
- A tall glass
- Tape
- A sink with a water tap and a drain
- A pin or needle
- Scissors

taped area where one straw is pushed inside the other
bendable straw that was cut short
uncut bendable straw

What you should do:

1. If you have flexible tubing that can be ruined, just bend it into a "U" shape and skip to step 6. Your tubing just has to replace the straws in the photo on the right.
2. Bend both straws so that they form an "L" shape.
3. Cut about two inches off the bottom of one straw. The bottom is the part that is farthest from the bendable section of the straw.
4. Push the end of one straw into the end of the other straw so that the bends are close together. You will have to crumple up the end of one straw to get it to fit into the end of the other straw. That's fine. Once it is crumpled and put inside the end of the other straw, push both straws together so that the inside straw goes deep into the outside straw. The two straws should form a "U" shape.
5. Use tape to seal where the one straw is pushed into the other straw. You want an airtight seal here.
6. Fill the glass with water and put it near the edge of the sink.
7. Put the shortest side of the U-shaped straws into the water, and allow the longer side to hang over the sink, as shown in the photo above.

8. Lift the glass so that you can put the end of the straw that is hanging outside the glass into your mouth. Try to keep it over the sink to avoid making a mess.
9. Suck water into your mouth through the straw. (If you hear air being sucked into the straws, the tape did not seal the place where the straws were pushed together. You need to make the U-shape again, trying to make an airtight seal.)
10. Once you get water in your mouth, pull the straw out of your mouth and let it dangle over the sink, making sure that the end of the cut straw never leaves the water. If you did everything properly, water should start pouring out of the end of the straw.
11. Set the glass back down on the counter, all the time keeping the end of the straw over the sink so that the water falls into the sink and the cut straw stays in the water.
12. Watch the water fall out of the straw until it stops. Note how much water was removed from the glass.
13. Refill the glass with water.
14. Repeat steps 8-10.
15. While the water is still flowing, use the scissors to cut the long straw (the one out of which water is flowing), right below its bend. What happens?
16. Repeat steps 1-7 so you have a new experimental setup, and then repeat steps 8-11 so that water is flowing into the sink again.
17. Use the pin to stick a couple of holes into one of the bends in one of the straws. What happens?
18. Clean up your mess.

The U-shape you made out of straw acted as a siphon. When you sucked water into it and then let the water flow, it pulled a lot of water out of the glass. Even though you supplied only a little bit of energy to get the water to fill both straws, water flowed freely for quite some time, which emptied a lot of water out of the glass.

How does a siphon work? Hero answered that question well. He realized that water starts falling out of the siphon simply because it is heavy. Like any other heavy thing, it falls. We now know that's because gravity attracts everything that has weight to the earth. Hero didn't know that, but he knew that anything heavy will fall unless it is held up.

Now, of course, he knew that in order for the water to fall out of the correct side of the siphon, there had to be more water in that side of the siphon than in the other side. What happened in step 15 of your experiment? When you cut the long straw, the water stopped flowing, didn't it? Why? Because after you cut the straw, there was more water in the *other* side of the siphon. This caused the water to be pulled back into the glass. So to get the water flowing properly, the weight of water in the open side of the siphon has to be greater than the weight of the water in the other side of the siphon.

Everyone already knew that, but Hero realized there had to be more to the story. It wasn't just the weight of the water that was causing the siphon to work. According to Hero, water keeps flowing in a siphon because as the water falls down one side of the siphon, *something must replace it.* The only thing available to replace it is more water, so water is pulled up one side of the siphon to replace the water that falls down the other side. Of course, it takes work to do that, because water is heavy. If something else could come in to take the place of water, it would. That's why you did the experiment a second time. When you poked a hole into the system, what happened? The water should have stopped flowing pretty quickly. Why? Because air came in to take the place of the water that was flowing out of the siphon.

We now understand a bit more than Hero understood, so we realize that this works because air exerts **air pressure** on anything that is exposed to it. The surface of the water in the glass has air pushing down on it. That doesn't push water up the straw, however, because there is also air in the straw, exerting the same pressure. However, when you filled the straws with water, there was no longer air in them. As the water poured out of one straw, that reduced the pressure inside the straws. Since there was reduced pressure inside the straws, the air pressure pushing down on the water in the glass forced water up the straw. When you poked a hole into the straws, air was able to leak in as the water rushed out. That meant there was never any reduction in pressure inside the straws, so the water didn't get pushed up anymore.

When you suck on a straw, you are removing the air in the straw, and air pressure pushing on the liquid in the glass pushes the liquid up the straw to replace the air you sucked out.

This is, in fact, the reason a straw works in the first place. You think that you are sucking liquid into the straw when you drink from a straw, but that's not really correct. At first, all you are doing is sucking the air out of the straw. Because of this, the pressure in the straw is reduced. The air pressure pushing down on the surface of the liquid then pushes liquid up the straw to replace the air that you sucked out, and that's what forces liquid into your mouth.

So Hero had the key to understanding a siphon, even though he didn't know about air pressure. He just knew there was something nature did to keep from having an area of nothingness, which he called a "void." He knew that if something didn't replace the water that fell down the one end of the siphon, there would be a void there. As a result, something had to push water to fill the void. We now know that in both a siphon and a straw, it's air pressure that pushes water to fill the void.

LESSON REVIEW

Youngest students: Answer these questions:

1. What does a siphon do?

2. When a hole is poked in a siphon that is working, what happens? Why?

Older students: In your notebook, draw a picture of a siphon working, and write an explanation of how it works. Be sure to explaining why the siphon can't have holes in it and why the end that has water flowing out of it cannot be too short.

Oldest students: Do what the older students are doing. Also, explain why a siphon wouldn't work in space, where there is no air.

Lesson 34: Hero and Steam

Hero's most important work was not about siphons; it was about steam. Hero understood that when you heat water to make steam, the steam has to expand and travel away from the water. He realized that if you designed things properly, you could use this to make steam work for you. Try this experiment to see what I mean.

Harnessing the Power of Steam

What you will need:

- A square of paper, about 13-cm (5-inches) by 13-cm (5-inches)
- Scissors
- A pencil that has never been used
- A straight pin
- A pan in which you can boil water
- Water
- A stove
- An adult to help you

What you should do:

1. Put some water in the pan and heat it on the stove so that the water will eventually boil.

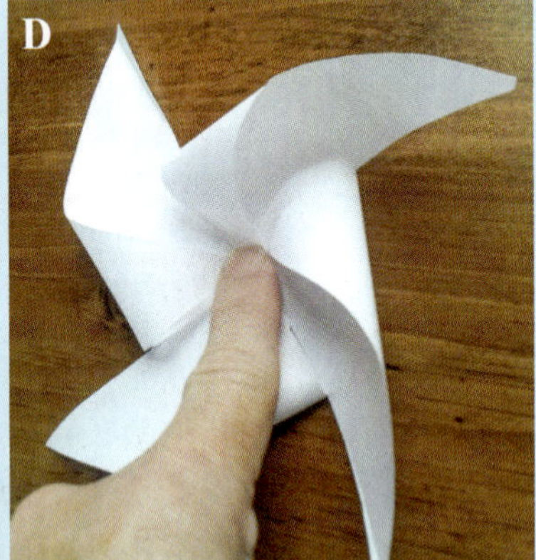

2. While the water is heating, fold the square of paper into a triangle by bringing two opposite corners together, as shown in the picture labeled "A."
3. Unfold it and fold it into a triangle again, using the two corners you did not use before.
4. Unfold it, and now you should have two diagonal creases, as shown in the picture labeled "B."
5. Use the scissors to cut along the diagonals from each corner towards the center, but stop when your cut is about a centimeter (½ an inch) from the center. You now have a square of paper with 4 cuts that go from the corners of the square to almost the center of the square.
6. Take one corner and fold it into the center of the square, as shown in the picture labeled "C."
7. Do that three more times, so that you have a pinwheel shape, as shown in the picture labeled "D."
8. Have an adult help you push the pin through the center of the pinwheel (where the finger is holding it in the picture labeled "D") and through the eraser on the pencil. The adult shouldn't push the pin all the way to the pencil. Instead, there should be some room between the pencil and the pinwheel, as shown in the picture labeled "E."

9. You should now have a working pinwheel. Blow on the pinwheel to see that it spins around on the pin.
10. If the water isn't boiling vigorously, wait until it is.
11. Holding the end of the pencil that you would normally sharpen, hold the pinwheel horizontally so that the pencil is on top and the pinwheel faces down.
12. Carefully hold the pinwheel so that it is over the pan and steam rises through the pinwheel. **BE CAREFUL.** Don't let your hand get close to the pan. I asked you to make this out of an unused pencil so that it would be long enough to hold the pinwheel over the steam without your hand getting too close to the steam. If your hand starts to get uncomfortably hot, put on an oven mitt and hold the pencil with the oven mitt.
13. The pinwheel should droop a bit as it gets wet, but that's not a problem. What happens after a few moments?
14. Clean up your mess.

What did you see in the experiment? When you made your pinwheel, you were able to get it to spin by blowing on it. Of course, that required you to do some work. What happened when you held your pinwheel over the steam? It should have started spinning, even though you weren't blowing on it and there was no wind. Why did it start spinning? Because the steam was rising out of the boiling water. The steam acted like wind, spinning the pinwheel around.

In the end, Hero of Alexandria realized that he could use the motion of steam to do all sorts of things. Probably his most famous use of steam is a steam engine called an **aeolipile** (ee ol' uh pahyl), which is shown on the right. It is a round container that has water in it. In addition, there are two bent tubes that lead to the inside of the container. Water is added to the container through the tubes, and then the container is place on a platform that allows it to spin. As the container is heated, the water eventually boils, and the steam has to expand and escape. The only way it can escape is through the tubes, and the force of the steam rushing out of the tubes causes the container to spin.

In a steam engine like Hero's aeolipile, the thermal energy from the fire is converted into the mechanical energy of motion.

How does Hero's aeolipile work? You don't know all the scientific concepts necessary to completely understand the process, but you know the basics. Remember that to get something moving, you need energy. In the case of Hero's aeolipile, where does that energy come from? It comes from the heat that is being used to boil the water. The thermal energy of the heat is being converted into the mechanical energy of the round container spinning. Obviously, steam plays a role in this conversion, and eventually, you will learn the details of how that works. Scientists didn't figure that out for quite some time, however, so you will have to wait to learn the details.

What good is a spinning ball on top of a fire? It's fun to watch, but it doesn't really accomplish anything. However, the principle that steam can be used to convert thermal energy into mechanical energy is incredibly important. For example, do you know how the first trains were powered? They were powered by steam, which is why they were called **steam locomotives**. The engine had a fire that had to be continually fed. The fire was used to boil water, and as steam was produced, it was used to make the engine's wheels turn. Lots of other things were also powered by steam. The first boats that were able to travel without sails or oars were powered by steam.

The steam pouring out of this steam locomotive is converting the thermal energy of fire into the mechanical energy of the engine's wheels turning.

Even though most boats and trains no longer use steam, there are still a lot of ways we use the principle of the steam engine. For example, do you know how we get the electrical power that runs your home? It is essentially the result of a steam engine. In order to produce electricity, electrical power plants need to move magnets. Guess how those magnets are moved? In most electrical power plants, they are moved by big turbines that are turned using steam. In fact, the main difference between a coal-powered electrical plant and a nuclear-powered electrical plant is how water is heated to make steam. In a coal-powered plant, coal is burned to boil water and make steam. In a nuclear-powered plant, a nuclear reaction is used to boil water and make steam. If it hadn't been for the discovery that steam can be used to convert thermal energy into mechanical energy, we might not have the electrical power that we enjoy today.

LESSON REVIEW

Youngest students: Answer these questions:

1. Fill in the blanks: In a steam engine, _______ energy is converted into __________ energy.

2. What powered the first trains?

Older students: In your notebook, explain what you did in the experiment and what you saw. Then explain how this shows that a steam engine converts thermal energy into mechanical energy. Write down the fact that we use steam to generate most of the electrical power we use today.

Oldest students: Do what the older students are doing. In addition, write in your notebook the following question and your answer to it: "If you didn't keep adding fuel to the fire under Hero's aeolipile, what would you see as you watched it for a long time? Why?" Check your answer and correct it if it is wrong.

Lesson 35: Hero and Flat Mirrors

Hero didn't limit himself to siphons and steam. He was interested in many aspects of God's creation, and like others before him, he studied the behavior of light. For example, he agreed with Euclid on the Law of Reflection (which you learned about in Lesson 23), but he used it to determine something that Euclid hadn't thought of. The best way to learn what Hero determined is to do the following experiment.

Where Is That Pencil's Reflection?

What you will need:

- Two pencils that are the same size
- A flat, hand-held mirror (It can't be a magnifying mirror.)
- A ruler or meter stick
- Some modeling clay, like Play-Doh
- A clear, flat surface

What you should do:

1. Place a small lump of clay on the flat surface close to you.
2. Stick one of the pencils in it so that the pencil easily stands straight up without any help.
3. Place the mirror on the other side of the pencil so that the pencil is centered in the mirror.
4. Lay the ruler down so that it is face up and you can read the distance between the pencil and the mirror.
5. Move the mirror around so that it is 10 centimeters (4 inches) from the pencil. Hold the mirror there with one of your hands.
6. Use your other hand to hold the other pencil directly behind the mirror so it is almost touching the mirror.
7. Lower your head so you can easily see the first pencil's reflection in the mirror and the other pencil sticking up above the mirror. At this point, your experiment should look something like the picture on the right.

8. With your head right behind the pencil that is in front of the mirror, move the pencil behind the mirror so that it is perfectly lined up with the image of the other pencil in the mirror and touching the back of the mirror. The pencil behind the mirror should look like it is sticking up straight above the image of the other pencil in the mirror, as shown in the drawing on the right.

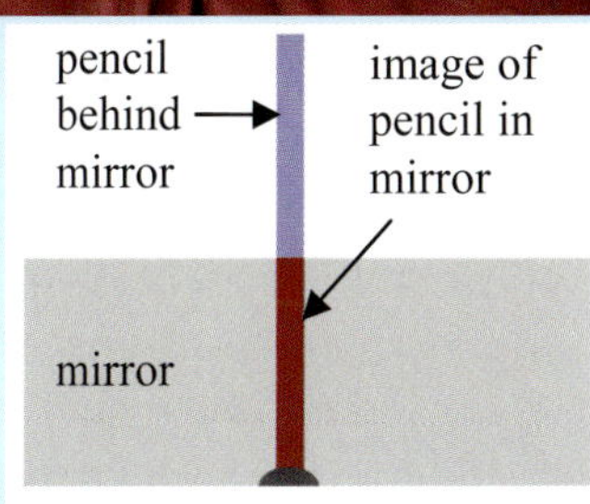

9. Move your head to one side so that it is lined up with one side of the mirror. Is the pencil behind the mirror still lined up with the image of the other pencil in the mirror? Probably not.
10. Move your head so that it is lined up with the other side of the mirror. Once again, the pencil behind the mirror should no longer look like it is lined up with the image of the other pencil in the mirror.
11. Move your head so that it is pretty much right behind the pencil that it is front of the mirror.

12. Move the pencil behind the mirror farther away from the mirror, but still keep it lined up with the image of the other pencil in the mirror.
13. Once again, move your head to one side of the mirror and check to see if it is still lined up with the image of the other pencil in the mirror.
14. Continue to repeat steps 12 and 13 until you get to the point that no matter where your head is (centered on the mirror, on one side of the mirror, on the other side of the mirror), the pencil behind the mirror looks like it is lined up with the image of the other pencil in the mirror.
15. Once you have succeeded, look at the distance between the mirror and the pencil that is behind the mirror. How does it compare to the distance between the mirror and the pencil that is in front of it?
16. Clean up your mess and put everything away.

Think about what happened in the experiment. When you looked in the mirror, you saw the front pencil's image in the mirror, right? Where was that image? It was in the mirror, I know, but *where was it in the mirror*? Was it right on the surface of the mirror? You might think it was. You might think that looking at a mirror is like looking at a picture. All the images are right there on its surface. However, your experiment shows that this is just not the case. Think about it. If the image was right on the surface of the mirror, the other pencil should have been very close to the image when it was almost touching the back of the mirror. In other words, the pencil behind the mirror should have been almost on top of the image in the mirror. However, when you looked at the mirror from a different spot, you saw that the pencil behind the mirror was no longer lined up with the image in the mirror. That's because the image in the mirror was *behind the mirror*!

When you look in a mirror, your image is not on the surface of the mirror. It is behind the mirror!

That might sound really strange to you, but think about it for a minute. If the pencil behind the mirror was right on top of the image in the mirror, it should be lined up with the image no matter where your head is when you are looking at it. If that's not the case, the pencil must not be on top of the image. As you changed the distance between the pencil and the mirror, however, you eventually got to the point that no matter where your head was, the pencil was always lined up with the image. That's the point where the pencil was right on top of the image!

So how far from the mirror was the pencil when it was actually sitting on top of the image? It should have been about 10 centimeters (4 inches) behind the mirror. It might not have been exactly 10 centimeters away, since there are a lot of ways you could make a small mistake in the experiment. However, it should have been just about 10 centimeters (4 inches) behind the mirror. Now let me ask you a question. How far was the other pencil in front of the mirror? You arranged things so that it was 10 centimeters (4 inches) in front of the mirror. This should tell you something:

When you put an object in front of a flat mirror, its image is the same distance behind the mirror as the object is in front of the mirror.

In other words, if you put an object 10 centimeters (4 inches) in front of a flat mirror, its image will appear to be 10 centimeters (4 inches) behind the mirror. If you put an object 25 centimeters (10 inches) in front of a flat mirror, its image will appear to be 25 centimeters (10 inches) behind the mirror.

Hero of Alexandria was the first to demonstrate this. He didn't use an experiment to show it like you did. Instead, he started with Euclid's Law of Reflection, and from that, he showed that it must be true. How he did that is a bit too complicated to explain, but the result is correct whether it is shown by experiment or by the way Hero showed it. What you see in a mirror isn't all on the surface of the mirror. Instead, the mirror shows the depth of the world that reflects in it. This makes what we see in a mirror more realistic than what we see in a picture. For example, if you see a picture of a boy holding a ball out in front of him, it is hard to tell exactly how far the ball is in front of the boy. However, if you look at your reflection while holding a ball in front of you, it is easy to tell how far the ball is in front of you. That's because the mirror's image has depth, while a picture does not.

Now you might wonder why I keep saying that the mirrors we are working with are "flat" mirrors. Aren't all mirrors flat? No. It turns out that there are curved mirrors. They don't reflect things the same way that flat mirrors do. As a result, they can actually magnify the things they reflect, depending on where that object is in relation to the mirror. You will learn more about curved mirrors later on in this course.

LESSON REVIEW

Youngest students: Answer these questions:

1. Fill in the blanks: If I put an object 12 centimeters in front of a flat mirror, its image will appear to be ______ centimeters _____ the mirror.

2. What law did Hero use to demonstrate where an object's image is in a flat mirror?

Older and oldest students: In your notebook, explain what you did in the experiment and what you saw. Explain how it demonstrated that when you put an object in front of a flat mirror, its image appears to be same distance behind the mirror as the object is in front of the mirror. Note how Hero demonstrated this same fact.

Lesson 36: Ptolemy (c. AD 90 – c. AD 168) and the Universe

This is how one artist imagines Ptolemy.

Shortly after most historians think Hero of Alexandria died, a very important philosopher was born. His name was Claudius (klaw' dee us) Ptolemaeus (tah' luh may' us), but he is generally known simply as **Ptolemy** (tah' luh me). Even though he did a lot to advance our understanding of science, he is best known for what he got wrong. He believed that the earth was the center of the universe and that everything (the sun, moon, planets, and stars) orbited around the earth. Now, of course, that view wasn't new in Ptolemy's day (he did most of his work between AD 127 and AD 141). You should remember that Aristotle had the same basic view of the universe, which is called the geocentric view. However, Ptolemy was considered a champion of this view. In fact, when scientists refer to the geocentric view of the universe, they often call it the **Ptolemaic system**.

If Ptolemy didn't come up with the idea, why is he remembered for it? There are really two reasons. First, he wrote a detailed book, called the *Almagest* (awl' muh jest), that compiled a huge amount of information regarding what was known about the moon, planets, and stars at that time. For example, it had a star catalog, and most (if not all) of the information was taken from Hipparchus. The information in the book was so valuable to **astronomers** (scientists who study the sun, moon, and stars) that it was used for more than 1,500 years after Ptolemy died.

In addition to compiling a lot of information about the heavens in his *Almagest*, Ptolemy also included his own ideas about how the universe was arranged. That brings us to the second reason Ptolemy is associated with the geocentric view of the universe. His book laid out a **geocentric model** of the universe that was in agreement with the many observations astronomers had made over the years. What do I mean by "model"? Well, what is a model of an airplane? It is something that represents an airplane but isn't really the airplane, right? A scientific model is something that represents a system in nature in an attempt to explain how it works. In his book, Ptolemy produced a representation of the universe that could explain everything that had (so far) been observed in the universe.

This was impressive, because there were some observations that had been made over the years that were hard to understand if the earth really was in the center of the universe and everything orbited around it. To see what I mean, let's use the model you made back in Lesson 21.

Epicycles

What you will need:

- The model of the universe you used in Lessons 21 and 22
- A paper plate
- Scissors
- Crayons, colored pencils, or markers

- A pushpin
- Someone to help you

What you should do:

1. When you last used your model, you made it look the way Aristarchus thought the universe looked. While his view was more correct than Ptolemy's, it didn't catch on in the ancient world. As a result, you need to make it look like it did when you first made it – with the earth at the center and the moon, a planet, and the sun orbiting the earth. In case you have forgotten, the picture on the left shows you what your model needs to look like now. Fix your model so it looks like that. If you have lost the circle that had the sun on it, just make a new one. Now remember, in this model, the ball of clay is the earth. The first circle has the moon on it. The second circle has a planet on it. The third circle has the sun on it, and the plate has the stars on it.

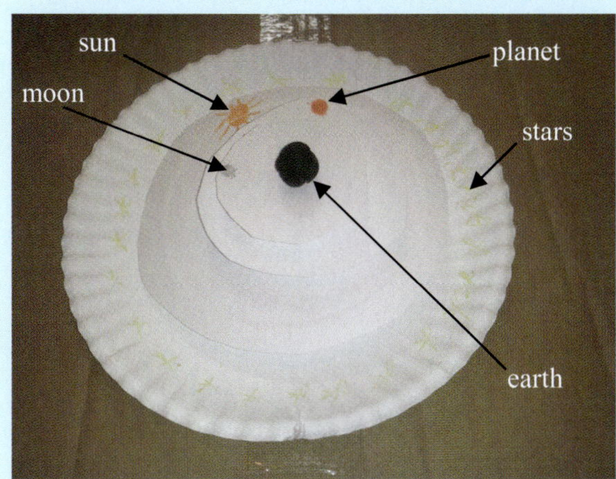

2. Turn the second circle (the one with the planet on it) so that the planet orbits the earth. As you do this, think to yourself what a person on earth would see if he were watching the planet in the sky over a period of many, many days. It would look like the planet was always moving in the sky, right? The important thing is that it would look like the planet was always moving in the same direction. Just as the sun always moves from east to west in the sky, the planet would always move in the same direction across the sky.
3. The problem is, that's not what astronomers in the ancient world saw. Instead, a planet would start moving in one direction across the sky (from west to east), but then for a few days, it would move in the opposite direction (from east to west)! Can you imagine what you would think if the sun did that? Suppose you watched the sun moving from east to west in the sky and then suddenly, it moved in the opposite direction for a few hours. That would be weird, wouldn't it? Ancient astronomers thought the motion of the planets in the sky was just as weird. Ptolemy, however, tried to fix the problem, and you are going to do the same thing he did. Remove the ball of clay.
4. Remove the circle that has the moon on it. Ptolemy's model didn't get rid of the moon, of course. We are just going to ignore it for now to make the other things easier to see.
5. Cut a small circle out of the paper plate. It should be small enough so its center can be pinned to where the planet is now but it won't come close to touching the nail.
6. Draw a planet on the edge of the circle.
7. Use the pushpin to attach the center of the circle to where the planet is drawn on your model.
8. Put the ball of clay back on the nail. Your model should now look like the picture on the right.
9. Have your helper spin the large "planetary circle" slowly so it orbits the earth, while you spin the small "planetary circle" quickly, so it spins as the large circle orbits. As you do that, think about what an astronomer would see if he watched how the planet traveled in the sky. He would see that it mostly moved in one direction, but for a brief time, it would move in the opposite direction. This makes the model agree with what astronomers saw.
10. Clean up your mess. You are now done with the model.

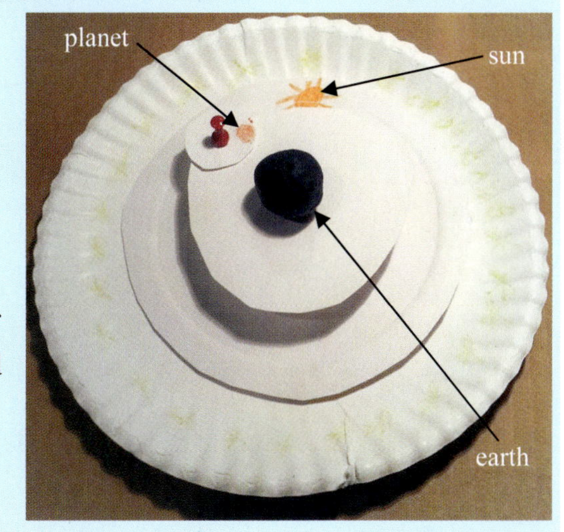

What Ptolemy's model did was explain something that puzzled ancient astronomers. If all the planets really did orbit the earth, they should appear to travel in the same direction across the night sky. However, the planets didn't do that. They would start out traveling in one direction, but then they would appear to stop and reverse direction for a while. Then, they would stop again and reverse direction again! This is called **retrograde** (reh' truh grayed') **motion**, and Aristotle's view of the universe couldn't explain it.

This is why Ptolemy had the planets travel in smaller circles, called **epicycles** (eh' puh sy' kulz), as they orbited the earth in larger circles. Using those epicycles, Ptolemy's model could explain retrograde motion. The planets' overall course across the sky was in one direction, determined by the large circles in which they orbited the earth. However, as they traveled in their epicycles, they would sometimes move in the opposite direction of their overall course. This was an ingenious way to get the geocentric model to at least be able to explain why the planets behaved as they did. As a result, Ptolemy was considered to be a great astronomer.

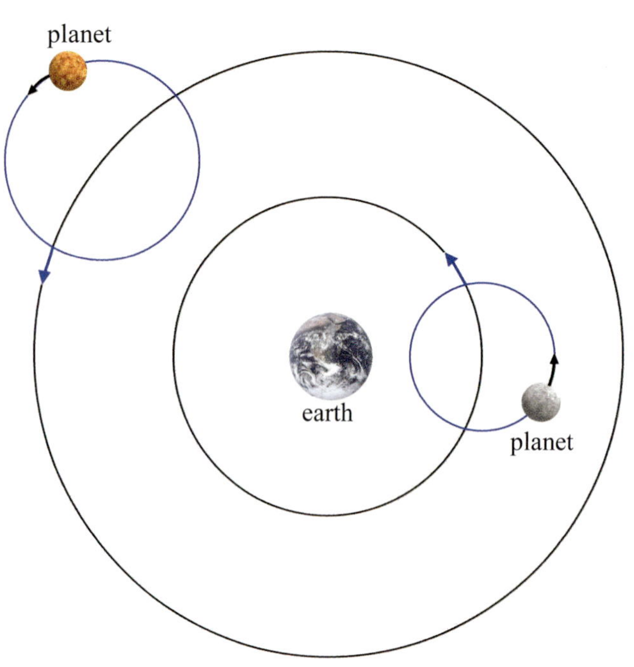

This shows the Ptolemaic system for earth and two planets. The blue circles are the epicycles. The planets travel in the epicycles, and epicycles orbit the earth along the black circles.

Of course, as time went on, more and more observations were made about planetary motion, and because of that, the geocentric model faced more and more problems. At first, following the example of Ptolemy, astronomers just added more epicycles to the model. Eventually, however, it had to be changed, and the heliocentric model was shown to be correct. Of course, I am getting ahead of myself here, because that doesn't happen for about 1,400 years!

LESSON REVIEW

Youngest students: Answer these questions:

1. What do astronomers study?

2. What did Ptolemy add to the geocentric model to account for the fact that planets sometimes reversed their course in the night sky?

Older students: In your notebook, draw the Ptolemaic system for a couple of planets, as shown in the drawing above. Label the epicycles and explain how the model has the planets moving. Define retrograde motion and explain how the epicycles account for it.

Oldest students: Do what the older students are doing. In addition, indicate which is faster: the motion of the planet in its epicycle or the motion of the epicycle around the earth.

Lesson 37: Ptolemy and Optics

Although he is best known for his work in astronomy, Ptolemy wrote about many other aspects of science. For example, he wrote a book called *Geographia* that compiled everything that was known about the geography of the world at the time. Of course, since travel was difficult back then, the known world was mostly Europe, some of western Asia, and some of northern Africa. Even though Ptolemy mostly compiled information that had been gathered by others, he did make one very important contribution to the study of geography. It was well known back then that the earth was a sphere. However, maps were printed on paper, which is flat. As a result, map makers had to practice the art of taking a round surface and drawing it on a flat page. This is called "projection," and Ptolemy used his mathematical skills to improve how it was done. As a result, his book contained the most accurate maps of the time. Nevertheless, there were still some obvious problems, as you can see from the map on the left. Notice how the shape of Africa is not at all correct.

This is a map from a 15th-century copy of Ptolemy's *Geographia*. It is thought that aside from the added color, the map is faithful to Ptolemy's original.

Not only did Ptolemy write about geography, but he also wrote about music. Do you remember studying Pythagoras a while back? Ptolemy analyzed music in a mathematical way. He discussed the mathematical relationships between different notes, starting with what Pythagoras had done and then adding more to it. In addition, do you remember the music of the spheres you learned about when you studied Aristotle? Ptolemy wrote about that as well.

I want to concentrate on something else that Ptolemy wrote about, however. Like many other natural philosophers, he was fascinated by light. He studied light quite a bit and ended up writing a book called *Optics*. In it, he includes the Law of Reflection, which had been discovered by Euclid. In addition, he made many observations about **refraction** (rih frak' shun), the process by which light bends when it starts traveling through a different substance. While you might know something about this process already, I want you to do the following experiment to learn more about what it is and how it works.

Refraction in Water and Oil

What you will need:

- A small container, like a Tupperware storage container (It needs to be one you can mark up.)
- A quarter
- A toothpick
- Modeling clay, like Play-Doh

- A flashlight
- Scissors
- A marker
- A dim room
- Water in a pitcher
- Vegetable oil
- Paper towels or a dishtowel
- Someone to help you

What you should do:

1. Mold the clay around the quarter so that the clay surrounds the quarter and has a bump near one edge of the quarter. Don't use too much clay. The quarter is supposed to give weight to the clay, and the bump of clay just needs to be enough to securely hold a toothpick upright.
2. Use the scissors to cut the toothpick in half.
3. Stick one half of the toothpick in the bump of clay.
4. Put the toothpick/clay/coin thing in the middle of the container. If the toothpick half is taller than the sides of the container, cut some more of it off. You need to be able to cover the toothpick half with water.

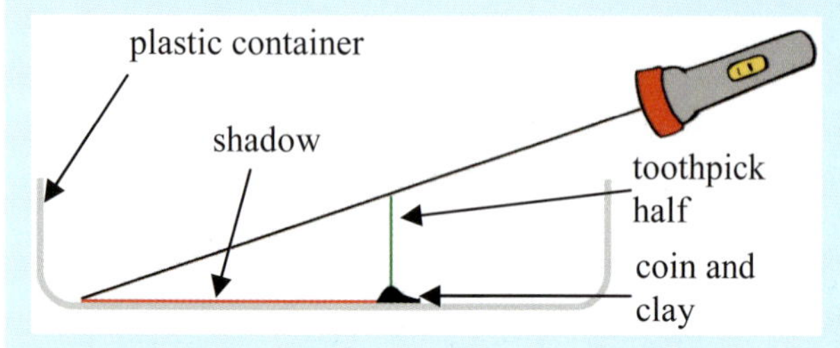

5. Take everything into the dim room.
6. Hold the flashlight up at an angle so it shines on the toothpick and the toothpick half casts a shadow into the container. Turn the toothpick/clay/coin thing so the shadow doesn't fall across the coin. Instead, it should fall onto the bottom of the container.
7. Change the angle of the flashlight to get a long, distinct shadow from the toothpick half. The shadow should be as long as possible but not extending beyond the bottom of the container.
8. Have your helper make a mark on the outside of the container to indicate the length of the shadow.
9. Have your helper *slowly* fill the container with water while you keep the flashlight exactly where it was when your helper marked the length of the shadow. Your helper needs to add the water slowly enough so that the toothpick/clay/coin thing doesn't move.
10. Once the toothpick half is covered with water, notice the length of the shadow. What happened?
11. Have your helper make a mark on the outside of the container to indicate the new length of the shadow.
12. Empty the container and dry it out with the paper towels or dishtowel.
13. Repeat steps 4-7 so that the setup is the same as before and the shadow's length is once again indicated by the mark your helper made in step 8.
14. Have your helper *slowly* fill the container with vegetable oil while you keep the flashlight exactly where it is. Your helper needs to add the oil slowly enough so that the toothpick/clay/coin thing doesn't move.
15. When the toothpick half is covered in oil, once again, note the length of the shadow. What happened?
16. Clean up your mess.

What did you see in the experiment? The shadow cast by the toothpick half should have gotten *shorter* as your helper filled the container with water. When your helper filled the container with vegetable oil, the shadow should have gotten even shorter! What caused this? Well, as you probably know, when light enters water from the air, its direction changes. It actually bends a bit. This is what we call refraction. Because the light bent when it hit the water, it was as if you were holding the flashlight at a different angle. As you already know, the angle at which light hits something affects the length of its shadow. As a result, the refraction of light as it hit the water caused the shadow's length to change. The light bent downward, casting a shorter shadow. What about the oil? Well, refraction is different in every substance. Some substances cause light to bend a bit, others cause it to bend even more. It turns out that oil bends light more strongly than water, so the shadow was even shorter when you filled the container with oil.

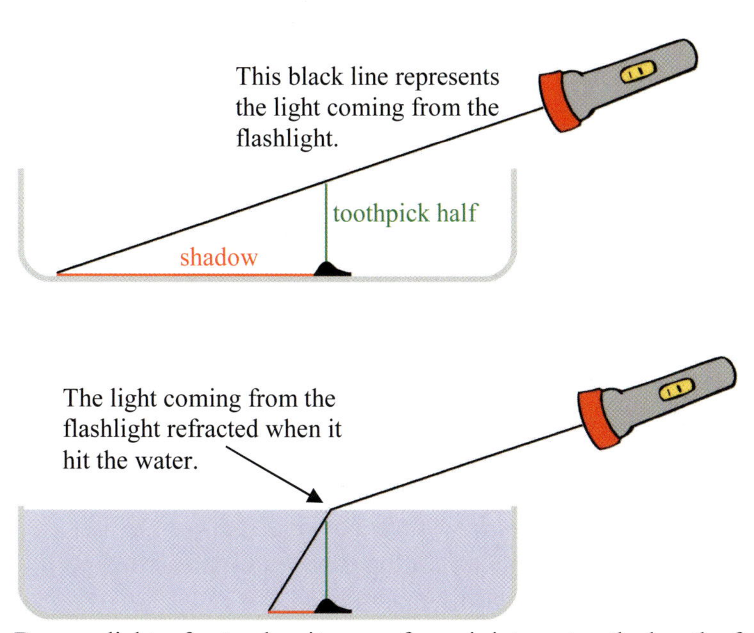

Because light refracts when it passes from air into water, the length of the shadow that was cast by your flashlight changed when you added water to the container. Since oil refracts light even more strongly than water, the shadow was even shorter when you filled the container with oil.

Ptolemy's book on optics detailed this fact. He discussed how some substances refract light weakly and others refract light more strongly. He thought that this was due to the fact that different substances resisted the passage of light differently. It turns out that he was essentially correct. Light travels more slowly in water than in air, for example, which is why water refracts light. We now know that different substances refract light differently because light travels at different speeds in different substances. The more light has to change speed going from one substance to another, the more it refracts.

Ptolemy made a device that measured how strongly light refracts when it enters a new substance, and he tried to figure out some mathematical way to analyze the results. Unfortunately, he couldn't do it. However, his measurements inspired others to study the effect, and eventually, a mathematical analysis was found. It requires more math than you have right now, so I won't go into it. When you study physics in high school, you will learn the relationship.

LESSON REVIEW

Youngest students: Answer these questions:

1. What is refraction?

2. Which refracts light more: water or vegetable oil?

Older students: In your notebook, draw three pictures that represent your experiment. The first should show your experiment without water, the second should show your experiment with water, and the third should show your experiment with oil. Explain the results below the drawings.

Oldest students: Do what the older students are doing. In addition, write down this question and your answer: "Does light travel faster in water or vegetable oil? Why?" Check your answer and correct it if it is wrong.

Lesson 38: Galen the Physician (AD 129 – c. AD 200)

About 40 years after Ptolemy was born, another very important figure in the history of science entered the world. His name was Claudius (klaw' dee us) Galenus (gay' luh nus), but today he is better known simply as **Galen** (gay' lun). We know that he was born in AD 129, because he says so himself in one of his books. We are not sure exactly when he died. Some historians think he died as early as AD 199, but most think he died around AD 216. Even though historians can't agree on the span of Galen's life, all agree that it was a very productive one! It is thought that he wrote about 600 different works, and almost half of all the literature that has survived from ancient Greece comes from Galen!

This is how 19th-century French artist Pierre Roche Vigneron imagined Galen.

What subject interested Galen enough to produce so many works of literature? It was the study of medicine. At the age of 16, his father sent him to study medicine. Three years later, his father died, leaving him a very wealthy man. While studying medicine, Galen read the works of Hippocrates (of course), and one of the things Hippocrates wrote was that anyone who wanted to study medicine should travel widely to learn as much as he could about different health problems and their treatments. Galen eventually attended a medical school in Alexandria. After a lot of study, he began working as a doctor and became so well known that he eventually became the personal physician for two different Roman emperors (Commodus and later Septimius Severus)!

While Galen was influenced strongly by the ideas of Hippocrates (he believed in the four humors, for example), he developed many of his own ideas that he put into his practice of medicine, and some of them were quite good. One of the most important things he did was to emphasize the importance of knowing human **anatomy** (uh nah' tuh me), which tells us about the **organs** in the human body and where they are located. Now, of course, in human anatomy the word "organ" doesn't refer to the musical instrument. Instead, it refers to a structure in the body that performs a function. For example, your stomach is an organ that performs a vital function when it comes to how your body processes the food that you eat. To get an idea of what anatomy is all about, do the following activity.

Trying Your Hand at Anatomy

What you will need:

- A photocopy of page A1 from the Appendix of your parent/teacher's *Helps and Hints* book
- Scissors
- Glue (A glue stick works best, but any glue will do.)
- Your notebook (a blank sheet of paper if you are doing the review exercises for the youngest students)

What you should do:

1. Use the scissors to cut out the outline of a human body that you find on page A1 of the Appendix.

2. Start a new page in your notebook and label it "A Little Human Anatomy."
3. Paste the human outline on that page of your notebook.
4. Use the scissors to cut out all the organs on A1 of the Appendix. You don't have to cut them out perfectly. If there is white space around the organ, don't worry about it. If you are having trouble, get an adult to help you.
5. Without using any books or resources, lay each organ in the outline of the body where you think it is in a real human body. Don't glue the organs onto the outline yet. Just lay them there. Some parts will lie on top of other parts. That's fine. Just put them where you think they should go.
6. When you are done, check your work against the drawing that your parent/teacher has in his or her *Helps and Hints* book. It is on the page where the answers to the review exercises can be found.
7. Fix any mistakes you made, and then glue the organs in their correct places.
8. Clean up your mess.

How did you do? It really doesn't matter. I just wanted to give you a feel for what the study of anatomy involves. If you found the activity a little difficult, just realize that you only had to place a handful of organs into the body. In fact, there are *a lot more* organs in the body, and someone who is really good with anatomy knows where all of them are!

One reason you might have had a bit of trouble putting the organs in their correct places is that you probably have never seen them, because they are on the *inside* of the body. If you knew the right place to put some of them, it is probably because you had seen a drawing in a book or read about the organ somewhere. But how did scientists figure out where the organs are in the body? Believe it or not, they actually cut open dead bodies and looked inside. As you already know, this is called dissection, and while it sounds pretty gross, it is an important part of learning to be a doctor. Nowadays, there are many people who donate their bodies to science. This means that when they die, they are giving permission to the scientific community to use their bodies to help people learn more science. Some of those bodies are sent to medical schools so that medical students can dissect them. Most medical schools require their students to dissect a dead person. It's just part of the training.

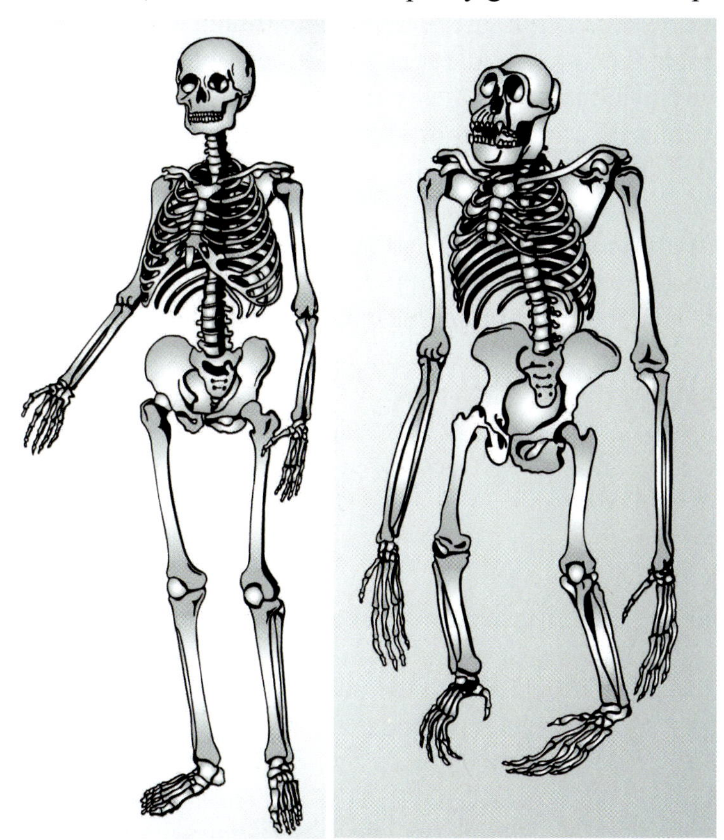

Notice how similar a person's skeleton (left) is to the skeleton of an ape (right). These similarities allowed Galen to learn a lot about human anatomy by dissecting apes.

For Galen, dissecting dead people was not an option, because it was against Roman law. As a result, he did the next best thing. He dissected dead animals (mostly apes) to learn about their organs. He then tried to guess how that applied to people. After all, apes and people look very similar on the outside, so it only makes sense that they should look very similar on the inside. As you can see on the left, for example, there are many similarities between the skeleton of a person and that of an ape. There are clear differences, of course, but there are also a lot of similarities. Galen used those similarities to help him understand the skeleton and many other organs in the human body. Because of the differences, he got some things wrong, but because of the similarities, he got a lot of things right.

Galen's knowledge of anatomy helped him a great deal when it came to treating injuries, especially the wounds suffered by soldiers. He also took the opportunity to study what he could of human anatomy when he treated a wound. In fact, he called wounds "windows into the body", and he looked into those windows to better understand human anatomy. He wrote all of his knowledge down in painstaking detail, and he became the recognized authority when it came to human anatomy. In fact, his books were considered the "final say" in matters of human anatomy for more than 1,000 years. As a result, he is often called the father of human anatomy.

While it is a testament to Galen's hard work that his books became so influential for so long, this did cause a problem. As I said before, Galen got many things right when it came to human anatomy, but he got a lot of things wrong. For example, the humerus (the arm bone that attaches to the shoulder) in an ape is gently curved, but in a human, it is pretty much straight. The same can be said about the femur, which is the leg bone that attaches to the hip. When later scientists were able to dissect dead people and show that the humerus and femur were straight, unlike Galen described them, some scientists refused to believe that Galen was wrong. They thought that since the time of Galen, people had gone through a change that straightened their humeruses and femurs! Some scientists blamed it on the tight clothing that became fashionable after Galen's time.

This gives me a chance to repeat a very important truth about science: *Even a very good scientist can be wrong about a great many things.* While Galen was a very good scientist for his time, as science progressed, a lot more was learned about human anatomy. Unfortunately, a lot of that knowledge was not taught, because it contradicted what Galen taught, and no one wanted to contradict such a great man. That's not the way science should be done. While it is good to respect those who have done great things, we can't be afraid to contradict them when the evidence shows that they are wrong. Science should be done not based on the opinion of the authorities, but based on what experiments and observations show us. Remember that!

While there is a whole lot to learn about human anatomy, I am not going to teach it to you now, because at this point in human history, there were still too many errors in the understanding of human anatomy. You will learn a lot more about human anatomy later on in this course.

LESSON REVIEW

Youngest students: Answer these questions:

1. What is anatomy?

2. What does the word "organ" refer to in anatomy? Give an example of an organ.

Older students: In your notebook, label the organs that are glued into the outline of the person. You can use your parent/teacher's *Helps and Hints* book to learn which organ is which. Below that, define the words "anatomy" and "organ."

Oldest students: Do what the older students are doing. In addition, explain why Galen's dissection of apes helped him learn about human anatomy but ended up causing him to be wrong when it came to some things.

Lesson 39: Galen and Blood

Galen contributed more to science than just his insights regarding anatomy. He also taught us a lot about **physiology** (fih' zee ah' luh jee), which is the study of the specific functions and activities of the organs in a living organism. Do you see the difference? Anatomy tells us what organs exist in a body and where they are, while physiology tells us what they do and how they work. For example, long before Galen, philosophers knew there were "tubes" in the human body. They even knew where some of those tubes were. However, until Galen's time, scientists thought that the tubes carried air. Well, it turns out that Galen showed that the tubes did not carry air – they carried blood. So before Galen's time, philosophers knew the anatomy of these tubes (which we now call blood vessels), but they had the physiology wrong. Galen corrected that.

How did Galen figure out that blood vessels carried blood? Because he actually opened up living animals and looked inside and saw the vessels carrying blood. Opening up a live animal to study how it works is called **vivisection** (vih' vuh sek' shun), and it sounds even worse than dissection. However, without it, Galen would have never figured out that the blood vessels in the body carried blood. He also wouldn't have figured out one of his most important contributions to the field of medicine. You can learn about that contribution by performing the following experiment.

Your Pulse

What you will need:
- Some modeling clay, like Play-Doh
- A toothpick
- A stopwatch or a watch with a second hand.
- An adult

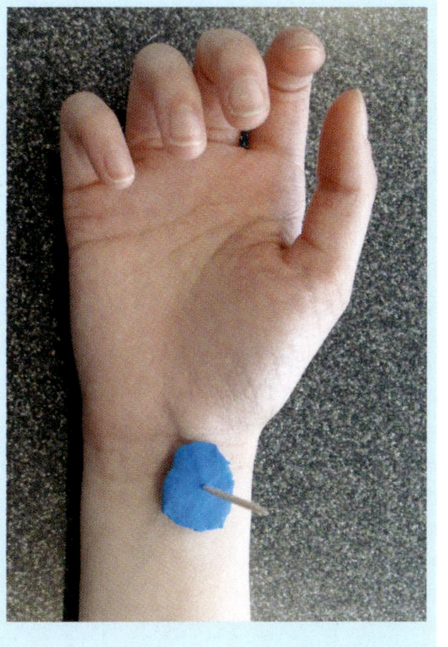

What you should do:
1. Lay your hand on a flat surface, palm up.
2. Place a dime-sized lump of clay on your wrist right below the base of your thumb (see the picture on the right).
3. Stick the toothpick into the clay so it stands straight up.
4. Watch the toothpick carefully. What is it doing?
5. Take the toothpick and clay off your wrist.
6. Put your forefinger where the clay was, and press down a bit. Do you feel something? You should. If you don't feel anything, press a bit harder and move your finger around. As long as you don't push too hard, you should eventually feel what was making the toothpick move. That's your **pulse**.
7. Start the stopwatch or wait until the second hand reaches an easy-to-read spot, and then count how many bumps you feel from your pulse for 30 seconds. Multiply that number by 2. That's your **pulse rate**.
8. Repeat steps 6 and 7 on the adult. How does the adult's pulse rate compare to yours?
9. Stand up and do 20 jumping jacks as quickly as you can.
10. Repeat steps 6 and 7 to measure your pulse rate. How did it change compared to the pulse rate you measured before?
11. Have the adult do 20 jumping jacks as quickly as he or she can.
12. Repeat steps 6 and 7 on the adult to measure the adult's pulse rate. How did it change compared to the pulse rate you measured before?
13. Clean up your mess.

The toothpick in the experiment should have allowed you to actually see your pulse. There is a blood vessel in your wrist that throbs as it sends blood to your hand, and that throbbing is your pulse. As it throbbed, it should have moved the clay, which in turn moved the toothpick. Thus, the toothpick should have moved back with each pulse.

When you count how many throbs that blood vessel makes every minute, you are measuring your pulse rate. What did you notice about the difference between your first pulse rate and the one you measured after doing 20 jumping jacks? You should have seen that after the jumping jacks, your pulse rate was higher. In the same way, the adult's pulse rate was higher after the jumping jacks, wasn't it? What about the difference between the pulse rates you measured for yourself and those you measured for the adult? They were different as well, weren't they?

The pulse rate obviously tells you something about a person's body. Your pulse rate is probably different from that of the adult you measured, because your body is different from that of an adult. In the same way, your pulse rate changes based on what you have been doing. Galen figured this out. He realized that the pulse tells a lot about a person's body, so he wrote an entire series of books about how a doctor should use the pulse rate (as well as other things about the pulse) to help him figure out what is wrong with a patient.

When a doctor is trying to figure out what is wrong with a patient, we say that the doctor is trying to **diagnose** (dye' uhg nohs) the patient's illness. This process is often long and complicated, but early on, the doctor or someone from the doctor's staff usually takes the person's pulse. Sometimes it is done with a medical instrument, and sometimes it is done the same way you just as you did it, by putting a finger on the patient's wrist and counting the throbs. That's what Galen did. He not only counted the throbs, he tried to find anything abnormal about the rhythm or feel of the patient's pulse. He realized that he could learn a lot about a patient through the patient's pulse.

This doctor is measuring his patient's pulse rate the same way you did in your experiment.

Why is a person's pulse so important? Galen had the overall picture right, he just had some of the details wrong. Based on his vivisections, he knew that a person's body contained all sorts of blood vessels that carried blood throughout the entire body. He therefore reasoned that the blood carried things that the body needed. He figured it carried nutrients that helped the various organs do their jobs, and he figured it probably carried other important things as well. Thus, he decided that the way blood was traveling throughout the body could tell you a lot about the body's condition and its needs.

In addition, Galen noticed the difference between veins and arteries. You've already learned that veins carry blood back to the heart, while arteries carry it away from the heart. Galen didn't know that, but he did discover that the blood in veins was colored slightly different from the blood in arteries. In each case the blood was red, but the blood in the arteries was brighter red, and the blood in the veins was darker red.

Galen actually thought the blood was made and then used up as it traveled through the blood vessels, but we now know that is not true. However, he rightly concluded that since blood was traveling to all parts of the body, it was supplying something important to the various parts of the body. He also reasoned that the throbbing of the artery in the wrist is related to how that blood is moving. As a result, he rightly concluded that the pulse was an important way of learning about how a person's body is using its blood. The faster the pulse rate, the more the body is using what is in the blood. If the pulse felt abnormal, the blood flow was abnormal. A weak pulse, for example, indicated weak blood flow.

Interestingly enough, even though Galen's vivisections showed him that the heart was related to how the blood flowed throughout the body, and even though he actually saw hearts pumping, he didn't think the heart's job was to pump blood through the blood vessels. He thought the arteries did that themselves. He thought that the heart's job was to suck blood in from the veins. So even though Galen helped us understand the basics of blood, arteries, and veins, he ended up having a lot of wrong ideas about them as well.

Of course, as time has gone on, we have corrected a lot of Galen's mistakes. For example, we now know that there is a third type of blood vessel, capillaries, and that they connect the veins to the arteries. In the end, then, blood is not continually made and used up, as Galen thought. Instead, it flows throughout the body in a completely closed series of vessels. We also understand that the color difference between the blood in the arteries and veins is just a reflection of how much oxygen the blood has in it. Most arteries carry blood that has a lot of oxygen in it, while most veins carry blood that has only a small amount of oxygen in it. You will learn more about that in another course.

LESSON REVIEW

Youngest students: Answer these questions:

1. What is a doctor doing when she diagnoses a patient's illness?

2. What does the word "physiology" mean?

Older students: In your notebook, draw a hand (palm up) and point out where you put your finger to feel a pulse. Describe what the pulse rate measures and how its rate is related to a person's activities.

Oldest students: Do what the older students are doing. In addition, do some research to find out how your pulse rate changes over the course of your life. Write what you learn in your notebook.

Lesson 40: Galen and Muscles

Your body is full of muscles. Some of them allow you to run and play, while others allow you to breathe. Some allow your stomach to churn so it can digest your food. Galen didn't know about all the muscles in the body, but he studied muscles a lot and ended up adding a lot to our knowledge of muscles and how they work. He spent a lot of his time studying what we call **skeletal muscles** today. Skeletal muscles are the ones that attach to your skeleton and allow you to move different parts of your body when you want to. The drawing on the right shows you many of the skeletal muscles in the body, but there are a lot more. In fact, there are over 600 different skeletal muscles in your body!

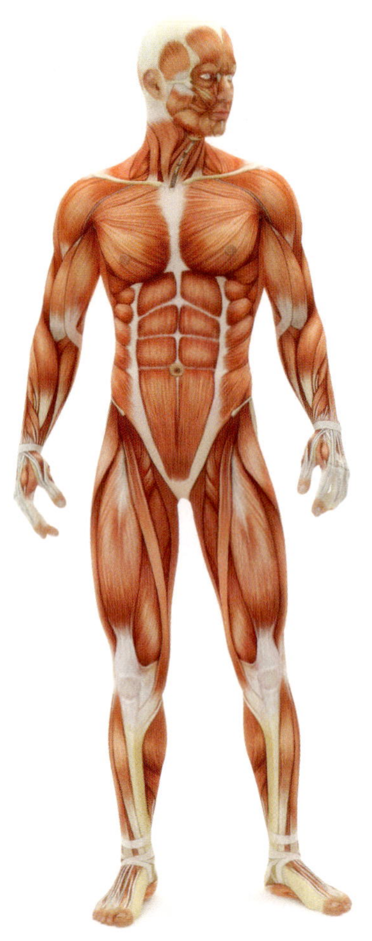

This drawing shows you some of the skeletal muscles (the red structures) in the body.

Why do you have so many skeletal muscles? Because your body needs to move in many different ways. As you will soon see, it usually takes at least two muscles to move a part of your body in one specific way. The more ways you need to move your body, then, the more skeletal muscles you need to have. Amazingly enough, your body has even more muscles than just your skeletal muscles. Galen, for example, recognized that the heart is made of muscle. It's not the same as a skeletal muscle, but it is definitely made of muscle.

Scientists say that the heart contains muscle **tissue**. Do you know what the term "tissue" means? It doesn't refer to the stuff you use when you blow your nose. Instead, it refers to the stuff that organs are made of. If you examine a stomach, for example, you will find that it is made of material that is quite different from the material that makes up bone. Bone is made of one kind of tissue, while the stomach is made from other kinds of tissue. The heart is made mostly of muscle tissue, and most of your organs have some muscle tissue in them.

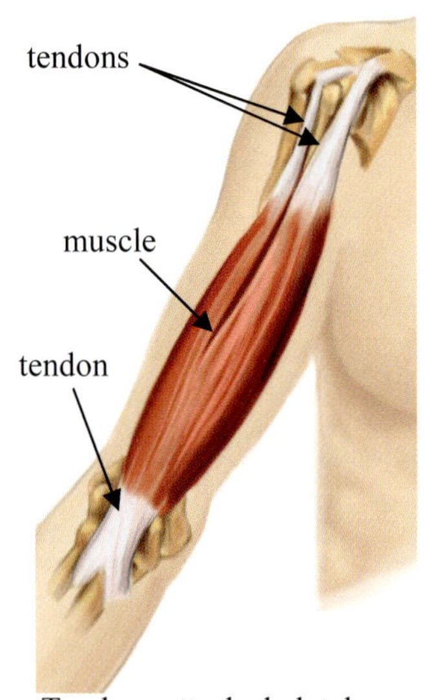

Tendons attach skeletal muscles to the skeleton.

You will learn a lot more about muscles when we get to the point in history where scientists better understood them. However, I do want to spend some time talking about one thing Galen figured out when it comes to muscles. First, Galen saw that skeletal muscles were not attached directly to the skeleton. Instead, he found that there was a different kind of tissue between the muscle and the bone. We call that tissue **tendon**. In the drawing above, you can see that there are white areas and red areas. The red areas represent muscles, and the white areas represent tendons. In the drawing on the left, I am showing you only one muscle, the **biceps** (by' seps) **brachii** (bray' kee eye'). As you will learn in a moment, this muscle helps you move your arm at the elbow. Notice the muscle is in the middle, and it is attached to the skeleton with tendons.

The real advance that Galen made when it came to skeletal muscles is that he showed the basics of how they work. The best way to see what Galen figured out is to do the following activity.

Skeletal Muscles Work in Antagonistic Pairs

What you will need:
- A photocopy of page A2 from the Appendix of your parent/teacher's *Helps and Hints* book
- A large piece of thick cardboard (It needs to be thick enough to push a thumbtack into.)
- String
- Scissors
- Tape
- Two pushpins or thumbtacks

What you should do:
1. Cut out the two illustrations from the photocopy of page A2 in the appendix.
2. Use a pushpin to attach the humerus to the top of the piece of cardboard.
3. Use another pushpin to attach the radius and ulna to the end of the humerus and to the piece of cardboard. Don't push the pin too hard, because the radius and ulna need to be able to rotate freely.
4. Cut two pieces of string that are long enough to attach to the radius and ulna and still reach a few centimeters (a couple of inches) above the top pushpin.
5. Use tape to attach one end of one string to the top side of the radius and ulna, right below the pushpin that is holding them to the humerus.
6. Use the tape to attach one end of the other string to the underside of the radius and ulna. It should be at about the same position as the end of the first string, but on the back side of the radius and ulna.
7. Position the radius and ulna so that the arm is stretched out as long as it can be, and stretch out the strings so that they reach above the top pushpin. Your setup should now look something like the picture on the right.
8. Pull on the string that is attached to the top side of the radius and ulna. The radius and ulna should pull up, as if the arm is being bent at the elbow. As you do that, notice how both strings behave. While you pull *up* on the one string, the other string has to move *down*.
9. Once the lower part of the arm has moved up so it forms an "L" shape, release the string you are pulling on.
10. Pull on the other string so that the radius and ulna are pulled back down to their original position. Once again, watch the strings as you do this. Notice that while the string you are pulling on moves *up*, the other string moves *down*.
11. Repeat steps 8-10 a couple of times so that you see the way the strings behave as you bend the arm at the elbow.
12. Clean up your mess.

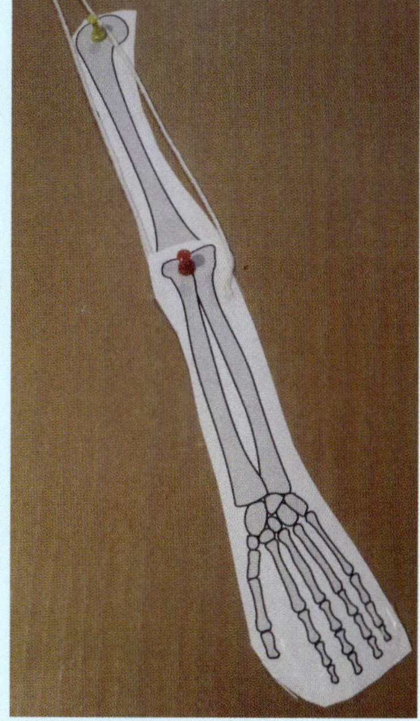

I doubt that what happened in the activity surprised you. In order to get the arm to bend at the elbow, the string you pulled on had to move up, while the other string moved down. Then, in order to get the arm to straighten back out, you had to pull on the other string, making it move up while the first string you pulled on moved down. So each string did the opposite of what the other string did: when one string moved up, the other one moved down.

Believe it or not, this is how your muscles work. Your muscles can do two things: get shorter or get longer. When a muscle gets shorter, we say that it **contracts**. It turns out that it takes energy to do that. However, when you stop using energy, the muscle **relaxes**. A relaxed muscle can be easily stretched. So, in order to move a part of your body, one muscle contracts. However, in order for the body part to actually move, another muscle needs to relax so that it can be stretched.

Let me explain how that works for the arm motion you played with in the activity. Stretch your arm straight out in front of you with your palm up. Place your other palm on your stretched-out arm, between your shoulder and elbow. Now bend the lower part of your arm at the elbow so that it comes up and touches the palm you placed between your elbow and shoulder. What does that palm feel? It should feel a little bump forming between your shoulder and elbow. That's your biceps brachii muscle contracting to bend your arm at the elbow.

However, that's not the end of the story. You have another muscle on the other side of your arm called your **triceps** (try' seps) **brachii**. In order for your arm to bend at the elbow, your triceps brachii must relax so it can be stretched out. If you want to straighten your arm, your triceps brachii must contract to pull your arm straight, but for that to happen, your biceps brachii must relax so it can be stretched out. In the end, these two muscles do opposite things. If your biceps brachii contracts, your triceps brachii relaxes and is stretched out, and your arm bends at the elbow. If your triceps brachii contracts, your biceps brachii relaxes and is stretched out, and your arm straightens.

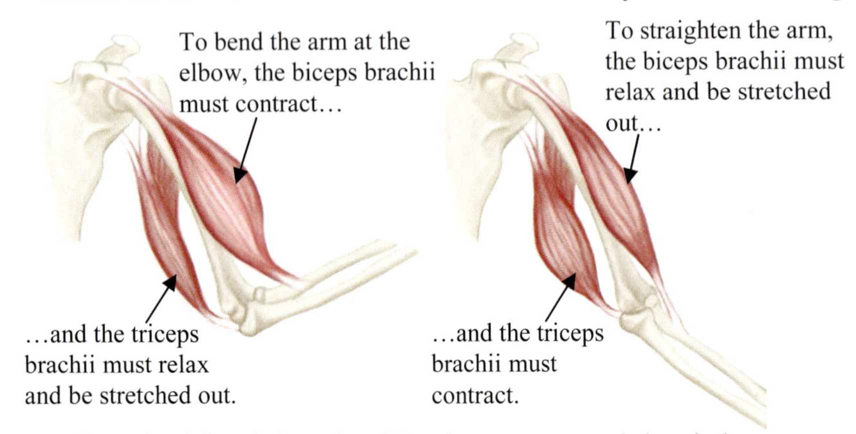

The biceps brachii and triceps brachii make up an antagonistic pair, because they do opposite things: when one contracts, the other must relax.

We call these muscles an **antagonistic** (an tag' uh nis' tik) **pair**, because they work against each other, just like the strings in the activity. When one muscle does something (like contracting), the other muscle does the opposite (like relaxing and getting stretched out). It turns out that all skeletal muscles work this way. They all have at least one antagonist, and Galen was the first to explain this important aspect of how skeletal muscles move your body.

LESSON REVIEW

Youngest students: Answer these questions:

1. What do tendons do?

2. Fill in the blanks: In order to bend your arm at the elbow, your biceps brachii _______ and your triceps brachii _________.

Older students: In your notebook, make a drawing like the one above and use it to explain how the biceps and triceps brachii work as an antagonistic pair. Also, explain the role of tendons in the body and point out the tendons in your drawing.

Oldest students: Do what the older students are doing. In addition, do some research to find out what "biceps," "triceps," and "brachii" mean and write that down in your notebook.

Lesson 41: Galen and Nerves

Have you ever heard someone say, "You've got a lot of nerve"? Well, everyone has lots and lots of nerves! They run all through the human body. What do these nerves do? Galen didn't give us the full answer to this question, but he gave us a really good start. Remember, Galen was dedicated to testing and observing things about the body. He couldn't use people in his experiments, but he used animals. In one of his most famous experiments, he showed that if he cut a specific nerve in a pig's body, the pig would no longer be able to squeal, even though it was clearly trying to. He traced that nerve back to the brain, and it made him realize that the brain controls how the pig makes noises.

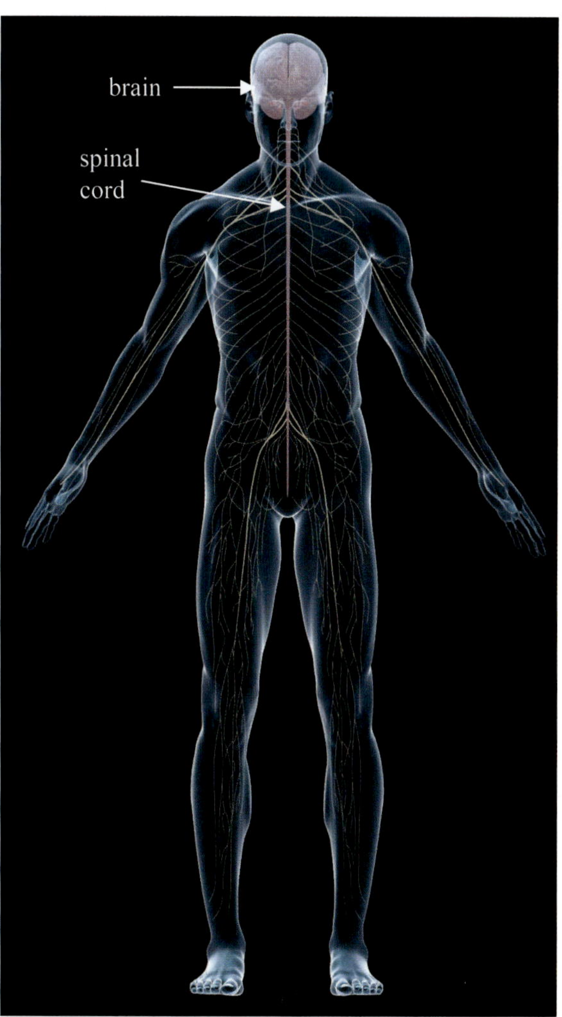

Nerves (the thin lines in this drawing) run throughout your body and are either directly connected to your brain or connect to your brain by your spinal cord.

This was revolutionary, because a lot of people thought that the heart was the main control center of the body. After all, the heart beats all the time, and when it stops beating, you are dead. Thus, it made sense to a lot of people that the heart was the main control center of the body. However, Galen showed that the muscles (like the ones that controlled a pig's voice) were not controlled by the heart. Instead, they were controlled by nerves, and those nerves could be traced directly to the brain or indirectly to the brain through the **spinal cord**, which is attached to the brain. In the end, then, Galen demonstrated that the brain controls the muscles by sending them "messages" through the nerves. Galen was wrong about most of the details regarding *how* that all happened. Nevertheless, he was the first to tell us that it does happen.

I would like you to perform an experiment that gives you some experience with how the nerves and spinal cord are necessary components for the control of your muscles.

The Patellar Reflex

What you will need:

- A chair that is high enough that your feet cannot touch the floor when you sit on it (If you don't have a chair that high, pile some large, hardcover books on the chair and sit on them.)
- Someone to help you

What you should do:

1. Sit on the chair so that your legs are bent at the knees but do not touch the ground.
2. Have your helper hold his palm flat like he is going to do a "karate chop."

3. Have your helper stand to one side of you and use the side of his hand to hit you right below the knee. So he is essentially doing a "karate chop" on your leg right below your kneecap. Your helper shouldn't hit you hard enough to hurt you, but he must hit you hard enough for you to feel it.
4. Does your leg kick out in response to the hit? If not, have your helper change where he is hitting slightly. Your helper should eventually find the spot where your leg kicks out when you are hit.
5. Once your helper finds the right spot, have him hit you there a couple of times to get an idea of how this works.
6. Concentrate with all of your might on holding your leg still.
7. While you are concentrating hard, have your helper hit you in that spot again.
8. Repeat this a few times, trying as hard as you can to keep your leg from kicking.
9. Have your helper move away, and kick your leg out just like you saw it happening when your helper hit you. Notice that you can make that happen when you want to make it happen, but you can't stop it from happening when you are hit in the right spot.

Did this activity remind you of anything? Sometimes, a doctor will hit you with a small hammer right below the kneecap, and your leg will kick out, just like it did in your experiment. Why does a doctor do this? The doctor is testing your **reflexes**, which are actions that your body automatically performs in response to something happening. Your reflexes are important, because they indicate how well your spinal cord is transmitting information.

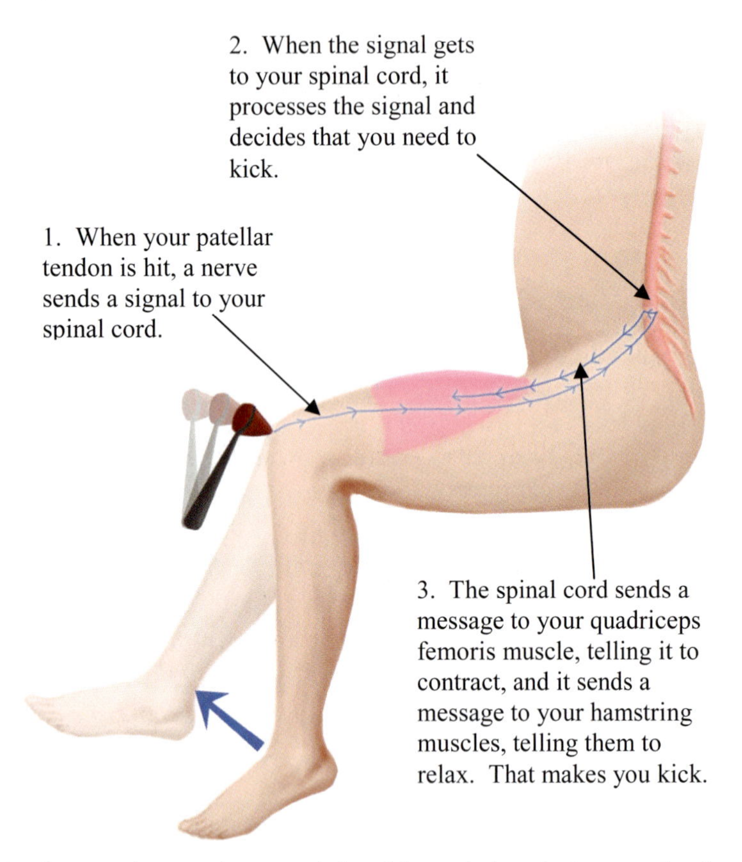

In the experiment, when your helper hit you below the knee, a signal was sent to your spinal cord, causing the spinal cord to send a message to one of your muscles making your leg kick forward. Your brain was not involved at all.

The scientific name for your kneecap is the **patella** (puh' tell uh). Right below your patella there is a tendon called the **patellar tendon**. When something hits that tendon, it causes a sensitive nerve to send a message to the spinal cord, telling the spinal cord that something just hit you below the knee. The spinal cord then sends messages to a muscle above your knee (the quadriceps femoris) telling it to contract and to the hamstring muscles on the underside of your leg, telling them to relax. When that happens, your leg kicks out automatically. Notice that in this entire discussion, I didn't mention the brain. That's because in this case, the brain is not involved. This is an automatic reaction your body has in order to protect itself. Essentially, if something hits your patellar tendon, your leg immediately kicks out to push away whatever just hit you.

Now at the end of the experiment, you were able to use your brain to make your leg kick out as well, right? That's because your brain can also use your spinal cord to send a message to the same leg muscles, telling them to make the leg kick out. So in the end, the spinal cord is the key. If you are going to use your brain to cause your leg muscles (or arm muscles or most muscles) to move, the brain must send a message down the spinal cord, which then sends a message to the muscle. However, that's not the only way to make a muscle move. Some muscles have a reflex action. That action doesn't involve the brain at all, but it still involves the spinal cord.

When you make a motion because you think about it and do it, that's called a **voluntary motion**. At the end of the experiment, for example, you thought about kicking your leg out, and you did it. That's a voluntary motion, because you controlled it with your brain. A reflex, however, is an **involuntary motion**. You can't control it. It just happens in response to something else that happens. The reflex you experienced in your experiment is just one of many involuntary motions that happen throughout your body. Have you ever touched something really hot? What happened? You pulled your hand back even before you realized it was hot. That's the result of a reflex. A nerve sent a message to your spinal cord saying "this is too hot," and the spinal cord relayed that information to the brain, but rather than waiting on the brain to decide what to do, it sent a message to some muscles in your arm telling them to contract and relax so that your hand would move away from the hot object.

Your brain is involved in voluntary motions, but in many involuntary motions, it is not involved at all. As you found in your experiment, you couldn't even control the involuntary motion when you tried. Now all of this was a bit beyond Galen, but his experiments showed conclusively that the brain is involved in voluntary motion, and that it controls muscles with nerves. Sometimes, those nerves come straight from the brain, and they are called **cranial** (kray' nee uhl) **nerves**. Otherwise, the brain sends messages down the spinal cord, and then the spinal cord sends messages out to the muscles through nerves that are called **spinal nerves**. This was the first step along the way to understanding how your brain, spinal cord, and nerves work together to produce muscle movements.

LESSON REVIEW

Youngest students: Answer these questions:

1. Fill in the blanks: When you want to move a leg muscle, your brain sends a message down your _____ ______, which then sends a message to the muscle using a ______ nerve.

2. What kind of motion can you control: voluntary motion or involuntary motion?

Older students: In your notebook, explain how your brain, spinal cord, and spinal nerves work together so you can move your legs the way you want to move them. Explain how a reflex like the one you experienced in the experiment is different. Also, explain the difference between voluntary and involuntary motions.

Oldest students: Do what the older students are doing. In addition, think about someone whose spinal cord is broken near his shoulders and cannot send messages across the break. Will the person be able to control his or her legs? Would the person have a reflex like the one you saw in your experiment? Discuss your answers in your notebook. Check your answer and correct it if it is wrong.

Lesson 42: Galen and the Voice

(Lessons titled in red are challenge lessons. If you are not doing challenge lessons, skip to Lesson 45.)

Since Galen learned about nerves by seeing that a pig couldn't squeal when a certain nerve was cut, it makes sense that he learned a lot about the voice as well. In fact, he was the first person who gave the proper explanation of how the voice works. He said that if you look at the lungs and the pipes attached to them, you see something that should look familiar. Look at the drawing and tell me what the main pipe coming up from the lungs looks like. To Galen, it looked like the kind of musical instrument people blew into. These days, we call them woodwind instruments. Galen said that since this part of the body looks like a woodwind instrument, and since it is attached to the lungs, which take in and blow out air, it only made sense that this is where the voice comes from.

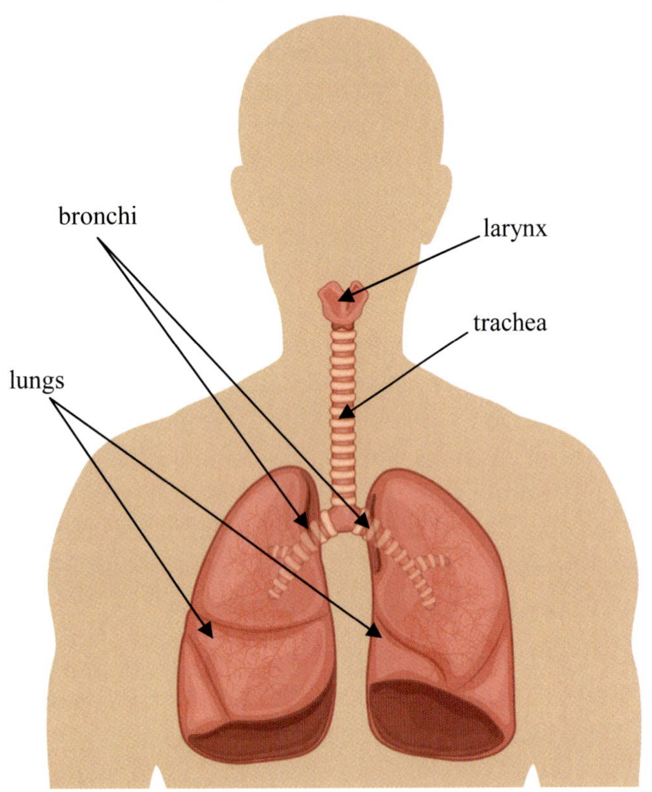

This drawing shows you the lungs, the tubes that lead to each lung (called the bronchi), the trachea, and the larynx.

As Galen looked closer at the anatomy, that reasoning made even more sense to him. As labeled in the drawing, the main pipe running up from the lungs is called the **trachea** (tray' kee uh). At the top of the trachea, we find the **larynx** (lar' inks), which is also called the **voice box**. If you look inside the larynx, you find tiny folds of tissue that are separated by a small distance. Using the muscles in this structure, we can change the distance between these folds of tissue. Galen argued convincingly that these folds are the source of the voice.

How could folds of tissue in the larynx be the source of the voice? Perform this experiment to find out.

A Model of the Larynx

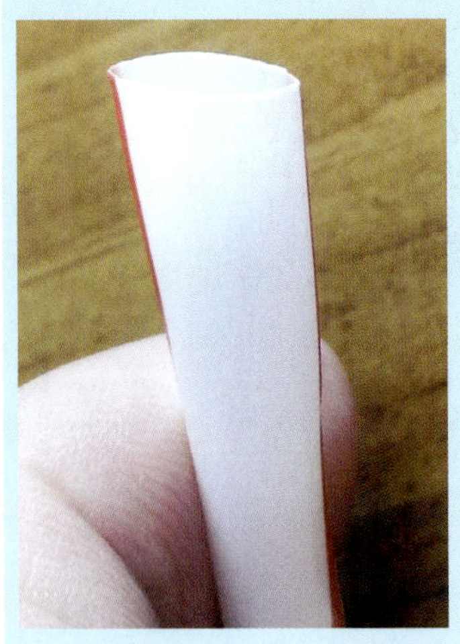

What you will need:
- A plastic straw (preferably one that can bend on one end)
- Scissors
- A spoon

What you should do:
1. Lay the straw on a smooth, flat surface and run the edge of the spoon over the last inch or so of the straw. Your goal is to flatten that end of the straw, as shown in the photograph on the left. If you have a bendable straw, flatten the end near where it bends.
2. Use the scissors to cut a diagonal on each side of the flattened end. You should start cutting about 1 centimeter (half an inch) from the end. You want to cut so that the flattened end of the straw is now a point.

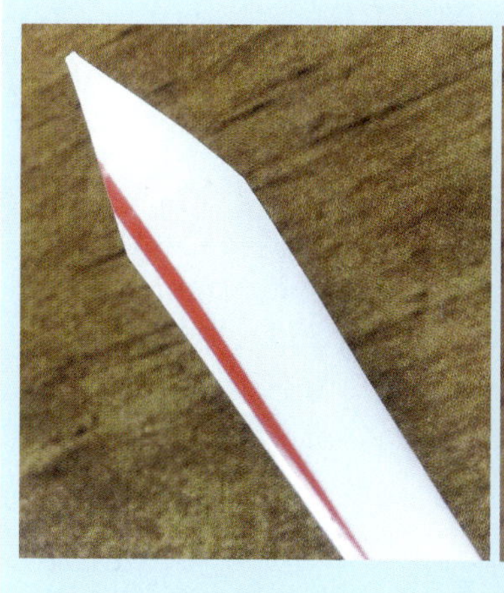

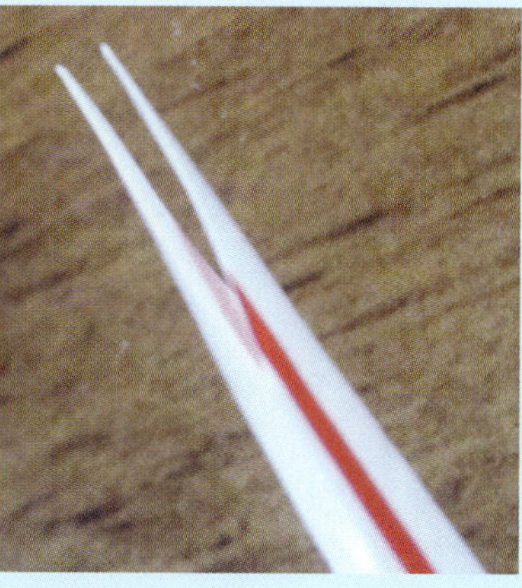

Your straw should now look like the picture on the far left.

3. Turn your straw so that you are looking at it from the side. You should see that you have two pointed ends close to each other but not touching, as shown in the picture on the near left.
4. Put the straw in your mouth so that the pointed end is completely in your mouth and your lips seal around the straw well beyond where you cut into it. Don't use your teeth to hold the straw. Hold it with your hands.
5. Blow through the straw as hard as you can. Use your lips to seal around the straw so that the only way air can escape is to go through the straw. Did you hear something?
6. If you didn't hear something, take the straw out of your mouth and use the spoon to flatten the edge again so that the two points come close together, as shown in the picture above and to the right. The points have to be close but not touching each other.
7. Once you can get the straw to make a sound, play with it a bit. Clench your lips slightly harder on the straw. Does that change the sound? Use as little force as possible with your lips. Does that change the sound? Try to blow harder and softer. Does that change the sound?
8. What's making the sound? Turn the straw around so that the pointed end is now outside your mouth.

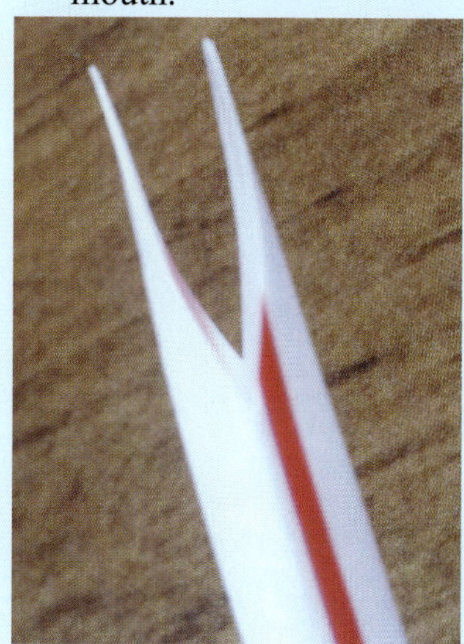

9. If you have a bendy straw, bend it so that you can look at the points from the side.
10. Suck in on the straw from the uncut end as hard as you can. If you suck hard enough, you should hear the sound, and you should see the points vibrating back and forth.
11. Use your thumb and forefinger to push the flattened end of the straw into a more circular shape again. Your goal is to make the points much farther apart, as shown in the picture on the left.
12. Try to make the sound again, either by sucking in on the straw with the points outside of your mouth or blowing through the straw with the points on the inside of your mouth. Can you still make the sound? If so, try to push the points farther apart. You should get to the point where you cannot make the sound anymore.
13. Clean up your mess and put everything away.

Do you remember back when you were studying Pythagoras and you learned about how vibrating rubber bands produced sound? Remember, sound is a wave that travels through air, so if something is vibrating in air, it can make a sound wave. In the "musical instrument" you just constructed, the pointed ends of the straw vibrated when you blew through the straw (or sucked in on the straw, depending on where the pointed ends were). That vibration, in turn, made a sound wave, producing the sound that you heard. You probably found that if you changed how your lips gripped the straw or how hard you blew, you could change the pitch of the sound as well as its volume. That's because making those changes caused the pointed ends to vibrate differently, producing a different sound. Also, you probably found that when the pointed ends got too far from one another, no sound was made at all. That's because they will only vibrate in response to air passing between them if they are fairly close to one another to begin with.

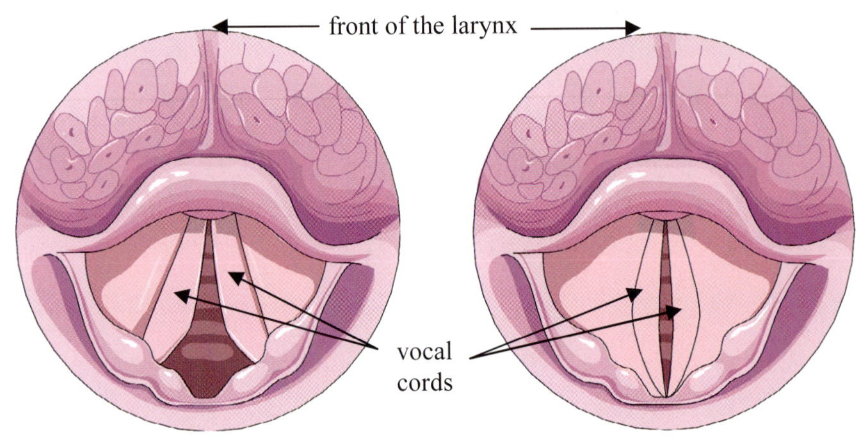

This drawing shows you what you would see if you looked down someone's larynx from the top. When the vocal cords are far apart (left), they don't make any sound. When they are close together (right), they vibrate to make sound.

It turns out that the larynx has something very similar to what the straw had. Look at the drawing on the right. It shows you what you would see if you looked straight down someone's larynx from the top. The two folds of tissue shown in the drawing are called your **vocal cords.** When your vocal cords are far apart, as shown on the left side of the drawing, air passes between them easily, and they do not vibrate. This is like what happened in the last part of the experiment, when you forced the points on the straw to get far apart from one another. When you are breathing without speaking or singing, your vocal cords are far apart, and no sound comes from your larynx.

When you want to talk, you contract some muscles in your larynx. This brings your vocal cords closer together. When that happens, air passing between them causes them to vibrate, and you start making sounds. By controlling the muscles in your larynx, your brain can change how close the vocal cords are to one another, how tight they are, and their length. This changes the sounds you make so that, in the end, you can talk, sing, hum, etc.

How do you control the volume of your voice? You do that by the amount of air you use. If you want to shout, you take a deep breath and use a lot of air. If you want to whisper, you use just a little bit of air. In the end, then, the muscles of your larynx mostly control the pitch of the sounds that you make, while the amount of air you use mostly controls the volume. Galen figured out most of this through the dissection of apes and vivisection of pigs.

LESSON REVIEW

Youngest students: Answer these questions:

1. What is another name for the voice box?

2. What do we call the folds of tissue in your voice box that vibrate to make sound?

Older students: In your notebook, make a drawing of the lungs, bronchi, trachea, and larynx, as shown on page 126. Label the parts, and indicate where the vocal cords are found. Explain how the vocal cords make it possible for us to talk and sing. Be sure to mention the difference between when the vocal cords are far apart and when they are close together.

Oldest students: Do what the older students are doing. In addition, think about what you have learned regarding how vibrating strings make sounds and write down whether you think a man's vocal cords are longer or shorter than a woman's. Check your answer and correct it if you are wrong.

Lesson 43: Galen and Breathing

In the previous lesson, you learned how the vocal cords vibrate to make sound when air passes between them. But where does that air come from? It comes from breathing, of course. You breathe air in, and when you breathe it back out, you can use it to vibrate your vocal cords. The scientific word for breathing is **respiration** (res' puh ray' shun), and Galen was the first to properly explain how it worked. To get an idea about what Galen figured out, try this experiment.

Respiration

What you will need:

- A tape measure (You could use a piece of string that wraps around your chest with plenty of string to spare and a ruler instead. You will be measuring the distance around your chest, so a tape measure is the easiest thing to use.)
- Your notebook and a pencil (a blank sheet of paper if you are doing the review exercises for the youngest students)
- Someone to use the tape measure to determine the distance around your chest

What you should do:

1. Start a new page in your notebook and title it "Breathing Measurements."
2. Sit on a stool or the edge of a chair so that your back is not leaning against anything.
3. Sit up straight and breathe out normally. Don't breathe out more than you usually do. Just breathe out normally, and right before you breathe in, hold your breath.
4. While you hold your breath, have your helper wrap the tape measure around your chest so that it is as high as possible on your chest while still passing under your arms. It should be wrapped snugly, but not uncomfortably tight.
5. Have your helper measure the distance around your chest and then remove the tape measure.
6. In your notebook, write "Distance around my chest while breathing out normally:" and then the measurement that your helper made.
7. Breathe normally again for a while.
8. After you have been breathing normally for a while, hold your breath right after you breathe in and right before you breathe out. Don't breathe in more than usual. Just breathe normally, and right before you let out your breath, hold it.
9. Have your helper measure the distance around your chest again, exactly the way he or she did before.
10. In your notebook, write "Distance around my chest while breathing in normally." and then the measurement that your helper made.
11. Is there a difference between the two measurements?
12. Repeat steps 2-11, but this time, breathe very deeply. Breathe out as much as you possibly can, and then hold your breath so your helper can measure the distance around your chest. Write down this measurement as "Distance around my chest while breathing out deeply."
13. Breathe in as much as you can and hold your breath so that once again, your helper can measure the distance around your chest. Write down this measurement as "Distance around my chest while breathing in deeply."
14. Is there a difference between the two measurements?

What did you see in the experiment? If things went really well, you might have seen that while breathing in normally, the distance around your chest is a bit larger than while breathing out normally. Unless things went really poorly, you definitely should have seen that while breathing in deeply, the distance around your chest was larger than while breathing out deeply. To understand why your measurements turned out the way they did, you need to know a bit more about the anatomy of your lungs.

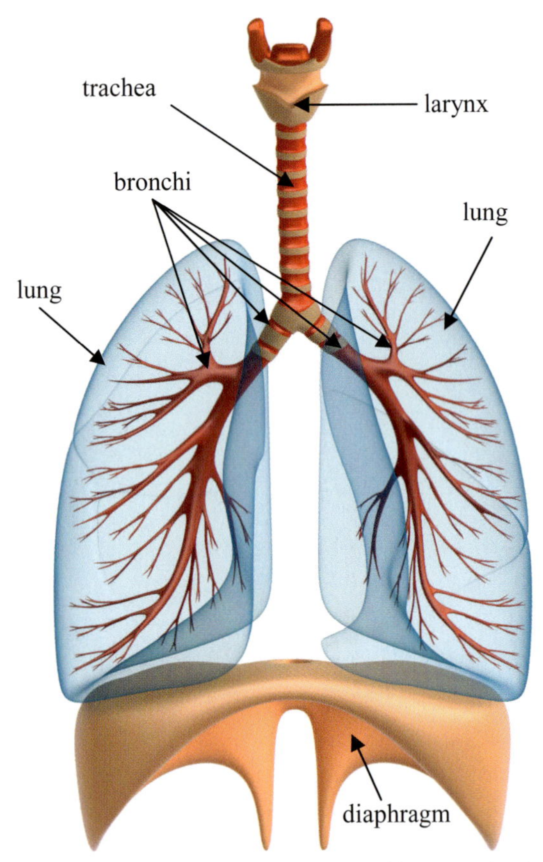

This drawing does not show the ribcage. That is shown in the drawing on the next page.

First, you need to know that your lungs are surrounded by bones. The bones are a part of your ribcage, and one of their jobs is to protect the organs that are inside, including your lungs. In the drawing on the left, the ribcage and other organs are not drawn so that you can better see the lungs and the **diaphragm** (dye' uh fram'), which is the most important muscle when it comes to breathing. Notice how large the diaphragm is. Also, the lungs have been drawn so that you can see through them. They aren't really see-through, but because they have been drawn that way in the picture, you can see all the tubes that branch out from the trachea. As you learned before, those tubes are called bronchi. They carry air into and out of the lungs.

Have you ever heard of a problem called **bronchitis** (brahn kye' tus)? When a person has bronchitis, it becomes very hard to breathe. That's because bronchitis tends to partially close down the bronchi. Since the bronchi are partially closed down, air can't travel easily into or out of the lungs, which is what breathing is all about. That should tell you how important the bronchi are to breathing. Without them, there would be no way for air to get into and out of the lungs.

The main thing I want to focus on, however, is the diaphragm. Remember, it is a muscle. It can contract and relax. What happens when the diaphragm contracts? It gets smaller, right? So that big muscle suddenly gets smaller. Now remember where the lungs and diaphragm are located. They are inside the ribcage, right? When the diaphragm gets smaller, there is suddenly a lot more room inside the ribcage. What happens as a result? The lungs have room to expand, so they do. This fills them with air. That's how you breathe in.

That's not all there is to breathing in, however. There are also muscles attached to your ribs. They can pull your ribs up and apart, making even more room for your lungs to expand. Well, what do you think happens when your ribs pull apart? Your chest gets bigger! So when you breathe in, what you are really doing is making room for your lungs to expand. When they do, they take in air.

When you breathe out, you do exactly the opposite. The diaphragm relaxes, which means it moves to its original size. That pushes against your lungs. Also, if the muscles in your ribs have lifted and pulled apart your ribs, they relax, and your ribs push against your lungs. What happens when you push on something filled with air (like a pool toy) when it is not sealed? Air rushes out, right? The same happens to your lungs. When your breathing muscles relax, air rushes out of your lungs.

When you breathe in, we say that you are **inhaling**. When you breathe out, we say that you are **exhaling**. The drawing on the right illustrates what happens during both processes. Notice that when you inhale, your diaphragm contracts, increasing the room in your chest. Also, your ribcage moves up and expands, making even more room. When you exhale, everything relaxes, pushing air out of your lungs. Now one thing that you probably know but I want to emphasize is that inhaling is the part that usually takes work. When muscles contract, they use energy, and muscles contract when you inhale. When muscles relax, they don't use energy. So breathing in is the hard part. Breathing out usually happens without using any energy, unless you are exercising, at which point some muscles contract so you can exhale more air than usual.

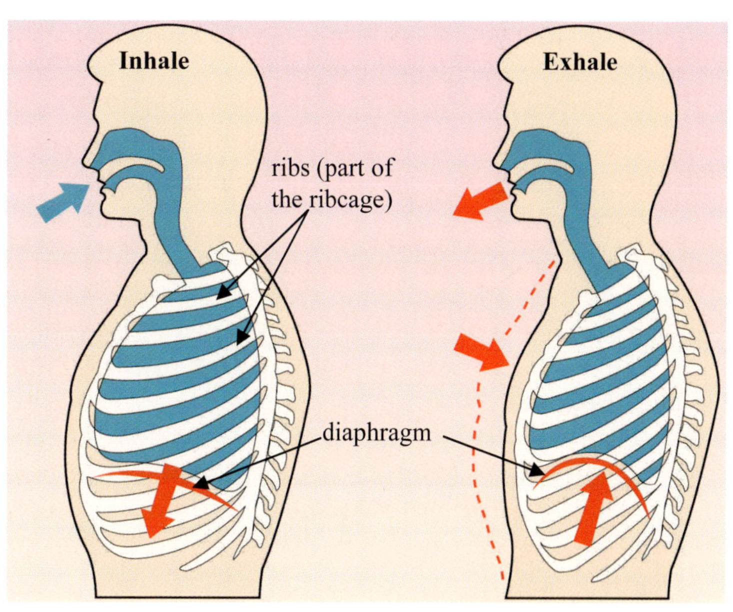

When you inhale (left), your diaphragm contracts and your ribcage gets bigger, allowing your lungs to fill with air. When you exhale, your diaphragm relaxes, and your ribcage gets smaller, pushing air out of your lungs.

Now what about the results in your experiment? If you noticed a difference in the distance around your chest in the first part of the experiment, it was very small. However, you should have noticed a much larger difference in the distance around your chest in the second part of your experiment. Why? Well, when you breathe normally, your diaphragm is doing most of the work. Your rib muscles don't come into play nearly as much. As a result, your ribcage doesn't expand much, so the distance around your chest doesn't expand much, either. However, when you breathe deeply, you use your rib muscles a lot more. This expands your ribcage more, which increases the distance around your chest. In your experiment, then, you saw the difference between what happens when your diaphragm is doing most of the work, and what happens when your rib muscles get more involved in respiration.

LESSON REVIEW

Youngest students: Answer these questions:

1. What is the main muscle used in breathing?

2. What do scientists call the process of breathing?

Older students: In your notebook, explain how inhaling and exhaling happen. Be sure to mention the diaphragm and the muscles attached to the ribs. Also, explain when the diaphragm is mostly involved and when the rib muscles come more into play. Finally, explain which process takes energy to perform.

Oldest students: Do what the older students are doing. In addition, compare a person sitting quietly with a person exercising vigorously. In which case do you expect the muscles attached to the ribs to be most involved in the respiration process?

Lesson 44: Galen and Medicines

Do you remember Pedanius Dioscorides? You studied him back in Lesson 32. He wrote a book discussing all the different things from plants and animals that could be used to treat sickness and disease. He did a great job with the information that he had, but Galen added something very important to his work. The best way to understand what Galen added to Dioscorides's work is to try an experiment.

Making Bubbles

What you will need:

- Vinegar
- A few antacid tablets (TUMS would be best, but anything that says calcium carbonate is the active ingredient will work.)
- Baking soda
- Dish soap
- A large metal spoon
- A spoon for stirring
- A paper plate or bowl
- A measuring teaspoon (If you have two measuring teaspoons, it would make things easier.)
- A ½-cup measuring cup
- Two glasses (They shouldn't be juice glasses. They should be taller than that.)
- A sink
- Paper towels

What you should do:

1. Put ½ cup of vinegar into each glass.
2. Add one teaspoon of dish soap to each glass.
3. Use the stirring spoon to *gently* mix the dish soap and vinegar in each glass. Don't mix so strongly that you make a lot of bubbles.
4. If you have only one measuring teaspoon, thoroughly rinse all the soap out of it and dry it with paper towels so it is completely dry.
5. Put one glass in the sink.
6. Put about three TUMS tablets in the paper plate or bowl and use the spoon to crush them. It is best to start by using the edge of the spoon on the edge of the tablet. That will break the edge down, making smaller bits of TUMS. Those bits should be easy to crush. You want to crush it all down to powder.
7. Use the spoon to scrape the powder up from the plate and pour it into a dry measuring teaspoon.
8. Once you have filled the measuring teaspoon with TUMS powder, pour the powder into the glass that is in the sink. What happens?
9. Watch what happens for a minute or two, and then use a paper towel to get all the TUMS powder out of the measuring teaspoon.
10. Dump the contents of the glass that is in the sink, rinse the glass out, and put it where you put your dirty dishes.
11. Put the other glass that has vinegar and soap into the sink.
12. Fill the measuring teaspoon with baking soda.
13. Pour the baking soda into the glass. What happens this time?
14. Watch what happens for a couple of minutes and then clean up your mess.

What happened in the experiment? If all went well, you should have seen some bubbles form in the first glass, but it probably wasn't very impressive. However, what happened in the second glass? Most likely, bubbles probably rose quickly up the glass. There were probably so many bubbles forming that they spilled out over the top of the glass. That's why I had you put the glass into the sink.

How did the bubbles form? Well, you know that the bubbles were produced by the soap, but why were they produced when you added powder to the glass? Soap bubbles form when soapy water is mixed with air (or any other gas). The soapy film stretches around the gas, holding it inside a little bubble. When you mix soap and water together strongly, you are also mixing in the air that exists above the surface of the water. That's why strongly mixing soap and water together makes bubbles. Well, when you added both powders to the vinegar, the powders reacted with a chemical in the vinegar called **acetic** (uh' see tik) **acid**. That reaction produced a gas called carbon dioxide. So by adding the powders to the vinegar, you were making a gas, which the soapy film stretched around to make bubbles.

Soap bubbles are made when a soapy film traps a gas, like air.

Now, both powders produced carbon dioxide gas when they mixed with the vinegar, so both powders made bubbles. However, if you wanted to make a lot of bubbles, which powder would you use? You would use baking soda, wouldn't you? After all, when mixed with soapy vinegar, baking soda made a lot more bubbles than the powdered TUMS did. In other words, when it comes to making bubbles out of soapy vinegar, baking soda is a lot better than powdered TUMS.

It turns out that we can find a lot of different chemicals in nature that do very similar things. For example, just as both baking soda and powdered TUMS will make bubbles in soapy vinegar, a chemical called lithium and another chemical called sodium will make hydrogen gas when mixed with water. However, if you add them both to water, you will find that sodium makes hydrogen gas much more quickly than lithium does.

Since medicines are made of chemicals, if a doctor wants to treat a patient using medicine, the doctor usually has a lot of choices, because there are often many medicines that do essentially the same thing. For example, in your experiment, you used TUMS or some other antacid tablet. Why do people use antacids? Because they experience something called "heartburn." Interestingly enough, heartburn has nothing to do with your heart. It is the result of too much acid being produced by your stomach. It turns out that there is a lot of acid in your stomach, but sometimes, too much of it is produced and it ends up rising into your esophagus (the tube that leads from your throat to your stomach). When that happens, a person usually feels a burning sensation near the middle of the chest, which is where the heart is. That's why it's called heartburn but has nothing to do with the heart.

Antacid tablets help relieve heartburn because they react with the acid in a person's stomach, turning it into chemicals that don't produce a burning sensation. Now go look at the baking soda box. Most likely, somewhere on the box, it says that it can be used to relieve heartburn. That's true. Both antacid tablets and baking soda can relieve heartburn. However, most people find antacid tablets to be more effective at it. Part of the reason that baking soda isn't as good for most people when it comes to relieving heartburn is because it is better at making carbon dioxide when it is mixed with an acid. That gives the person a lot of gas!

So in the end, when it comes to medicine, you might be able to find several medicines that treat a particular illness. However, some will probably be better than others. Galen figured this out by trying many different medicines on people and seeing their effects. As a result, he developed a way to measure a medicine's effect on a particular problem. He listed the "degrees" of effectiveness of each medicine he used for each problem he treated. If the medicine was really effective, it was given the highest degree, a "4." If it was not very effective at all, it was given the lowest degree, a "1." That way, when a doctor wanted to treat a problem, he could look at all the medicines he had and choose the one with the highest degree. According to Galen, that would produce the most effective treatment.

Now remember, a lot of what Galen believed was wrong. In fact, he still followed Hippocrates's idea that the body contained humors and that many illnesses were the result of an imbalance in those humors. This meant that a lot of his ideas about medicines were also wrong. As a result, we don't use his degrees or his advice on medicines anymore. However, we do recognize that he was right when it came to the fact that although many medicines might treat a problem, some are more effective than others. For example, if you are in a little pain, a doctor might tell you to take a couple of aspirins. However, if you are in serious pain, the doctor might prescribe a better pain reliever, such as codeine. That's because while both aspirin and codeine relieve pain, codeine is more effective. This kind of thinking is partly the result of Galen's work.

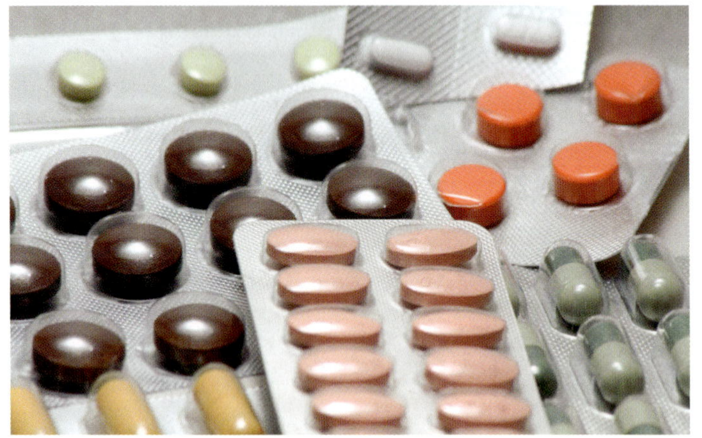

Often, many medicines will treat the same condition, but they usually have different levels of effectiveness.

LESSON REVIEW

Youngest students: Answer these questions:

1. What gas is made when baking soda reacts with the acetic acid in vinegar?

2. Why did Galen assign degrees to medicines?

Older students: In your notebook, explain why Galen assigned degrees to medicines. Use your experiment's results to illustrate the reason.

Oldest students: Do what the older students are doing. In addition, write down the following question and your answer in your notebook: "Vinegar is mostly water; it contains very little acetic acid. What results would you expect from your experiment if you used nearly pure acetic acid instead of vinegar?" Check your answer and correct it if it is wrong.

Lesson 45: Anicius Manlius Severinus Boëthius (c. AD 480 – C. AD 525)

Galen died around AD 200, and history records very few scientific works for the next 300 years, at least in Greece and the surrounding areas. This doesn't necessarily mean there weren't people working on science during that time. It just means that we don't find any books that contain original scientific ideas during this period, and we don't find any references to such books in history. One problem with history is that you don't always get a complete picture of the time that you are studying. You only get a picture of what survived from that period. It may be that later on, someone will discover an incredible natural philosopher who lived in Greece and the surrounding areas during this 300-year period. For right now, however, history doesn't record one.

However, in about AD 480, a man named Anicius (an uh see' us) Manlius (man lee' us) Severinus (sev uh ree' nus) Boëthius (boh ee' thee us), was born in Rome. Thankfully, most historians call him **Boethius**, and that's what we'll call him as well. One interesting thing we know about Boethius is that he was a Christian. He is actually the first Christian natural philosopher I have talked about in this book. He was almost certainly not the first Christian natural philosopher. However, he is the first one that history says much about.

This illustration comes from a 14th-century copy of one of Boethius's books. It is how the artist imagined Boethius teaching students.

Obviously, I couldn't talk about Christians in the two sections that dealt with science before the birth of Christ, but now that I am discussing the science that was done after Christ was born, are you surprised that this is the first Christian I have discussed? You shouldn't be. As I said before, history rarely gives us a complete picture of the time period that we are studying. We learn about natural philosophers in ancient history from their own books or from books in which other people discuss them. Thus, we tend to learn about the people whom other people thought were important. There is a lot we know about Christian theology from the earliest centuries AD, and this makes sense. After all, theology was the most important thing to early Christians, so that's what they preserved and wrote about. As Christianity grew, however, the importance of Christian natural philosophers became obvious, so their books started being preserved, and other natural philosophers began writing about them.

Boethius wrote on a great many topics, including theology, mathematics, logic, and philosophy. What's important to us in this course, however, is what he wrote about mathematics and music. Like many natural philosophers, Boethius studied the mathematics of music and tried to understand how it actually works. Boethius tried to understand how sound works, and he was pretty successful.

Do you remember from Lessons 3-5 that sound can be described as a wave that travels through air? The wave makes clumps of air called "crests" and spread-out regions of air called "troughs." The frequency with which those crests and troughs hit your ear determines the pitch of the sound. The more crests and troughs that hit your ear each second, the higher the pitch. The larger the crests are, the higher the volume. Well, Boethius was the first to come up with this idea. He gave a wonderful description of why this was the best way to think about sound. To understand his description, perform the following experiment:

Making Waves

What you will need:
- A bathtub that can be filled with water
- A medicine dropper
- Someone to help you

What you should do:
1. Put the plug in the bathtub and start filling it with water. Continue to add water until it is about 2 centimeters (1 inch) deep in the tub. Then, turn off the water.
2. Have your helper fill the medicine dropper with water from the tub.
3. Wait for water to settle so that it is still.
4. Have your helper hold the medicine dropper near the center of the tub at least 1 meter (3 feet) above the water.
5. While you are watching the water in the tub, have your helper squeeze the medicine dropper so that a single drop of water falls into the tub.
6. Watch the ripples that result.
7. Have your helper add a single drop of water to the tub each second for 10-20 seconds.
8. Watch the ripples that result. Specifically, watch how they hit the sides of the tub.
9. Have your helper refill the medicine dropper.
10. Let the water become still again.
11. Have your helper hold the medicine dropper where it was in step 4, and have him or her add drops as fast as possible. They need to be individual drops, but they should be added to the water as quickly as possible – much faster than one each second.
12. Watch the ripples that result, once again concentrating on how the ripples hit the side of the tub. Do you notice that a lot more waves hit the sides of the tub now than they did in step 8? That's what frequency is all about. In step 8, the frequency was low, because the ripples hit the side of the tub about once every second. In this step, the frequency was high, because lots of ripples hit the sides of the tub every second. If these were sound waves, the pitch in this step would be higher than the pitch in step 8.
13. Have your helper refill the medicine dropper.
14. Let the water become still again.
15. Have your helper hold the medicine dropper near one end of the tub and about 2 centimeters (1 inch) above the water.
16. Have your helper drop one drop into the water, and try to see how far the ripples travel towards the other side of the tub before they get too small to see.
17. Repeat step 16 so you get an idea of how far the ripples travel in the tub.
18. Have your helper hold the medicine dropper near the same end of the tub, but this time, have him or her hold it at least 1 meter (3 feet) above the water.
19. Have your helper drop one drop into the water, and try to see how far the ripples travel towards the other side of the tub before they get too small to see.
20. Repeat step 19 so you get an idea of how far the ripples travel in the tub. Did you notice a difference? You should have. The ripples should have traveled much farther in step 19. That's because the ripple started off much larger when the drop was released from higher up. As a result, the ripples lasted longer.
21. Drain the bathtub, empty the medicine dropper, and put it away.

Boethius said that we should think of sound as a ripple, like what is made when something is dropped in water. When the sound is made, it produces waves that travel out in all directions, just like the ripple travels outward, forming a larger and larger circle. The pitch of the sound depends on how fast the ripples hit your ear, and the volume of the sound depends on how high the ripples are.

What was wonderful about this explanation is that it shows why sounds get softer the farther you are from the source, but they do not change pitch. If I stand right next to someone who blows on a loud whistle, the sound will be annoyingly loud in my ear. If I move far away from the person and he then blows on the whistle again, I will hear the same pitch, but it will be much softer. If you think about the experiment, that's exactly what you saw: The frequency of the waves depended on how fast water was dropped into the tub. However, the distance that they could travel depended on their initial height. The higher they were to begin with, the farther they could go.

Why is that? Well, think about how the ripples traveled. They spread out from the point where the water was dropped. As they spread out, the circle that they formed got larger and larger, didn't it? That means the wave had to spread its energy out over a larger and larger area. Eventually, the area got so large that there just wasn't enough energy to keep the ripple large enough for your eyes to see, so it disappeared. The same can be said for a sound wave. As a sound wave spreads out, its energy must be spread out over a larger and larger area. As a result, the wave gets smaller and smaller. This means that the farther you are from where the sound originates, the lower the sound's volume will be.

Boethius said to think of sound like ripples that travel in water. Notice that the height of the wave crests gets smaller as the waves spread out in larger and larger circles.

Boethius's description of sound behaving like ripples on water was the first example of a scientist thinking of sound as a wave. While Boethius wrote about a great many things, this was his main contribution to science.

LESSON REVIEW

Youngest students: Answer these questions:

1. Fill in the blank: When you drop something in water, the ripples spread out in __________.

2. Fill in the blank: The ripples in water get _______ as they form larger circles.

Older students: In your notebook, explain why Boethius's view of sound as a wave traveling like a ripple in water explains why a sound gets softer the farther you are from what is making the sound.

Oldest students: Do what the older students are doing. In addition, use Boethius's view of sound traveling like a ripple through water to explain why the pitch of the sound is not affected by how far you are from what is making the sound.

Science in the Early Middle Ages

GEOMETIA·MAGISTRI·TH
OME·BRDVARDINI·INCIPIT
EOME
TRIA
ASSE
CVTI
VA·Ē
ARIT
HMETICE·NAM
POSTERIORIS
DINIS·EST·ET·
PASSIONESNVMÆ
RORVM·IN·MAGNI
TVDINIBVS·DE
SERVIVNT·PROPTER·

This is a page from *De Geometria Speculativa*, a book about geometry written by Thomas Bradwardine.

Lessons 46-60: Science in the Early Middle Ages

Lesson 46: John Philoponus (c. 490 – c. 570) and Creation

John Philoponus was born just ten years after Boethius, so they lived and pursued natural philosophy at roughly the same time. We know very little about his early life, but we know a lot about what he did, because he wrote at least 40 works on subjects like philosophy, theology, mathematics, and science. We also know that he was a Christian.

One thing Philoponus is famous for is arguing against the idea that the earth is eternal. Nearly all natural philosophers at the time (at least the ones we know about) agreed that the earth had always existed and would continue to exist forever. However, this idea goes against what the Bible teaches, since Genesis 1:1 says, "In the beginning God created the heavens and the earth." That means earth had a beginning, and since Philoponus believed the Bible, he decided that the idea held by most natural philosophers was wrong. As a result, he argued against it.

This is a view of earth from outer space. You can see North and Central America under the clouds. In Philoponus's time, natural philosophers thought the earth was eternal.

How did Philoponus make his case? He didn't argue from the Bible, because he wanted to use the kind of argument that natural philosophers of the time were using, so they would appreciate it. As a result, he used a scientific argument to show that the earth had a beginning. To get an idea of what his argument was all about, do the following activity:

Don't Lose Your Marbles

What you will need:
- Several marbles (If you don't have marbles, get several balls, like baseballs, soccer balls, etc. If you use balls instead of marbles, you will need a big area to work in. If you have a pool table, you can use the pool table and pool balls.)
- A large, flat area that will allow you room to roll the marbles or balls around

What you should do:
1. Put a marble (or ball) in the center of the flat area.
2. Roll another marble at the first marble so that it hits the first marble as it is rolling. In essence, you used one marble to get the next marble rolling, didn't you?
3. Now set up two marbles about 2 centimeters (about an inch) away from one another.

4. Roll a third marble into one of the two marbles so that it starts moving and then hits the *other* marble. That way, the marble you rolled hits *only one* marble, and then that one marble moves to hit the other marble.
5. Try to repeat the process using one more marble. Three marbles should sit near the middle of the surface. The first one should be about 2 centimeters (about an inch) from the second one, and that should be about 2 centimeters from the third one. You then need to roll a fourth marble so that it hits one marble, which then hits the next marble, which then hits the next marble.
6. See how many marbles you can get moving in this way. As the number of marbles grows, how does the time over which it all happens change?
7. Clean up your mess.

How many marbles could you get rolling in that way? I couldn't get any more than seven rolling. After that, I had to roll the first marble so hard that I couldn't aim well, and one of the marbles would end up flying off in a direction that made it miss the next marble. Depending on how patient you are (and how good you are with such things), you might be able to do more. That, of course, would make the rolling last longer. Do you think you could do 10 marbles? What about 100 marbles? What about 1,000 marbles? At some point, not even the most patient, skillful person would be able to add more marbles. So even the most patient, skilled person couldn't keep this process going very long. Even with 1,000 marbles, the rolling would eventually come to a halt.

What does this have to do with Philoponus and his argument against an eternal earth? Well, Philoponus noticed that in order for earth to continue, certain things needed to be replaced. For example, new animals had to be born to replace the animals that had died. New plants had to grow to replace the plants that had died. In each case, however, two animals had to exist already in order to be able to have baby animals. At least one plant would have to exist in order to produce a baby plant. For earth to be eternal, such things would have needed to happen for all eternity. New animals would have to continually be made to replace dying animals. New plants would have to continually be made to replace old plants.

However, that presents a problem. If I see a plant or animal, I know that it had to be made by parent plants and animals. But where did *those* parent plants and animals come from? They came from parent plants and animals that lived before them. But where did *those* parent plants and animals come from? They came from parent plants and animals that lived before them. Since plants and animals have to come from other plants and animals, and since they live for only a certain amount of time, we know that *all* plants and animals that exist on the earth had to come from other plants and animals that existed before them.

If the earth is eternal, Philoponus argued, that means it has existed for an **infinite** (in' fuh nuht) number of years. Do you know what "infinite" means? It means limitless or endless. An infinite number, then, is a limitless number. Suppose you started counting. You start with 1, move on to 2, go to 3, and just keep counting. Suppose you counted every second of your life, even while you were asleep. When you finally died, you would have gotten to a *huge* number, wouldn't you? If you lived 80 years and counted every second of your life from the time you were born to the time you died, you would reach 2,524,608,000. That's a big number, but it is not an infinite number, because no matter how big the number is, all you have to do is add 1, and you get to a bigger number. If you add one to 2,524,608,000, you get 2,524,608,001, which is bigger. So there is no such thing as an infinite number, since there is no limit to the numbers you can make.

If the earth had existed for an infinite number of years, there would have to be an infinite number of plants and animals. This would be like trying to make your experiment last forever. To make the rolling eternal, you would need an infinite number of marbles. The problem is, you can't have an infinite number of plants, animals, or marbles. No matter how many plants, animals, or marbles you have, if you add one more, you now have a larger number of plants, animals, or marbles. If the earth had existed for an infinite number of years, there would also have to be an infinite number of plants and animals. Since it is impossible to reach an infinite number, it is impossible for earth to be eternal. As a result, earth had to have a beginning.

This drawing shows a computer that is displaying a computer that is displaying a computer that is displaying a computer, and so on. Even though there are a lot of computers being displayed, it is not an infinite number of computers, because it is impossible to have an infinite number of computers.

Even though the reasoning Philoponus used is not used in modern science, all modern scientists agree that the entire universe (and therefore the earth as well) had to have a beginning. This brings up an important point. I am sure there were scientists back in Philoponus's day who said that you shouldn't believe what the Bible says about creation, because natural philosophy (science at that time) had shown that the earth was eternal. As a result, natural philosophy had shown that the Bible was wrong about the earth having a beginning. As it turns out, however, it was natural philosophy that was wrong, not the Bible. That's something you will see again and again if you study science from a historical point of view. Throughout history, science has said that this part of the Bible can't be true, or that part can't be true, because it conflicts with science. However, time and time again, science has been shown to be wrong, and the Bible has been shown to be correct.

LESSON REVIEW

Youngest students: Answer these questions:

1. What does the word "infinite" mean?

2. Why did Philoponus believe that the earth is not eternal? I am not asking how he argued against the idea. I am asking *why* he didn't believe the idea.

Older students: In your notebook, explain the reasoning Philoponus used to argue that the earth could not be eternal. How does the argument relate to the activity you did?

Oldest students: Do what the older students are doing. In addition, write in your notebook why Philoponus didn't believe that the earth was eternal, and explain why he didn't use that reason when arguing with natural philosophers.

Lesson 47: John Philoponus and Aristotle

Do you remember Aristotle? Aristotle had a lot of correct ideas, but he also had a lot of incorrect ideas. For example, he was one of the philosophers who believed the earth is eternal. Obviously, John Philoponus disagreed with him on that point. In fact, one of the reasons Philoponus made his argument against an eternal earth was to challenge Aristotle's teaching on the matter. Not surprisingly, Philoponus disagreed with Aristotle on other things as well. This set him apart from many natural philosophers, because Aristotle was so highly respected that few were willing to challenge his teachings. They thought that if Aristotle said it, it must be true. Philoponus recognized the great work that Aristotle did, but he was also willing to disagree with Aristotle when he thought Aristotle was wrong.

Do you remember what Aristotle believed about falling objects? He thought that heavier objects fell faster than lighter objects. In fact, he thought that if one object weighed twice as much as another object, it would fall twice as fast. You already did an experiment to learn that's not true. Guess what? Philoponus wrote about just such an experiment. He wrote that while some heavy objects fell slightly faster than some light objects, the difference wasn't nearly as large as what Aristotle thought. Even though a lot of natural philosophers disagreed with Philoponus on this point, as you found in your experiment, he was correct.

To learn about another one of the things Philoponus disagreed with Aristotle on, perform the following experiment.

Projectiles and the Medium Through Which They Travel

What you will need:
- A rubber band (It should be short enough to comfortably stretch between your index finger and thumb, but long enough to allow you to use it to launch a pebble into the air. It also needs to be wide enough so a pebble can fit into it.)
- A pebble that fits comfortably in the rubber band.
- An open space outside
- A bathtub (A swimming pool would be even better, if you have access to one and it is a good time of year for a swim. If so, do steps 6 and 7 in the pool with the pebble and rubber band underwater.)

What you should do:
1. Loop the rubber band around your thumb and index finger and stretch it tight.
2. Use your other hand to put the pebble in the rubber band, on the side facing away from you.
3. While holding the pebble in the rubber band, pull back on the rubber band and then release so that the pebble launches into the air. Notice how far it goes.
4. Get the pebble (or more pebbles like it) and repeat steps 2-3 several times, observing how the pebble travels through the air and how far it goes.
5. Go inside and start filling the bathtub with water. It doesn't have to be very full. It just has to have enough water so that you can put your entire hand underwater.
6. Repeat steps 2-3, but launch the pebble underwater from one end of the bathtub to the other. Aim it so that it stays underwater the whole time. Notice how far it travels this time.
7. Repeat step 6 a few more times, getting an idea of how the pebble's motion is different under water than it was in the air.
8. Drain the bathtub and put everything away.

The results of the experiment probably didn't surprise you much. The pebble couldn't travel nearly as far underwater as it could in the air. Why is that? To fully understand what is going on here, let me define a couple of terms. First, when I launch an object that travels through the air without its own source of power, it is called a **projectile** (pruh jek' tyl). The pebble you launched in your experiment, for example, was a projectile. The rubber band gave it an initial push, but once it left the rubber band, there was nothing powering its flight. It was just flying through the air because of the push that the rubber band gave it.

Second, when a projectile is traveling, it usually travels through something. In the first part of your experiment, the projectile traveled through air. In the second part of the experiment, it traveled through water. In science, when an object travels through something, we say that it is traveling through a **medium**. I know you think of the word "medium" as meaning something between big and small. In science, though, it can also mean something through which an object travels. In the first part of your experiment, then, the medium through which the projectile (the pebble) traveled was air. In the second part of your experiment, the medium through which it traveled was water.

Aristotle was familiar with the motion of projectiles, but he believed that for something to move, it must be continually pushed or pulled. In his mind, objects "wanted" to be at rest. For an object to move, then, it must be continually pushed or pulled. However, this idea didn't make sense in terms of projectiles. For example, Aristotle observed arrows being shot from a bow. The bow obviously gave the arrow an initial push, but after that, the arrow would fly through the air without anything Aristotle could see pushing or pulling on it. As a result, Aristotle actually thought that the air was pushing the arrow. He thought that as the front of the arrow pushed air out of its way, the air traveled to the back of the arrow and pushed the arrow along. So Aristotle thought that the medium through which a projectile travels actually helps it to continue to travel.

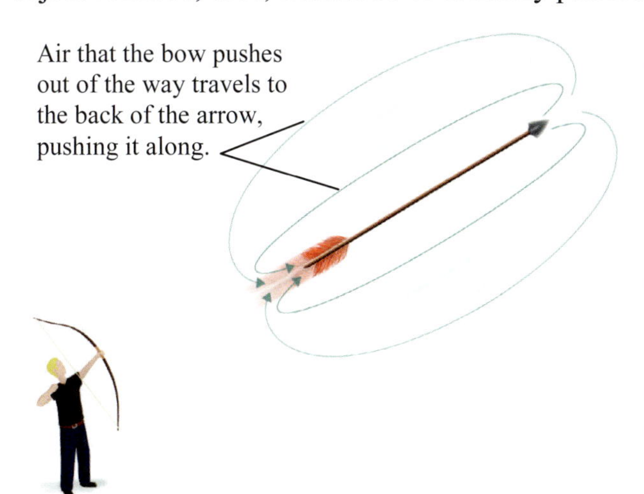

Aristotle thought that as an arrow flew, the air that it pushed out of its way would travel back to the rear of the arrow and push it along.

Philoponus didn't believe that. Instead, he thought that the medium through which a projectile travels actually slows it down. Based on your experiment, who do you think is right? The water is a thicker medium than air, right? Because of that, you would think that if the medium pushed the projectile along, water should do a better job of pushing than air. Your experiment showed, however, that a projectile doesn't travel nearly as far in water as it does in air. That's because the medium through which a projectile travels resists the motion. When a projectile travels through air, it is met with air resistance, which you already learned about. When a projectile travels through water, it is met with **water drag**, which resists the projectile's motion even more.

So in the end, Philoponus was right. The medium through which a projectile travels doesn't help the projectile; it hinders the projectile. However, that doesn't answer the question of why the projectile stays in motion long after it stops being pushed. Your pebble traveled a long distance in air without the rubber band pushing on it. An arrow travels an even longer distance. Why? Philoponus thought that there must be some lingering force that the initial push gave the projectile, and while that force lingered, the projectile could continue to move. Now it turns out that Philoponus was also wrong. You will learn in a later course that you don't need a continual force to produce motion. Instead, if an object is moving, it will continue to move until it is acted on by another force. However, natural philosophers won't learn this principle for about a thousand more years!

Since I mentioned water drag, I thought I should mention how some of the best swimmers on the planet deal with it. Dolphins are mammals that live in the ocean. As you probably know, mammals are warm-blooded animals that breathe air and give birth to live young. Well, it turns out that dolphins can swim really quickly, even when they are completely underwater. Some scientists have observed them swimming as fast as 35 kilometers per hour (22 miles per hour).

This is an impressive feat, since water drag resists motion strongly. As a result, scientists have been studying dolphins for years, trying to find out exactly how they overcome water drag and swim so quickly. Interestingly enough, scientists haven't completely figured it all out, but they do know a few things. First, a dolphin's shape is important. You probably know that certain shapes can reduce the resistance of a medium better than others. It turns out that a dolphin's shape is ideal for reducing water drag. In addition, dolphins have ridges on their skin. These ridges are arranged in a particular way, and it seems that they interact with the water as it flows by the dolphin. The result of that interaction is to make the water travel more smoothly across the dolphin's skin, which reduces water drag. In addition, when dolphins swim, their skin vibrates with very tiny vibrations. Some scientists think that this also helps reduce water drag.

Dolphins are designed to reduce water drag so they can swim quickly even underwater.

We don't know all the details about how dolphins reduce water drag, but then again, that shouldn't surprise you. Dolphins were designed by God to live underwater, and many of God's designs can be amazingly complicated!

LESSON REVIEW

Youngest students: Answer these questions:

1. Which is a projectile: an airplane flying in the sky or a ball that has been thrown in the air?

2. When a projectile travels through a medium, what does the medium do?

Older students: In your notebook, explain what a projectile is and how Aristotle thought projectiles flew through the air. Then explain why your experiment shows that his idea is incorrect.

Oldest students: Do what the older students are doing. In addition, write down the following question and your answer in your notebook: "Airplanes will often fly very high when they are traveling for long distances, because there is less air at those elevations. Why would that be important?" Check your answer and correct it if you are wrong.

Lesson 48: Robert Grosseteste (c. 1175 – 1253)

History presents us with another "hole" when it comes to scientific discovery from the late 500s to the early 1200s. This isn't because natural philosophers stopped thinking, of course! Quite a lot of philosophy was written during this time period. However, not a lot of *new* ideas were discussed in the field of natural philosophy. However, all that changed in the early 1200s, when **Robert Grosseteste** (grohs' test) started writing.

He was a teacher at the University of Oxford in England, but eventually became the Bishop of Lincoln, a very important position in the church at that time. Prior to his appointment as bishop, he wrote about philosophy, theology, and the great thinkers that came before him, such as Aristotle. However, he is best remembered for emphasizing the importance of observations and evidence when it came to studying nature. He thought that no explanation about the natural world and how it works should be accepted unless evidence could be brought forth that supported the explanation.

Because of this, he is considered by some to be the father of the method that is used by modern science, which emphasizes experiments and observations when it comes to explaining the world around us. Does it surprise you that someone who was a theologian and ended up in an important church position is considered by some to be the founder of the method by which modern science is done? It shouldn't.

You see, science assumes that the universe is understandable and is controlled by certain laws. It also assumes that in general, there isn't anything that gets in the way of these laws. There's no reason to think these things are true unless you think that the universe was made by a rational Lawgiver who set the laws up and then allows them to work in nature. Of course, this is exactly what Christians think, so it is not surprising that modern science grew out of the Christian church.

This stained glass window from a church in the United Kingdom honors Robert Grosseteste.

Now, of course, the kinds of evidence that scientists could collect back in the 1200s was limited, so what Grosseteste could conclude based on the evidence was limited. Nevertheless, he was able to solve one puzzle that scientists had argued about for some time: Exactly how does the sun warm the earth? Does it warm the earth like a fire warms a room, giving heat directly to everything in the room? After all, the sun looks like a big ball of white fire in the sky. It would make sense that it could warm the earth like a fire. However, it was also known that the sun was pretty far from the earth, and everyone knows that the farther you are from a fire, the less it is able to warm you. Since the sun warms the earth so well, many natural philosophers thought that it could not be acting like a fire. Grosseteste was able to use evidence to answer the question once and for all.

How the Sun Heats the Earth

What you will need:
- A magnifying mirror (It is easiest if the mirror is a small, hand-held magnifying mirror. However, it can be a larger one, such as one that sits on a vanity. The larger the mirror, the more dramatic the effect.)
- Some newspaper
- Kitchen tongs or pliers
- A sunny day (It can be partly cloudy – just wait until the sun comes out from behind the clouds.)

What you should do:
1. Tear off a small part of a page of newspaper. The piece should be about the size of your hand.
2. Use the tongs or pliers to hold onto the outer edge of the piece of newspaper.
3. Take the mirror and the piece of newspaper (which is being held by the tongs or pliers) outside.
4. Stand so that you are facing the sun and it is shining down on you.
5. Point the mirror at the sun. **NEVER LOOK IN THE MIRROR DURING THIS EXPERIMENT!**
6. Hold the piece of newspaper (which is being held by the tongs or pliers) in front of the mirror far from it. If the paper casts a shadow on the mirror, move the paper down so that it doesn't block the sun from the mirror.
7. Using the hand that is holding the mirror, tilt the mirror until light from the sun reflects off the mirror and lands on the piece of newspaper (see the photo on the right). The spot of light on the newspaper should be reasonably circular or oval-shaped. If it has a different shape, tilt the mirror to get an oval or circular shape. 
8. Once you have the light shining on the piece of newspaper, start bringing it closer to the mirror. You should eventually see the light concentrating into a spot on the newspaper.
9. Move the piece of newspaper back and forth until the spot of light is as small as possible.
10. Wait. What happens?
11. If nothing happens after a few minutes, drop the newspaper and tongs and do the next three steps. **Do not do those steps if something happened to the newspaper**!
12. Hold the hand that you were using to hold the tongs like you were holding the newspaper before, so that the light reflects off the mirror and hits your hand. **Your hand should be far from the mirror**.
13. **Slowly** move your hand closer to the mirror. When it gets uncomfortable, pull your hand away.
14. Go back inside and put everything away. Remember where the magnifying mirror is, however, because you will use it again in the next lesson.

The magnifying side of the mirror is pointed towards the paper.

What did you see in the experiment? If you live in most parts of the world and had a reasonably-sized magnifying mirror, you should have seen the light burn a hole in the piece of newspaper. If that didn't happen and you ended up using your hand, it should have gotten uncomfortably warm as your hand moved closer to the mirror.

This is essentially the experiment Grosseteste did to show that the sun does not heat the earth the same way a fire heats a room. A fire produces a lot of heat, and that heat travels outward. Anything near the fire that is cool will absorb the heat, and as a result, it will get warmer. However, this experiment shows that the sun doesn't do that to the earth. Why? Think about it. If you put a mirror near a fire, would the mirror reflect the heat? Of course not! It would absorb the heat like anything else, becoming warmer.

Your experiment, however, clearly showed that the sun was warming the newspaper because the mirror was reflecting the sun's *light* onto the newspaper. Now a regular mirror would not have warmed the newspaper so much, because it just reflects the light that hits it. A magnifying mirror, however, does something else. It reflects the light so that the light becomes *concentrated* at a specific distance away from the mirror. This distance is called the mirror's **focal point**.

In the experiment, when you held the piece of newspaper at the point where the spot of light was the smallest, you held it at the focal point. At that point, all the light that was hitting the mirror was shining on that small spot. This concentrated the light, putting a lot of radiant energy on the newspaper, and as that radiant energy was absorbed by the newspaper, it was converted into thermal energy, which heated up the newspaper until it burned.

Using this experiment (and some other observations), Grosseteste demonstrated that the sun heats the earth with its light. The light shines down on the earth, and when it is absorbed, it heats up whatever absorbs the light. Using a simple experiment, then, Grosseteste was able to determine that the sun did not act like a fire. Instead, it was a source of a lot of light, and that light provides heat to the earth.

If you are worried about how a magnifying mirror concentrates the light that it reflects, don't be. You will learn about that in the next lesson, provided you are doing the challenge lessons.

LESSON REVIEW

Youngest students: Answer these questions:

1. What is the big difference between the way a normal mirror reflects light and the way a magnifying mirror reflects light?

2. How does the sun warm the earth: with its heat or with its light?

Older students: In your notebook, explain why the experiment shows that the sun doesn't warm the earth like a fire warms the things around it, and how, instead, it warms the earth with its light.

Oldest students: Do what the older students are doing. In addition, explain why it isn't at all surprising that someone who was a very important person in the Christian church is considered by some to be the person who came up with the method by which modern science is done.

Lesson 49: A Curved Mirror

In the previous lesson, you used a magnifying mirror to reflect light onto a newspaper, hopefully setting the newspaper on fire. If nothing else, you found that the mirror reflected light in a way that warmed what the light shined on. How does the mirror do that, and why does it magnify things, when most mirrors do not? The answers to these questions are based on the *shape* of the mirror. To see what I mean, do the following experiment.

Curved and Flat Mirrors

What you will need:

- The magnifying mirror you used in the previous experiment
- A hand-held mirror that does not magnify (Often, a mirror will magnify on one side and not magnify on the other. You can use such a mirror in this experiment. Just use the magnifying side when the experiment calls for a magnifying mirror and the non-magnifying side when it calls for a non-magnifying mirror.)
- A person to help you

What you should do:

1. Hold the non-magnifying mirror close to your face and look at your reflection.
2. While you look at your reflection, pull the mirror away from your face slowly until it is as far from your face as you can hold it. How did your reflection change as you moved the mirror away?
3. While you are still holding the mirror out, have your helper take the mirror from you.
4. Have your helper hold the mirror the same way, so you can still see your reflection, and then have your helper slowly move the mirror even farther from your face. You may have to change position a bit to keep your face's reflection in the mirror. How does your reflection change?
5. Now repeat steps 1-4 with the magnifying mirror. You should notice a remarkable change in how your reflection behaves as your helper moves farther away from you.
6. Put the mirror or mirrors away.

What did you see in the experiment? If things went well, you should have seen that your reflection didn't change much when you moved the non-magnifying mirror away from your face. It got smaller, but that's about it. However, with the magnifying mirror, you should have seen your face go blurry and then turn upside down! When a magnifying mirror is held close to your face, your face looks larger, and it is not upside down. However, when the magnifying mirror is far from your face, it appears upside down in the mirror. Why?

Remember back in Lesson 23 when you learned the Law of Reflection? What did it say? In case you have forgotten, the Law of Reflection says, "When light reflects from a smooth surface, the angle of reflection will be equal to the angle at which the light hit the surface." In other words, when light hits a mirror, it will bounce off the mirror with an angle equal to the angle at which it hit the mirror. When you learned about this law, you saw how it worked with a regular, non-magnifying mirror. You shined a beam of light on the mirror, and you saw that the angle that the reflected beam made with the mirror was the same as the angle that the beam made when it hit the mirror.

The Law of Reflection is the same for non-magnifying mirrors and magnifying mirrors, but the mirrors themselves are different. A non-magnifying mirror is flat. However, a magnifying mirror is curved. Why does that make a difference? Well, look at the drawing below, which shows how light reflects off both a flat mirror and a curved mirror.

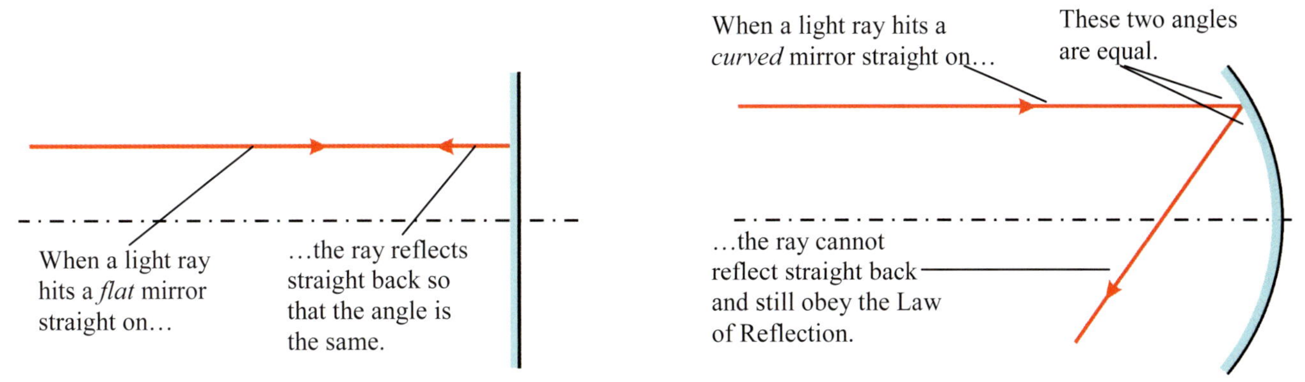

When light hits a flat mirror straight on, it reflects straight back, because of the Law of Reflection. Because of the curve, when light hits a curved mirror straight on, for the angle of reflection to be the same as the angle that the light ray hit the mirror at, the ray cannot be reflected straight back. Instead, it is reflected off at an angle.

Notice that since a flat mirror is flat, when a beam of light hits it straight on, it reflects straight back. That way, the angle at which the light hits the mirror equals the angle at which the light reflects from it. However, for a curved mirror, things are different. In order for the reflected angle to equal the original angle on a curved mirror, the light cannot reflect straight back. It is reflected at an angle. Now look what happens when we add a few more beams of light.

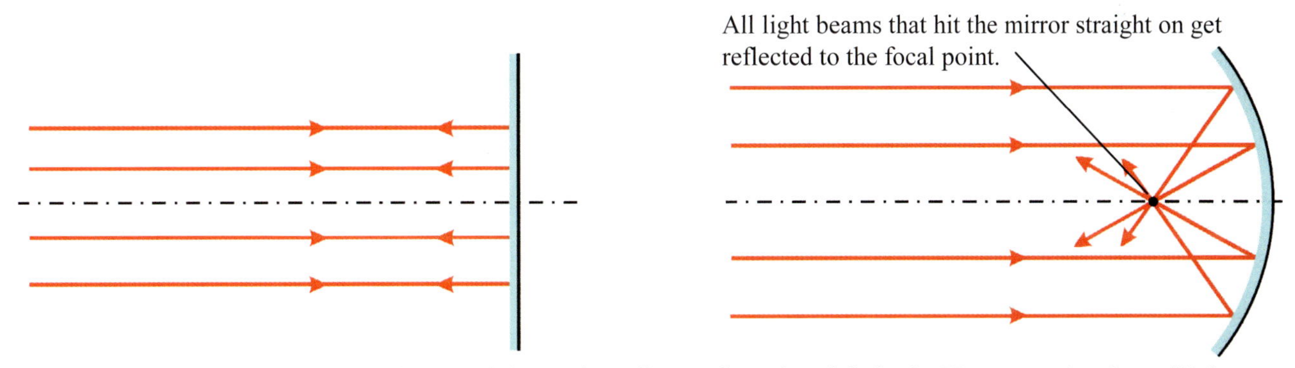

When lots of light beams hit a flat mirror straight on, they all get reflected straight back. However, when lots of light beams hit a curved mirror straight on, they get reflected towards the same point, which is called the focal point.

This is why the magnifying mirror was able to set the newspaper on fire, or at least warm your hand quite a bit. Light coming from the sun hit the mirror straight on. So all the beams of light that hit the mirror got reflected toward the same point – the focal point of the mirror. When you moved your newspaper back and forth to get the smallest spot of light on the newspaper, you were trying to find the focal point. When you got near the focal point, the light was concentrated well enough that it got the newspaper (or your hand) very hot.

In the end, then, the difference between a non-magnifying mirror and a magnifying mirror is the shape. A non-magnifying mirror is flat. As a result, your image is not changed when you look at it in such a mirror. A magnifying mirror is curved. Because of that, it focuses the light that hits it straight on towards the focal point. That changes the image you are looking at. When you are close to the mirror, it makes the image seem much larger than it is.

But what about the experiment? Why did your face get blurry and then turn upside down when you moved the magnifying mirror back far enough? Well, think about it. When you look at yourself in a mirror, you are looking at light that first reflects off your face, then travels to the mirror, then reflects off the mirror, and then hits your eyes. The image you see comes from the light that has reflected off the mirror. Think about what happens in a curved mirror. Look at the drawing on the left. There are two beams of light coming into the mirror. The dashed green line is drawn through the focal point. Anything to the right of the dashed green line is closer to the mirror than the focal point, and anything left of the dashed green line is farther from the mirror than the focal point.

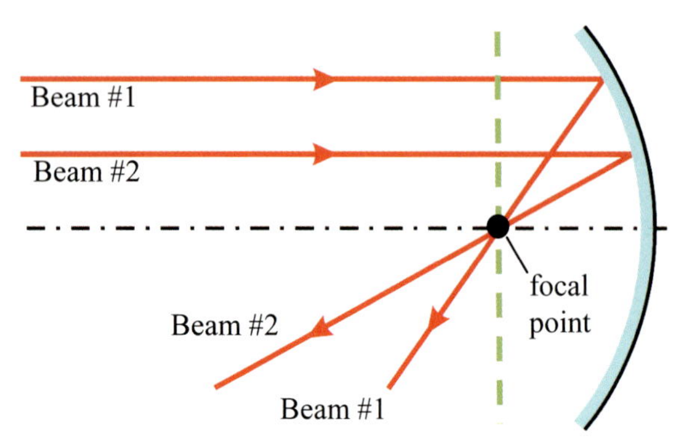

A curved mirror causes things to appear upside down when they are farther from the mirror than the focal point.

Now look at the two beams of light. Beam #1 is above Beam #2. Now look at the beams after they reflect off the curved mirror. The reflection of Beam #1 is above the reflection of Beam #2 when both beams are to the right of the dashed green line. However, what happens once the reflected beams pass the dashed green line? Beam #2 ends up above Beam #1.

So think about what happens when your face is closer to the mirror than the focal point. In other words, think about what happens when your face is to the right of the dashed green line. The reflection of Beam #1 will be above the reflection of Beam #2. As a result, you will see things the way they really are: right-side up. However, if you back away so that you are farther from the mirror than the focal point (in other words, left of the dashed green line) now the reflection of Beam #1 is *under* Beam #2. That means you will see things upside down!

In your experiment, when you saw your face right-side up, you were closer to the mirror than the focal point. However, as you pulled the mirror farther away, you eventually got to the point where you were farther away from the mirror than the focal point. This caused your face to flip upside down. This never happened with the non-magnifying mirror, because it was flat and had no focal point.

So a curved mirror magnifies things because it concentrates light that hits it straight on at its focal point. However, that causes a magnifying mirror to turn things upside down when they get farther away from the mirror than the focal point.

LESSON REVIEW

Youngest students: Answer these questions:

1. Which kind of mirror is curved: a magnifying mirror or a non-magnifying mirror?

2. What is the focal point of a curved mirror?

Older students: In your notebook, draw a flat mirror and a curved mirror. Each mirror should have two rays of light hitting the mirror straight on and reflecting back the way they are supposed to. Explain why the Law of Reflection causes a curved mirror to concentrate the light that reflects off of it and the name of the point to which all straight-on light beams are reflected.

Oldest students: Do what the older students are doing. In addition, explain why your face appears upside down if you are too far from a curved mirror.

Lesson 50: Roger Bacon (c. 1214 – 1294)

While Robert Grosseteste is the one who came up with the method that is used by modern science, **Roger Bacon** often gets the credit for it. As a result, many consider him to be the father of the method that is used by modern science. This is because Bacon, who did most of his work in the late 1200s, was a student of Grosseteste and took his teachings to heart. As a result, he did not rely on the word of others. He had to verify things for himself. In one of his books, he wrote, "For there are two modes of acquiring knowledge, namely, by reasoning and experience. Reasoning draws a conclusion and makes us grant the conclusion, but does not make the conclusion certain, nor does it remove doubt so that the mind may rest on the intuition of truth, unless the mind discovers it by the path of experience…" [R. B. Burke (trans.), *The Opus Majus of Roger Bacon*, Vol. 2, 1928, p. 583].

If you don't understand the quote, don't worry – I'll explain it. A lot of philosophers, both before and after the time of Bacon, thought that all you had to do to figure something out was think about it and reason it through. Once you did that, you had to be right. There was no reason to test your conclusion. So when Aristotle's reasoning, for example, concluded that light objects fall more slowly than heavy objects, there was simply no reason to doubt him on that.

Roger Bacon disagreed. He didn't think you could believe anything unless it was confirmed by experience. Do the following experiment to get an idea of what Bacon meant by "experience."

Candles and Air

What you will need:
- A small candle that stands up on its own (This can be a birthday-cake candle stuck in some clay, or it can be a candle that is wide enough to stand up on its own but still not very tall.)
- A glass that is taller than the candle
- A glass jar or other glass container that is significantly larger than the glass listed above
- A glass bowl or other glass container that is larger than the glass or jar listed above
- Matches or a lighter
- An adult to help you

What you should do:
1. Set the candle on a counter so that it stands on its own.
2. Have an adult light the candle.
3. Watch the candle for a minute.
4. Turn the small glass upside down and use it to cover the candle. What happens in a while?
5. Pull the glass away and have the adult relight the candle.
6. Turn the jar or larger glass upside down and use it to cover the candle, like you did with the glass. Did you notice any difference between what happened this time and what happened in step 4?
7. Pull the jar away and have the adult relight the candle.
8. Turn the bowl upside down and use it to cover the candle. Once again, try to see any difference between what happens now and what happened in step 6.
9. Clean up your mess and put everything away.

Roger Bacon did the first part of the experiment you just did because he read that a candle will stop burning when covered with a jar. This made sense to him, but he didn't believe it until he *experienced* it for himself. So what is experience, as far as Roger Bacon is concerned? It's an *experiment*! In the end, Roger Bacon would not believe something about nature until he confirmed it by experiment. This is why Bacon is often given the credit for coming up with the method we use in science today. In science, we confirm things by experiment, and Roger Bacon was a champion of that approach.

So what did the experiment tell Bacon? It told him that there was something in air that is necessary for things to burn. The candle used up this "something," and since no new air could come in through the glass container, it eventually got used up, and the candle could no longer burn. Bacon couldn't explain the process of burning any better than that, and it would take some time for natural philosophers to understand what the process of burning is all about, but Roger Bacon gave us the first step – he showed that there was something in air that is necessary to keep a candle burning. Now, he could have just believed what he read, but he knew that a lot of what you find in books is wrong. As a result, he had to test it for himself.

Your experiment was a bit more detailed than Bacon's. You used coverings of different sizes. Did you see any differences among them? You should have noticed that the larger the covering, the longer the candle was able to burn before it went out. If air is necessary for burning to happen, this should make sense. After all, the larger the container, the more air it holds. As a result, the larger the container, the more air the candle could use, so the longer it could burn. As we progress through history, you will eventually learn more about the process of burning, so I won't explain it to you here. For now, just understand that oxygen in the air is necessary to keep the burning process going. Once the oxygen is used up, the burning stops.

Have you ever seen someone who is trying to get a campfire started? What does he sometimes do? He sometimes blows on the fire, as shown in the picture below. What usually happens? The fire actually gets brighter, and usually, more flames are produced! So blowing on a fire actually helps the fire burn. Why?

Blowing on a fire can make it burn brighter, because you are replacing old air with newer air.

Believe it or not, the answer is found in your experiment! The process of burning needs oxygen that is in the air. As the wood in the fire burns, it uses up the oxygen. Since there is a lot of air around the fire, new air will eventually come in, and it will carry new oxygen to replace what the fire used up, but that takes time. As a result, the fire might not burn as well as it could, because it has to "wait" on new air to replace the oxygen it has used up. When you blow on a fire, you are forcing new air into the fire. That new air has more oxygen, so the fire is able to burn more brightly.

In the end, then, if you want to burn something, you need air. Air contains oxygen, which is absolutely necessary for burning. Now remember, at this point in history, natural philosophers didn't know that fire needed oxygen. They only knew that there was something in the air that was required for things to burn.

You might be surprised to learn that you are burning things in your body right now! Your body gets energy from the food you eat by first breaking it down into individual molecules and then burning some of those molecules. Chemically, we call the process of burning **combustion** (kum bust' yun), and there are all sorts of combustion processes going on in your body. Now don't get the wrong idea. There aren't little fires going on in your body. Your body has been so incredibly well-designed that it does combustion in a *controlled* way. So while there aren't any fires in your body, your body is doing all sorts of combustion to keep you alive.

Since your body is doing combustion all the time, what does it need? It needs the same gas that a fire needs – oxygen. In fact, that's why you breathe. You breathe so you can get oxygen from the air. When your body gets oxygen, it uses it to burn molecules from your food so that you can have the energy you need to live. If your body can't get enough oxygen, it can't do combustion. As a result, it can't convert the energy in your food into the energy you need to survive.

Being underwater is fun, but you can't stay there long, because your body will force you to breathe air so that it can do combustion.

That's why you can't hold your breath for very long. Your body knows that it needs a constant supply of new air to replace the oxygen that it needs for combustion. So no matter how hard you try, your body will eventually force you to stop holding your breath so that you can take in new oxygen.

LESSON REVIEW

Youngest students: Answer these questions:

1. When Roger Bacon said that experience is necessary in order to be certain, what did he mean by "experience?"

2. Why did the candle go out in the experiment when you covered it?

Older students: In your notebook, rewrite Bacon's quote on page 105 in your own words. In addition, explain what you did in the experiment and why the candle went out. Be sure to explain why the candle burned longer in the larger containers.

Oldest students: Do what the older students are doing. In addition, explain in your notebook why pouring water on a fire will usually put the fire out. If you don't know, ask your parent for the hint that is found with the answer.

Lesson 51: A Bit More About Fire

Remember that since this title is in red, it is a challenge lesson. You can skip it if you like, but the experiment is pretty interesting, so even if you don't typically do challenge lessons, you might want to do this one!

In the previous lesson, you learned that when you blow on a fire, you are pushing new air into the fire, and that usually makes the fire burn more brightly. This is because fire needs oxygen, and it pulls the oxygen out of the air that is close to it. By pushing new air to the fire, you are giving it more oxygen, and as a result, it burns more brightly. Now that ought to make good sense, but if you were thinking hard, you might have been a bit confused.

When you blow on candles, they typically go out. They don't burn more brightly.

After all, in your own experience, you know that sometimes, blowing on a flame will put it out. After all, what do you do to the candles on your birthday cake? You blow them out, don't you? Why doesn't blowing on those candles make them burn more brightly? You are pushing new air to the candles, and the candles need new air. Why does blowing on a fire make the fire burn more brightly but blowing on a birthday candle make it go out?

There's a big difference between a fire and a candle. To see that difference, perform the following experiment.

Relighting a Candle

What you will need:

- A birthday candle (It could be any candle that is tall and thin. A candle that is wide enough to stand on its own will probably not work well for this experiment.)
- Some matches
- Something to hold the candle upright (If it is a birthday candle, you can just use a lump of clay. If it is another kind of tall, thin candle, you might need a candleholder that is made for it.)
- An adult to help you

What you should do:

1. Set the candle on a counter so that it stands on its own.
2. Have the adult light the candle.
3. Allow the candle to burn for about 30 seconds, and then blow it out. Watch the candle carefully when it goes out. Do you see the stream of "smoke" that rises up from the wick of the candle?
4. Have the adult light the candle again.
5. Allow the candle to burn for about 30 seconds, and have the adult light another match.
6. Blow out the candle.

7. As soon as the candle is out, have the adult bring the burning match close to the wick of the candle and into the trail of "smoke" that is rising up from the candle. The match should not touch the wick. It should just be close to the wick and in the trail of "smoke." What happens?
8. If nothing surprising happens, try steps 4-7 a few more times. Remember, the idea is that as soon as you blow out the candle, the adult should bring the burning match close to the wick and into the stream of "smoke" that is coming off the wick. The match and its flame should not actually touch the wick.
9. Clean up your mess.

What did you see in the experiment? You should have seen that the candle would start burning again, even though the match and its flame didn't touch the wick! How in the world does that work? How can you relight the candle if the flame of the match never touches the wick?

The answer lies in how a candle actually burns. Some people think that a candle burns the wick that you light. That's only true for a few moments, however. If the wick of the candle is the only thing that burned, a candle wouldn't last very long, because the wick burns fairly quickly. A candle actually burns the wax out of which it is made. However, the way that it burns that wax is pretty surprising.

When you light a candle, the flame you are using to light the candle heats up the wick. The wick has wax on it, so the wax melts into liquid. The flame that you are using to light the candle is very hot, however, so the wax that is in the wick actually *boils*. In the end, then, when you light a candle's wick, you are first melting the wax to make liquid wax, and then you are boiling it to make wax vapor (wax gas). The wax vapor is what actually burns. So while a candle does burn its wax, it doesn't burn it in its solid phase or its liquid phase. The candle burns the wax in its gas phase.

When you light a candle, you are getting the wax in the wick hot enough to melt and then boil. The wax vapor produced by the boiling is what actually burns.

When you blew out the candle, you saw a stream of "smoke" coming from the wick. Believe it or not, that wasn't really smoke. It was *wax vapor*! You see, even though the flame was out, the wick was still hot. As a result, the wax in the wick continued to boil for a while, until the wick cooled off. Since there was nothing to burn the wax vapor, it simply started rising in the air, and you saw it as something that looked like smoke. Once again, however, it wasn't smoke. It was wax vapor. What does wax vapor do? It burns! When the adult put the match's flame into the wax vapor, then, it started burning. Since the vapor was coming up from the wick, the vapor near the wick started burning, and that relit the wick!

The trail of "smoke" coming from a blown-out candle is really wax vapor. If you light the vapor, the flame will travel back to the wick, relighting the candle.

So now you should be able to understand why blowing on a candle will put the flame out. When you blow on a candle, you are pushing the wax vapor away from the candle. At the same time, you are cooling the wick down to the point where it is not hot enough to light the vapor again. As a result, the candle goes out. Note how this is different from a fire. A fire burns something solid, like wood. When you blow on the fire, you aren't blowing the wood away! You are just bringing newer air close to the fire. As a result, the fire burns more brightly. When you blow on a candle, you are blowing away the stuff that is burning (wax vapor). That puts out the candle.

Have you ever seen trick birthday candles that relight after they are blown out? Those candles have wicks that contain small bits of a chemical like magnesium. When the candle is blown out, the wick is hot, but it isn't hot enough to light the wax vapor. However, it is hot enough to light the magnesium, so the magnesium burns, and it burns hot enough to light the vapor that is coming from the wick, so that starts the flame again. If you watch a trick candle closely, you will see it spark from time to time. That spark is the result of the magnesium in its wick burning, and that spark is hot enough to light the wax vapor when the candle is blown out.

LESSON REVIEW

Youngest students: Answer these questions:

1. What is the fuel that a candle actually burns?

2. In a candle, what is the phase (solid, liquid, or gas) of the fuel?

Older students: In your notebook, draw a candle that is burning, and then draw a candle that has just been blown out. Make sure there is a trail of vapor rising from the candle, as shown in the picture above. Use those drawings to explain what a candle really burns and why your experiment worked the way it did.

Oldest students: Do what the older students are doing. In addition, explain how you would extinguish a trick birthday candle. Check your answer and correct it if it is wrong.

NOTE: The next lesson has an experiment that requires gelatin. It is best to start the experiment the night before the lesson, so the gelatin is ready when you start the lesson.

Lesson 52: Roger Bacon and Lenses

One of Bacon's favorite things to study was light. He probably picked up that interest from Grosseteste, who was also fascinated by light. Bacon spent a good deal of his time studying the refraction of light. Do you remember that term from Lesson 37? When you studied Ptolemy, you learned that light bends when it starts traveling through a different medium. In Lesson 37, you used this fact to shorten the shadow cast by a toothpick.

Like Ptolemy, Bacon was interested in the refraction of light, but he was interested in it for a specific reason. Perform the following experiment to see what I mean.

Lenses and Magnification

What you will need:

- Unflavored gelatin (This kind of gelatin is clear when it sets.)
- A small dessert bowl that is made of glass or a small glass
- A round cookie cutter or a glass
- A butter knife
- A spoon
- A spatula
- A pie pan, preferably with a completely flat bottom
- Two white paper plates
- A piece of white paper that has print on it, such as a newspaper
- A flashlight
- Black construction paper
- Tape
- Scissors
- A dark room

What you should do:

1. Make the gelatin almost according to the instructions, but use ½ **cup** <u>**less**</u> **boiling water** than instructed, because you want the gelatin to be very firm.
2. Pour some of the gelatin into the small dessert bowl, so that there are at least a couple of inches of liquid in the bowl.
3. Pour the rest of the gelatin into the pie pan.
4. Put both the bowl and the pan in the refrigerator so the gelatin can set.
5. Once the gelatin has set, lay the piece of paper with print on it flat on a table.
6. Set the dessert bowl with gelatin down on the paper so that it is on top of some of the words.
7. Look through the gelatin at the words. Since the gelatin is clear, you should be able to see them.
8. Turn the bowl on its side so that the gelatin inside the bowl is over some words, and now look through the gelatin from the side of the bowl and read the words. What difference do you notice?
9. Use the butter knife to cut a square out of the gelatin. The square needs to be a bit bigger than the cookie cutter or opening of the small glass you are using in place of the cookie cutter.
10. Use the spatula to lift the square of gelatin out of the pie pan and put it in the middle of one paper plate.

11. Repeat steps 9 and 10 so that you have one square of gelatin on each plate.
12. Press the cookie cutter or opening of the small glass into one of the squares so that it cuts a circle out of the gelatin.
13. Use the spoon and knife to clean away all of the gelatin around the cookie cutter or small glass, so that the only gelatin left is what's inside the cookie cutter or jar lid.
14. Lift the cookie cutter or small glass (do not twist – just lift) so that now there is a circle of gelatin where there used to be a square of it.
15. Use the black construction paper, scissors, and tape to make a covering for the flashlight like you did in Lesson 23.
16. Take the two paper plates with gelatin on them and the flashlight into a dark room.
17. Shine the flashlight so that the beam of light comes out of the flashlight and hits one side of the gelatin circle straight on, as shown in the picture on the right. Notice how the beam exits the other side of the circle.
18. Move the flashlight up and down so that its beam hits the circle at different places on its side. Don't change the angle, however. The beam needs to hit the side of the gelatin straight on, just at different locations on the circle's side. Notice how that changes the way the light comes out the other side of the circle.
19. Repeat steps 16 and 17 using the square of gelatin. How did the square change the direction of the light beam compared to the circle?
20. Clean up your mess.

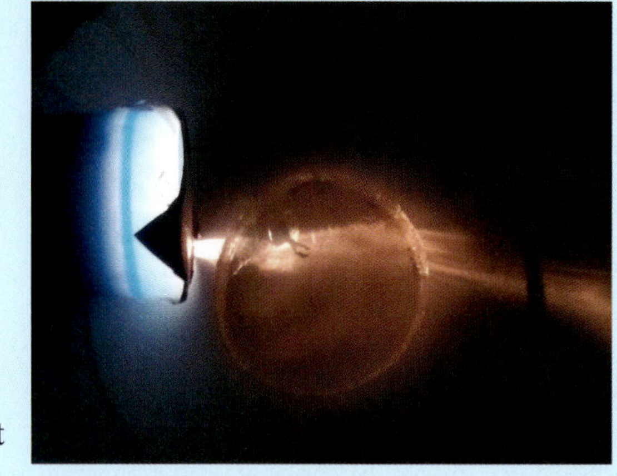

In the first part of the experiment, you should have seen that when you read the words through the flat surfaces of the gelatin, the words looked the same as those on the paper. However, when you looked through the side of the glass at the words, they were much bigger. In other words, curved gelatin magnified the words, but flat gelatin did not. Why? Well, the second part of the experiment explains that. Here's what you probably saw in that part of the experiment:

When light hits a square of gelatin (or glass) straight on, the light isn't bent (left). As a result, it passes straight through the square. However, when light hits a circle of gelatin (or glass) straight on, the curve of the circle causes the light to be bent so that no matter where the light hits the circle, it is directed to a specific point, called the focal point (right).

When you shined the beam of light into the square of gelatin, you shouldn't have seen much bending. The beam of light should have exited the square going the same direction in which it entered the square. Compare that to what happened with the circular piece of gelatin. When you moved the flashlight, the beam bent differently, depending on where it hit the circle. In the end, if you noticed, the circle of gelatin seemed to direct the beams to hit the same spot on the other side of the circle. That spot is called the focal point.

Bacon used glass and water in his experiments, but the results were the same. He found that when you looked at things through curved glass or water, they were magnified. He reasoned that this happens because of the way the curved glass or water bends light. Since it tends to focus light to a specific point, it increases the amount of light you see, so everything looks bigger. This led him to realize that we can change the size of things just by bending the light that is coming from them in a specific way. As a result, Bacon predicted that one day, we would learn enough about bending light that we could make pieces of curved glass that would help people with poor eyesight to read better. In other words, he predicted that we would eventually make eyeglasses. Indeed, the first eyeglasses were invented 18 years after Bacon wrote about curved glass and how it affects light. We use the word **lens** to refer to a curved piece of glass that bends light in a specific way.

The first eyeglasses were made only 18 years after Bacon's work on how light behaves in curved glass.

In addition, Bacon suggested that we would eventually be able to bend light well enough to make things that were far away appear to be very close. In other words, he predicted that we would eventually be able to make telescopes. Some historians even think that Bacon made his own telescope, but there's not enough evidence to be certain of that. The first telescopes of which we are certain don't appear in history until 1608, more than 300 years after Bacon wrote his book on optics.

Regardless of whether or not he made a telescope, Bacon knew the power of bending light by his own experiments (and those of Grosseteste), and he was able to reason through what could be done with it.

LESSON REVIEW

Youngest students: Answer these questions:

1. (Is this statement True or False?) If you look at things through a flat piece of glass, they will be magnified.

2. Fill in the blanks: A circle of glass (or gelatin) directs light that hits it straight on to a point called the _____ ________.

Older students: In your notebook, make a drawing that represents what happened in your experiment. Show both a square piece of gelatin and a circular piece of gelatin. Draw several straight lines that represent light hitting each, and draw what happened to the light after it passed through. For the circular piece of gelatin, point out the focal point.

Oldest students: Do what the older students are doing. In addition, answer the following question in your notebook: "Is the glass in a magnifying glass flat or curved? Why?" Check your answer and correct it if it is wrong.

Lesson 53: Petrus Peregrinus (lived c. 1250) and Magnets

During the same time when Roger Bacon was performing his experiments (the late 1200s), **Petrus** (peh' trus) **Peregrinus** (pehr' uh grin' us) was working as an engineer for the army of the king of Sicily. No, he didn't drive trains. An engineer is someone who designs and builds machines or at least keeps machines that have already been built running. In those days, armies had machines, like catapults, that threw heavy boulders or burning oil. Those machines had to be designed, built, and kept running, and that's one thing Petrus Peregrinus, also known as "Peter," did.

We don't know when he was born or when he died, but we do know that in his spare time, Peregrinus did experiments. The thing that seemed to interest him the most was **magnetism**. Philosophers had known about magnetism for a long time, because there are naturally magnetic rocks that you can find called **lodestones**. Philosophers already knew that somehow, certain metals were attracted to these lodestones, while other metals were not. They also knew that lodestones were attracted to one another. However, most philosophers considered this attraction to be very mysterious, and some even thought it was magical!

Peregrinus couldn't explain how these lodestones worked, but he did study them enough to learn some basic rules about them. These rules are sometimes called the Basic Law of Magnetism. To get an idea of what Peregrinus determined, perform the following experiment.

Magnets

What you will need:
- Two magnets (You cannot use refrigerator magnets for this experiment, because they are made of composite materials that behave differently than normal bar magnets.)
- Two metal sewing needles

What you should do:
1. Lay one magnet on a table.
2. Drop the other magnet on top of the first one.
3. Try to pull the two magnets apart. It will take some work, because the magnets are attracted to one another.
4. Once you pull the magnets apart from one another, turn one of them around so that it is facing opposite of the way it was when it was stuck to the other magnet.
5. Hold each magnet tightly so that it cannot move in your hand, and try to force the magnets to stick together. What happens?
6. Hold the magnets very close to one another, and then loosen your grip on one of the magnets so that it can turn freely in your hand. What happens?
7. Put the magnets off to one side and place one sewing needle on the table.
8. Touch the sewing needle that is lying on the table with the other one. Are the two needles attracted to one another?
9. Move one of the needles away, but keep the other one on the table.
10. Bring one of the magnets close to the needle. Slowly bring it closer and closer until something happens. What happens?
11. Let the needle stay on the magnet, and lay the magnet down. We will come back to it near the end of the lesson.

What happened in the first part of the experiment? The magnets stuck together at first, because they were attracted to one another. However, when you turned one of the magnets around, they didn't want to stick to each other anymore, did they? In fact, it was hard to push them close to one another, wasn't it? So, sometimes magnets are attracted to one another, and sometimes, they **repel** (push away) each other. Why?

Peregrinus decided that every magnet has two sides, or poles. He called one the **north pole** and one the **south pole**. He said that magnets are attracted to one another when the south pole of one magnet is close to the north pole of the other magnet. In other words, opposite poles attract one another. In the same way, magnets are repelled from each other when both of their south poles are close to one another or both of their north poles are close to one another. In other words, like poles repel one another. This is sometimes called the Basic Law of Magnetism:

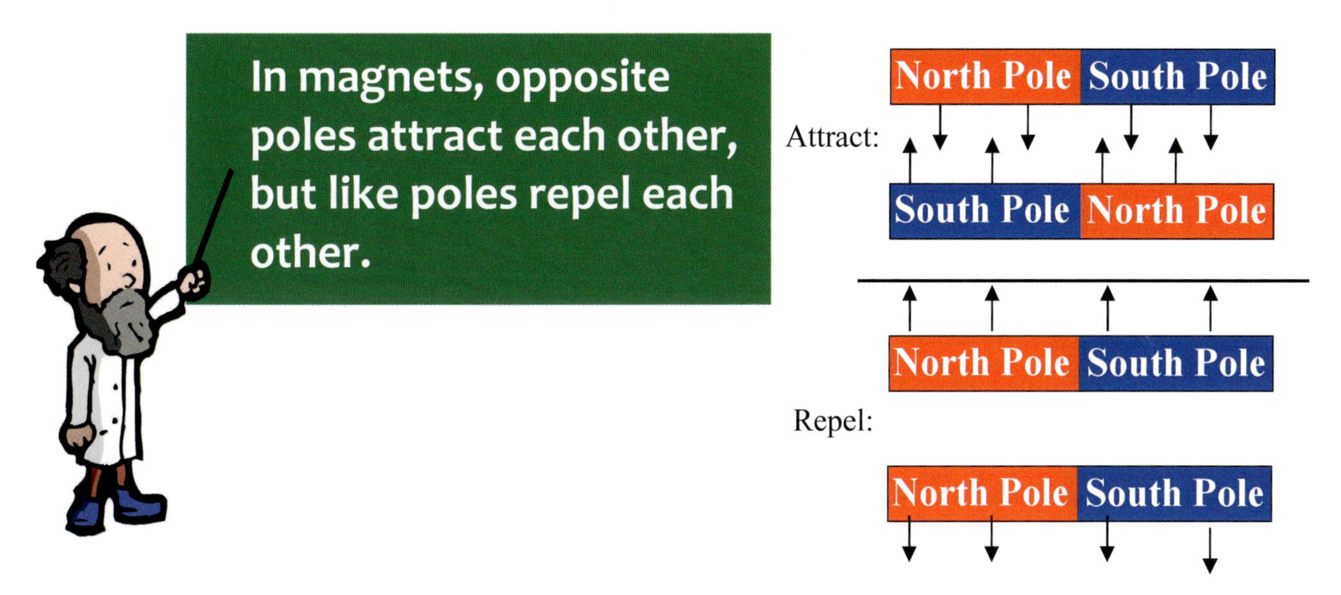

In your experiment, when the magnets were attracted to one another, it was because their opposite poles were close to each other. However, when you twisted one of the magnets around, you made it so that their north poles were close to each other or their south poles were close to each other. Since like poles repel, that made it hard to push the magnets together, because their like poles were repelling each other.

What happened when you released your grip on one of the magnets? It twisted around in your hand, didn't it? Why? Because opposite poles attract one another. Even though they weren't close together, the north pole of one magnet was attracting the south pole of the other magnet, and vice-versa. So when you released your grip, the magnet was free to move in the direction of that attraction, and it turned around so that the opposite poles of the magnets were lined up again!

What happened in the second part of your experiment? The needles weren't attracted to each other, because they weren't magnets. However, the magnet attracted a sewing needle, didn't it? In fact, the needle should have moved to the magnet on its own when the magnet got close enough. Why did it do that? Because the metal of the sewing needle can become magnetic, if it is placed near a magnet. When the magnet came close to the needle, then, it was able to turn the needle into a magnet. Then, it behaved like any other magnet, with its opposite poles being attracted to the magnet's opposite poles.

To see that the needle really did become a magnet, let's go back to the experiment for a moment.

More on Magnets

12. Pull the needle off the magnet.
13. Touch the needle you just pulled off the magnet to the other needle, and slowly try to move it away. What happens? If nothing interesting happens, try touching different parts of the needle that came off the magnet to the other needle. Eventually, you should see that the two needles stick together.
14. If you are doing the challenge lessons, stick one of the needles to the magnet and leave it stuck there. Put the needle and magnet somewhere out of the way so that you can get them the next time you do science. You will use the needle in the next experiment, but it needs to be stuck to the magnet for a while first.

Why did the needles stick together? Well, remember that the magnet attracted the needle to itself by first making the needle a magnet. So even though the needle wasn't initially a magnet, it *became* a magnet when it was put near a magnet. Once you pulled the needle away, it was still a magnet, so it could attract the other needle, just like the first magnet did! So the magnet really did make the needle into another magnet. The needle will eventually lose its magnetism, because it is only a weak magnet. Nevertheless, for a while, it will be a magnet.

That's something else Peregrinus figured out. He decided that the way a magnet attracts a piece of metal is to first make that metal another magnet. That way, the metal would have north and south poles, and the opposite poles would attract one another. So in the end, a magnet can impose its magnetism on certain metals. Not all metals can become magnets, so not all metals are attracted by magnets. However, if a magnet can, it will impose its magnetism on a piece of metal, giving the metal poles that are closest to its opposite poles. That way, the metal is attracted to the magnet.

LESSON REVIEW

Youngest students: Answer these questions:

1. Fill in the blanks: ________ poles of a magnet attract one another, but ________ poles repel each other.

2. How does a magnet attract a piece of metal that is not a magnet?

Older students: In your notebook, write down the Basic Law of Magnetism and use it to explain why the magnets were first attracted to each other in the experiment but were later repelled by each other. Also, explain what a magnet has to do to a piece of metal in order to attract it.

Oldest students: Do what the older students are doing. In addition, answer the following question in your notebook: "Suppose you have a very strong magnet and a very weak one. You put them together so that their like poles are closest to each other, and you hold them in place for a very long time. What will eventually happen to the poles of the weaker magnet?" If you are having a hard time answering the question, ask your parent/teacher for the hint that is in the answers.

Lesson 54: Petrus Peregrinus and the Compass

One very useful aspect of magnetism is that it allows us to find out which way is north and which way is south. Perform the following experiment to see what I mean.

Making Your Own Compass

What you will need:

- The sewing needle that has been stuck to the magnet since the last time you did science
- A large bowl
- Scissors
- A Styrofoam cup
- A thimble (or something else you can use to push on a needle without getting hurt)
- Water

What you should do:

1. Fill the bowl ¾ of the way with water and let it sit on a table or counter.
2. Cut a thin piece of Styrofoam from the cup. It needs to be about half as long as the needle.

3. Pull the needle off the magnet, and set the magnet down far away from the bowl.
4. Use the thimble to push the needle through the thin piece of Styrofoam. The needle doesn't need to go through all of the Styrofoam. It just needs to be solidly attached to the Styrofoam and lying along the length of the piece, as shown in the picture on the left.
5. Place the Styrofoam/needle contraption in the bowl so that it floats on the water.
6. Watch the needle for a few moments. Notice which way it ends up pointing.
7. If the needle ends up floating to the edge of the bowl and hitting the side, just use your finger to gently move it back so that it floats freely in the water.
8. Once you see where the needle is pointing, use your finger to spin the needle so that it is pointing in a completely different direction. Then, let go of the needle and see what happens. Once again, if it ends up hitting the side of the bowl, just use your finger to gently move it back.
9. Grab the bowl with both hands and spin the bowl slowly. Watch the needle. What happens?
10. Bring the magnet close to the needle, above the bowl. What happens to the needle?
11. Move the magnet around, keeping it close to the needle. How does moving the magnet affect the needle?
12. Clean up your mess and put everything away.

What did you see in the experiment? If things went well, you should have seen that no matter how you spun the needle, it always went back to pointing in the same direction. If you know the **cardinal directions** (north, south, east, and west), you might have noticed that one end of the needle always pointed north, and the other end always pointed south. In other words, you made a **compass**, which can allow you to know which way is north and which way is south no matter where you are.

Modern compasses, like the one pictured on the right, are easy to use and are a big help when people are trying to figure out which way they are headed. After all, if you know that you need to move in a specific direction, all you have to do is pull out your compass. Since the compass tells you which way is north, you now know that east is to the right of where the compass needle is pointing, west is to the left, and south is the exact opposite direction.

The blue end of this compass needle (the north pole of the magnet) always points north. This allows you to know all the directions, no matter where you are.

In the compass you made, you didn't know which end of the needle was pointing north and which was pointing south, but that's just because you didn't know enough about how the needle became magnetic. However, if you know which way is north at some location with which you are familiar (like your house), you could look at the needle, see which end pointed north, and then mark that end with some paint. After that, as long as you carried the needle, bowl, Styrofoam, and some water around, you would always know which way is north, because the marked end of the needle would always point in that direction.

So why does one end of the compass needle always point north? That's because the earth itself acts like a big magnet. Just like any other magnet, it has a north pole and a south pole. In fact, that's why Petrus Peregrinus used the terms "north pole" and "south pole" when describing magnets. It was well known before his time that if you floated a magnet in water like you did in your experiment, one end of the magnet would always point to the earth's North Pole, while the other end would always point to the earth's South Pole. So Peregrinus said that when a magnet floats on water, the north pole is the end that always points north, and the south pole is the end that always points south.

Since the earth acts like a magnet, it attracts one pole of a magnet to its South Pole, and it attracts the other pole to its North Pole. As a result, the magnet in a compass turns so its poles are as close as possible to the poles that attract them. However, earth is a fairly weak magnet. You should have seen that in the second part of your experiment. Even though the needle floated so that one end pointed north and the other end pointed south, you could change that by bringing a magnet close to the needle. When you did that, the magnet in your hand attracted the needle more than the earth did, so the needle moved in response to the magnet in your hand rather than the magnet of the earth.

This was the problem with using compasses back in ancient times. According to many historians, Chinese inventors had come up with compasses long before the late 1200s, which is when Peregrinus was doing his experiments. The problem is that because the earth is such a weak magnet, it can't force most magnets to point in the directions of the North and South Poles. For example, if you laid the needle you used in your experiment on a table, it wouldn't point north and south. It would point in the direction that you laid it. That's because it's hard to turn a needle that is lying on a table. However, it is very easy to turn a needle that is floating on water. The earth is such a weak magnet that it is not strong enough to turn a needle sitting on a table, but it is strong enough to turn a needle that is floating on water.

Up until the time that Peregrinus started doing his experiments, scientists didn't really understand this. Peregrinus realized that in order to make a useable compass, you had to make it so that the magnet in the compass would be very easy to spin. That way, it would be able to respond to the earth. Obviously, using a compass like the one you made wouldn't work well on a ship, for example, because the ship would rock back and forth, sloshing the water in the bowl around. Even if you were walking somewhere, carrying around a bowl, water, and a floating magnet would be a lot of trouble.

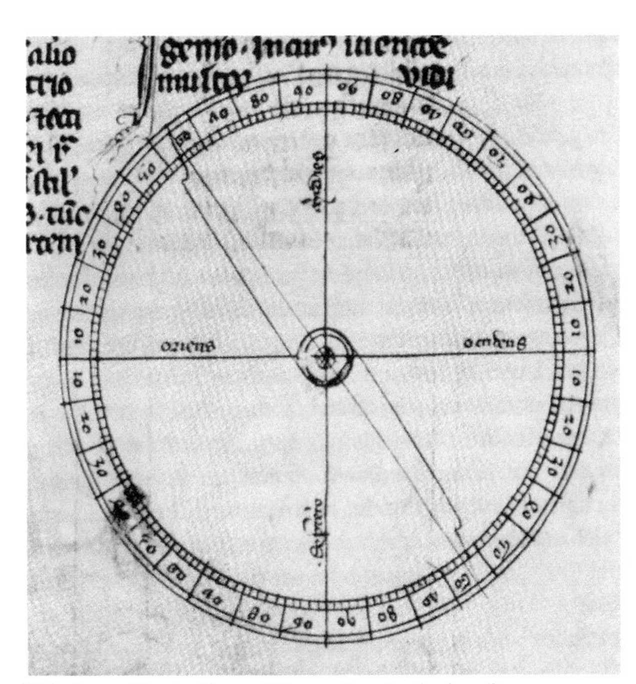

This is an illustration of Peregrinus's design for a compass that does not use water. It comes from a copy of his book. The copy was made in the 1300s.

To fix this, Peregrinus described in his book a way to make a compass without the use of water. Being an engineer, he was used to figuring out new designs, so he described a way to make a compass whose needle would spin easily without floating in water. The design was obviously very successful, because around 1300, compasses started becoming popular with sailors, allowing them to know which direction to steer their ships. Up until then, most sailors used the sun or the stars to determine direction, but as its design improved, the compass became a more reliable tool for them.

Now please don't misunderstand what I am saying here. Compasses were used by sailors prior to when Peregrinus did his work. It's just that they weren't very easy to use, so they weren't very popular. Peregrinus's design allowed people to make easy-to-use compasses, which made them significantly more popular. Because of his contributions to our understanding of magnets and compasses, the European Geosciences Union (EGU) gives the Petrus Peregrinus Medal to recognize modern scientists who are making important contributions to our current understanding of magnets and how they work.

LESSON REVIEW

Youngest students: Answer these questions:

1. Fill in the blank: The marked end of a compass needle always points __________.

2. Why did Peregrinus describe magnets as having a north pole and a south pole?

Older students: In your notebook, explain why a compass needle always points north. Also, explain why most magnets don't point north when you lay them on a surface, like a table.

Oldest students: Do what the older students are doing. In addition, remember that the north pole of a compass magnet always points north. Now think of the earth as a magnet. Which of the Poles (North or South) would be labeled as its *magnetic* south pole? Explain your answer in your notebook. Check your answer and correct it if it is wrong.

Lesson 55: Giles of Rome (c. 1243 – 1316)

Near the end of the thirteenth century (about 1280 or so), **Giles of Rome** started writing about philosophy and theology. He was a very important person in the church at the time. For example, he was in charge of an entire order of monks (the Augustinians), and he was eventually appointed to be the Archbishop of Bourges (boorzh), which meant he was in charge of all the priests in a certain region of France. Pope Benedict XIV called him "Doctor Fundatissimus" (fun duh tis' ih mus), which means "Best-Grounded Teacher" in Latin. He was clearly well-respected within the church.

This is how the 17th-century Italian artist Oliviero Gatti imagined Giles of Rome.

While he spent most of his time writing about theology, he was keenly interested in natural philosophy. As a result, he wrote a lot of commentaries on the works of Aristotle. If you don't know what a commentary is, think about reading a book. When you get done with it, someone might ask you what you thought about it. Was it good? Did you enjoy it? When you answer the person's questions, you are giving him your comments about the book. All of those comments together would be your commentary on the book.

Since all priests in the Roman Catholic Church were required to read the works of Aristotle, it was important for those who were considered great teachers to write their thoughts about his works. That way, students would not only learn what Aristotle taught, but they would also learn what current philosophers thought about those teachings.

Giles also came up with some original ideas, and one of them can be illustrated in the following experiment.

The Spaces In Between

What you will need:

- Two plastic one-liter bottles (Actually, anything that is transparent, tall, thin, and able to hold water will do. Glasses don't work very well because they tend to be wider at the top than the bottom, and that makes it harder to see the effect.)
- Water
- Alcohol (The container needs to indicate that it is at least 90% pure. Rubbing alcohol is usually 91% pure, and denatured alcohol, which is sold in hardware stores, is usually 95% pure. Either will work.)
- A ½-cup measuring cup (You need the kind that is used to measure out solids, because you will fill it to overflowing rather than filling it to a certain line.)
- A funnel
- A sink
- You may need someone to hold the funnel while you pour.

What you should do:

1. Set the bottles on the counter next to the sink, and put the funnel in one of the bottles. If the funnel or bottle isn't steady, have someone hold them steady.
2. Fill the ½-cup measuring cup full with water, holding it over the sink. Fill it all the way to the very top, and even allow it to spill over. You want the measuring cup to be completely full.
3. Carefully pour the water in the measuring cup into the funnel. Try not to spill *any* of the water.
4. Repeat steps 2 and 3 again, pouring the water into the same bottle as before.
5. Once the water has completely drained out of the funnel and into the bottle, remove the funnel and put it into the other bottle. If the funnel and bottle aren't steady, have someone hold them steady.
6. Repeat steps 2 and 3, pouring the water into the other bottle.
7. Now fill the measuring cup with alcohol. Hold it over the sink as you do it, because once again, you want to fill the measuring cup completely full. Allow some to spill out just to make sure it is full.
8. Add the alcohol to the bottle that has the funnel in it.
9. As things are settling, think about what you just did. You added ½ cup of water to the first bottle, and then you added another ½ cup of water to the first bottle. How much water should be in the first bottle? One cup, right? You added ½ cup of water to the second bottle, and then you added ½ cup of alcohol to the second bottle. How much solution should be in the second bottle? Well, let's take a look.
10. Put the bottles right next to each other so that you can compare the levels of the liquids in each. What do you see? Does this surprise you?
11. Clean up your mess and put everything away.

If you could magnify it enough, you would see that water is not a smooth liquid. Instead, it is made up of a bunch of individual water molecules with spaces in between them.

What did you see in the experiment? If everything went well, you should have seen that the bottle that held the solution of water and alcohol did not have as much liquid in it as the bottle that contained water! In other words, ½ cup of water plus ½ cup of water makes a full cup of water. However, ½ cup of water plus ½ cup of alcohol does not make a full cup of an alcohol/water solution.

How can that be? Well, you probably already know that water is made up of tiny things called molecules. If you could magnify water enough, you wouldn't see a smooth liquid. Instead, you would see individual little water molecules, and there is *space in between those molecules*. So even though a glass of water looks like there is only water in it, there is actually water and empty space in between the individual water molecules!

What does this have to do with the experiment? Well, the space in between water molecules is determined by the properties of the water molecules. When you added more water to the ½ cup of water that was already in the first bottle, nothing could change the distance between the water molecules, so adding another ½ cup just turned it into a full cup of water.

However, when you added ½ cup of alcohol to the ½ cup of water that was in the second bottle, you mixed two liquids that were made of two different kinds of molecules. When the molecules started mixing, the alcohol molecules could squeeze in between the water molecules, because they were attracted to the water molecules. This allowed the solution to take up a little less space. As a result, the solution of ½ cup water and ½ cup alcohol took up less than one cup of volume.

If you are having trouble understanding this, think about adding ½ cup of ice cubes to ½ cup of tea. Would the result be one full cup of iced tea? Of course not! There are lots of spaces in between the ice cubes, so when you add tea to the glass, the tea will slip in between the ice cubes, and the total volume will be much less than one full cup. Now, of course, in your experiment, the alcohol molecules couldn't fit nearly as well in the spaces between the water molecules as tea can fit in between ice cubes, but the ideas are the same.

Just as tea can slip into the spaces between ice cubes, the alcohol molecules in your experiment slipped into the spaces between water molecules.

What does all this have to do with Giles of Rome? He believed that all substances (like water and alcohol) contain empty space. He said that if you could magnify something like water enough, you would see that there were small regions where nothing existed and larger regions where water existed. In other words, while water appears smooth, when it is magnified enough, you will eventually see patches of water and patches of nothing in between the patches of water.

Even though Giles didn't use terms like "atoms" or "molecules," this thinking added to the ideas of Democritus (whom you learned about in Lesson 6) to help produce our modern idea of atoms and molecules.

LESSON REVIEW

Youngest students: Answer these questions:

1. When you add 1 cup of a liquid to 1 cup of another kind of liquid, will the volume always be 2 cups?

2. Fill in the blank: In between the molecules of a substance, you find _________ __________.

Older students: In your notebook, explain why ½ cup of water plus ½ cup of alcohol did not result in 1 full cup of alcohol/water solution.

Oldest students: Do what the older students are doing. In addition, write down this question and answer it in your notebook: "If you add 1 cup of one liquid to 1 cup of another liquid, can you get a solution that has a volume of *more than* 2 cups? Assume the molecules of the liquids don't change when the two liquids are mixed." Check your answer and correct it if you are wrong.

Lesson 56: Dietrich von Freiberg (c. 1250 – c. 1310) and the Rainbow

Note to the parent/teacher: This lesson requires a sunny day. If it is not sunny, skip this lesson and come back to it when it is sunny.

Throughout most of recorded history, rainbows have always fascinated people. In Genesis 9:12-17, the Bible tells us that God made the rainbow as a sign that the world will never again be destroyed by a flood. However, it doesn't tell us how a rainbow actually works. As a result, many scientists tried to puzzle out the details of why rainbows sometimes form after it rains and sometimes don't. Robert Grosseteste spent a lot of time trying to figure this out, specifically because he thought it was an important thing for Christians to understand. Roger Bacon tried as well. He could make rainbows using tiny balls of glass, but he could not explain how they formed.

A rainbow represents God's promise that He won't destroy the world by flood again.

In the early 1300s, a German monk named **Dietrich** (dee' trick) **von Freiberg** (fry burg') finally figured it all out. Building on the work of both Grosseteste and Bacon, he came up with the correct explanation. Perform the following experiment to get an idea of how a rainbow forms.

Making Your Own Rainbow

What you will need:
- A sunny day
- A spray bottle that can produce a fine mist
- A shady area outside

What you should do:
1. Fill the spray bottle.
2. Find a shady area, but stand outside of it. You should be in the sun, but you should be able to see the shady area. Stand with your back to the sun.
3. Spray water in front of you at eye level. The spray also needs to be in the sun, and you should be looking through the spray into the shady area. What do you see? If you don't see anything interesting, change the direction in which you are spraying so that you end up spraying all around the area in front of you. Eventually, you should see something interesting.
4. Continue spraying water in front of you so that something interesting is happening, but tilt your head to one side and then the other as you are spraying. What happens?
5. Go back inside and put everything away.

What did you see in the experiment? If all went well, you should have seen a rainbow in the mist that you sprayed. The rainbow should have appeared in different places in the mist, depending on how you tilted your head. Believe it or not, the same process that formed the rainbow in your experiment also forms the rainbows you see after it rains.

One of the keys to forming a rainbow is that there must be water drops in the air. You used the spray bottle to make sure that happened. Also, I told you to stand where you needed to stand in order to see the rainbow: with your back to the sun. That's one of the keys to seeing a rainbow: you need to stand with your back to the sun. Why? Well, remember that light can do two things when it hits a transparent substance: it can reflect or it can refract. In order to see a rainbow, *both* have to happen.

When light first hits a drop of water, some of it reflects off the surface of the water droplet, and some of it refracts into the water drop. The light that refracts into the water drop continues to travel inside the water until it reaches the other side. You should remember that white light contains all the colors of the rainbow. Well, each one of those colors refracts in the water a bit differently. As a result, while the light is traveling through the water, it starts to separate into its many colors.

When the light that is already separating into its colors reaches the other side of the water drop, it once again can either reflect or refract. The light that refracts back into the air leaves the water droplet, and it has been partially separated into its colors. If that were all, you wouldn't be able to see them, because they haven't been separated very well yet.

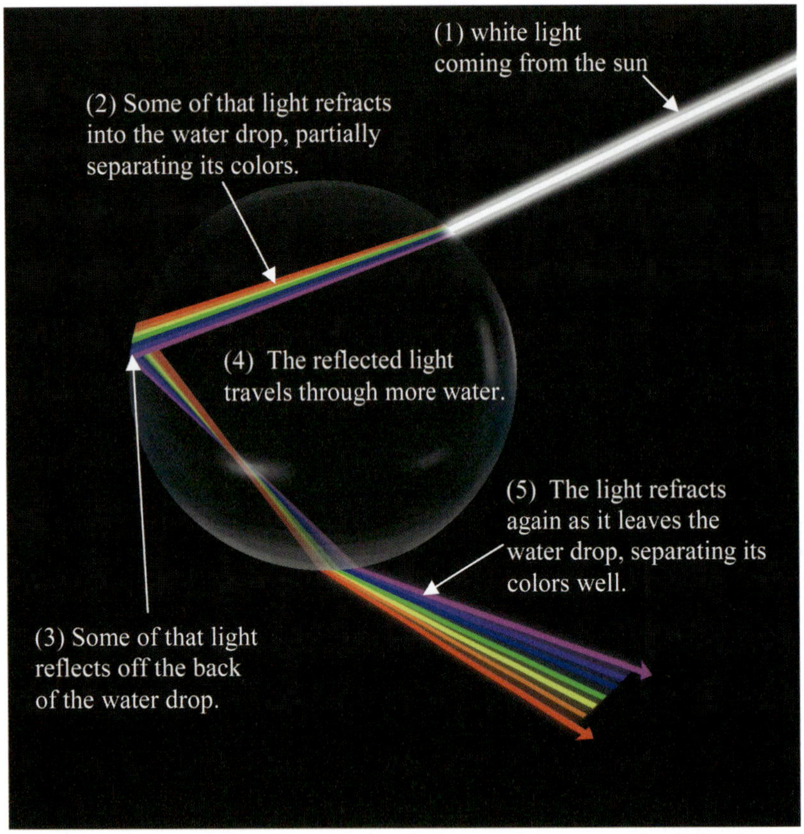

In order to see a rainbow, you need to be in a position to see the light that enters the water drop, reflects off the other side of the water drop, and then leaves the water drop.

However, the part that reflects back into the water drop continues to have its colors separated. As it travels back to the other side and refracts out of the drop into the air, the separation becomes pretty good. At that point, then, the light has its colors well separated, and you can see them clearly. Since there are lots of water drops in the air after it rains, you see the separate colors of light as a rainbow. To see that rainbow, however, the sun has to be at your back so that its light can refract into the water drop, reflect off the other side, and then refract into the air and hit your eyes.

Now please don't get confused here. You don't see the rainbow coming from every drop of water! Each color exits each drop at a specific angle, and you won't see them all. In fact, you will see only one color coming from each drop. For example, when you look at a rainbow in the sky, you see red on the top and violet (or maybe just blue) on the bottom. Why? Look at the illustration on above. When the light leaves the water drop, red is on the bottom, and violet is on the top. In other words, the red light is aimed low, while the violet light is aimed high. If the water drop isn't high enough, the red will be aimed below your eyes, and you won't see it. As a result, you see only red coming from the drops that are high in the sky, and you see only violet coming from the drops that are lower.

Each drop, then, sends a specific color to your eyes, and the color you see depends on the position of the drops. If the drops are high, you will see red coming from them. If they are low, you will see blue, indigo, or violet coming from them. Because of this, the colors of a rainbow are always ordered the same: red is on top, orange is next, yellow is next, green is next, blue is next, indigo is next, and violet is on the bottom. You might not see all those colors, depending on how good the rainbow is, but the order is always the same.

If you have ever been really fortunate, you might have seen a **double rainbow**, like the one pictured below. How does a double rainbow form? Well, in order to see a double rainbow, you have to be standing so that you see a normal rainbow (called the **primary rainbow**), but then you also have to be standing so that you see another set of light coming from the water drops. This light has to reflect *twice* inside the water drops before it leaves. Because the light reflects a second time to make this second rainbow (called the **secondary rainbow**), the order of colors is reversed. As a result, violet (or maybe just blue) is on top, and red is on the bottom.

The secondary rainbow in a double rainbow happens when light can reflect inside a water drop twice and still hit your eyes.

Isn't it wonderful that God gives us rainbows to tell us that He will never again destroy the world with a flood, and isn't it wonderful that men of God like Dietrich von Freiberg figured out how He does it?

LESSON REVIEW

Youngest students: Answer these questions:

1. Fill in the blank: To see a rainbow, the sun must be _________ you.

2. Why do rainbows usually form after it rains?

Older students: In your notebook, make a drawing like the one on the previous page, and use it to explain how a rainbow forms and why the sun must be behind you in order for you to see it.

Oldest students: Do what the older students are doing. In addition, explain why the colors in a primary rainbow are always the same. (Red is on the top, while violet is on the bottom.).

Lesson 57: Thomas Bradwardine (c. 1290 – 1349) and Motion

About 25 years after Dietrich von Freiberg figured out how rainbows form, **Thomas Bradwardine** (brad' war deen) started making important advances in the study of how things move. Like most natural philosophers of this time period, Bradwardine had been trained as a priest and served in various positions in churches around England. At one time, he was even chaplain to King Edward III and the Archbishop of Canterbury. However, he spent much of his free time studying natural philosophy.

As a part of his education, he learned Aristotle's ideas about how things moved, but he found errors in them. As a result, he tried to improve on our understanding of motion by correcting some of Aristotle's mistakes. As he studied the process of how things move, he made an important distinction that is still used today in modern physics. Perform the following experiment to understand what it is.

Separating Movement from the Mover

What you will need:
- A small ball, like a baseball, tennis ball, or golf ball
- Masking tape (or another type of tape that is easy to see)
- A board that is at least 60 centimeters (2 feet) in length
- Some books
- A hallway or a room that has a long, open space on the floor

What you should do:
1. Pile up the books at one end of the hallway until they are about 15 centimeters (6 inches) high.
2. Put one end of the board on top of the book pile so that the board forms a ramp sloping down.
3. Put a small piece of masking tape on the floor to mark the end of the ramp (in other words, where the board meets the floor).
4. Hold the ball at the top of the ramp, and then release it so that it rolls down the ramp and down the hallway.
5. Use a piece of masking tape to mark where the ball stops.
6. Remove the books and ramp so that there is now open space behind the piece of tape that marked the end of the ramp.
7. Use your hand to roll the ball down the hallway. Make sure the ball leaves your hand before it reaches the piece of tape that marks where the end of the ramp was. Your goal is to roll it just right so that it stops at (or at least very near) the piece of tape that marked where the ball stopped in step 5. Keep trying until you get the ball to stop close to the piece of tape. How hard was that?
8. Put everything away.

How hard was it for you to roll the ball so that it ended up stopping at the same place where the ball that rolled down the ramp stopped? Some people find it fairly easy, while others find it really frustrating. Whether or not you liked the experiment, it demonstrates a very important point: A scientist should be able to separate the *cause* of motion from the motion itself.

Think about it. In your experiment, the first cause of motion was gravity, which worked as the ball rolled down the ramp. The ball was motionless when you held it at the top of the ramp, but as soon as you let go, it started rolling down the ramp. If the board you used hadn't been propped up by the books, the ball wouldn't have rolled at all. However, the fact that the board was propped up on one side turned it into a ramp, which allowed gravity to cause the ball to start rolling. After it left the ramp, it continued to roll until it finally stopped somewhere down the hallway.

In the second part of your experiment, the ramp was gone, so it had no effect on the ball. Instead, the ball rolled down the hallway because you pushed it. Nevertheless, after a while, you were able to figure out how to roll it so that it stopped at the same place as when it rolled down the ramp. In other words, even though the *cause* of the motion in the second part of your experiment was completely different, the *result* ended up being the same.

Now imagine that a person was watching your experiment, but he only saw what happened between the two pieces of tape. He couldn't see what caused the motion (the ramp or your hand). All he could see was the ball rolling from one piece of tape to the other. What would he see? In part 1 of the experiment, he would see the ball rolling down the hallway. In part 2 of the experiment, he would see the ball rolling down the hallway *just as it did before*. If he wasn't careful, he might conclude that in both cases, the cause of the motion was the same.

Bradwardine made it clear that you cannot make such conclusions. In the end, different *causes* of motion can lead to exactly the same *results*. Because of this, he said that we should be able to separate the cause of motion from the motion itself. There are lots of ways to get a ball rolling, for example, but once it is rolling on its own, its behavior should not depend on what got it rolling in the first place.

In modern science, we say that this is the distinction between **kinematics** (kih' nuh ma' tiks) and **dynamics** (dye na' miks). When we study kinematics, we are studying *only* the motion. We don't care how it got going – we just want to know what will happen as the motion continues. In dynamics, we study what caused the motion to begin with. One of the most important things that Bradwardine taught was that the motion itself (kinematics) should be understandable without knowing what actually caused the motion (dynamics).

In bocce ball, the players try to roll their balls (the red, green, blue, and yellow ones) so they stop as close as possible to the jack (the white ball).

In your experiment, then, someone who just observed what happened between the two pieces of tape would be studying the kinematics of the experiment – how the ball rolls down the hallway. If the person instead studied what happened behind the tape that marked the end of the board, he would be studying the dynamics of the ball's motion.

By the way, if you enjoyed the challenge of trying to roll a ball so that it ended up stopping at a particular place, you might enjoy playing a game called **bocce** (bah' chee) **ball**. In this game, a player throws a small ball (usually called the "jack") on a long, flat surface. Once it stops, everyone tries to throw some larger balls so that they come as close as possible to the jack without touching it.

The Oxford Calculators were able to advance our understanding of motion by applying mathematics to it.

Probably the greatest thing that Bradwardine did was to emphasize mathematics in the study of motion. He thought that natural philosophers should be able to use mathematics to completely describe how things move. In fact, he was part of a group of philosophers that were called the **Oxford Calculators**. They were called that because most of them belonged to Merton College (a part of the University of Oxford), and they thought that calculations (math) were very important in the study of nature. It turns out, of course, that they were right. Today, mathematical formulas are at the core of how we understand motion.

As Bradwardine tried to apply mathematics to his study of motion, he saw that much of what Aristotle taught was simply wrong. He tried to develop his own mathematical description of motion, but he wasn't very successful. Most of his mathematical formulas didn't work very well when compared to experiments, but they worked a little better than Aristotle's ideas. As a result, many of those who study the history of science credit Bradwardine's work as another important step in the long process by which natural philosophers started to question Aristotle's teachings.

Questioning Aristotle was a very important step along the road to understanding motion and many other things about nature. A great many natural philosophers regarded Aristotle so highly that they simply refused to question what he taught. This was a problem, because a lot of what he taught was wrong. With the help of John Philoponus, Thomas Bradwardine, and others, however, natural philosophers slowly began to admit that there were many things wrong with Aristotle's teachings. As a result, our understanding of motion and other aspects of the natural world began to improve remarkably.

LESSON REVIEW

Youngest students: Answer these questions:

1. Fill in the blank: Bradwardine taught that different causes of motion can lead to the same ________.

2. Fill in the blank: Bradwardine and the other Oxford Calculators thought that ________ was very important in the study of science.

Older students: In your notebook, explain the difference between kinematics and dynamics. Also, explain what Bradwardine and the Oxford Calculators thought about science and mathematics.

Oldest students: Do what the older students are doing. In addition, do a bit of research to find the name of another Oxford Calculator. Note: They are called the Merton Calculators by some.

Lesson 58: Jean Buridan (c. 1300 – c. 1360) Builds on the Work of John Philoponus

We don't know exactly when he was born or when he died, but **Jean** (zhahn) **Buridan** (ber' uh duhn) was writing about philosophy just a few years after Bradwardine. The best guess is that Buridan lived from about 1300 until about 1360. He was born in France, studied and then taught at the University of Paris, and like most of the great scientists of this time, he was a priest. He believed that because God created the universe, it must act in a logical way, and he wanted to learn more about God by studying His creation.

Buridan had read a lot of books in the course of his education and his teaching career, and he was strongly influenced by John Philoponus. In Lesson 47, you learned that Aristotle thought projectiles (objects like arrows or thrown rocks flying through the air) must have something pushing on them in order to keep them moving. As a result, he thought that the air itself was somehow pushing the projectiles along. Philoponus disagreed; he actually showed that the air slows projectiles down.

Buridan built on that fact and decided that how well a projectile travels through the air depends on certain properties that the projectile itself has. To see what I mean, perform the following experiment.

Projectile Range

What you will need:
- A normal sheet of paper (writing paper, printer paper, etc.)
- Some toilet paper
- A small rock (It needs to be about the same size as a wadded-up sheet of paper.)

What you should do:
1. Lay the sheet of paper down on a table or desk.
2. Cover the sheet of paper with squares of toilet paper. In the end, you want to have essentially the same amount of toilet paper as the paper that is in the sheet. So, if your sheet of paper is 8.5" x 11", you want 8.5" x 11" worth of toilet paper as well.
3. Wad up all those squares of toilet paper into a tight ball.
4. Wad up the sheet of paper into a tight ball. The two balls should be about the same size.
5. Check your rock. It should be about the same size as each of the balls of paper. If not, find another rock that is the right size.
6. Go outside and find a nice, large area that is far away from anything breakable.
7. Throw the rock as far as you can. Make sure you don't throw the rock at someone or towards something that could be damaged by it.
8. Note where the rock landed.
9. Throw the wadded sheet of paper as far as you can.
10. Note where it landed.
11. Throw the wadded up toilet paper as far as you can.
12. Note where it landed.
13. Repeat steps 7-12 a couple more times. Do you notice a pattern in how far each projectile went?
14. Put the rock back where you found it, go back inside, and recycle or dispose of the paper.

What did you see in the experiment? You should have seen that overall, you could throw the rock farther than you could throw the wadded sheet of paper, but you could throw the wadded sheet of paper farther than you could throw the wadded-up toilet paper. Before I tell you what this means, let me define an important word for you. The distance that a projectile travels from where it is thrown to where it lands is called its **range**. Imagine throwing a snowball:

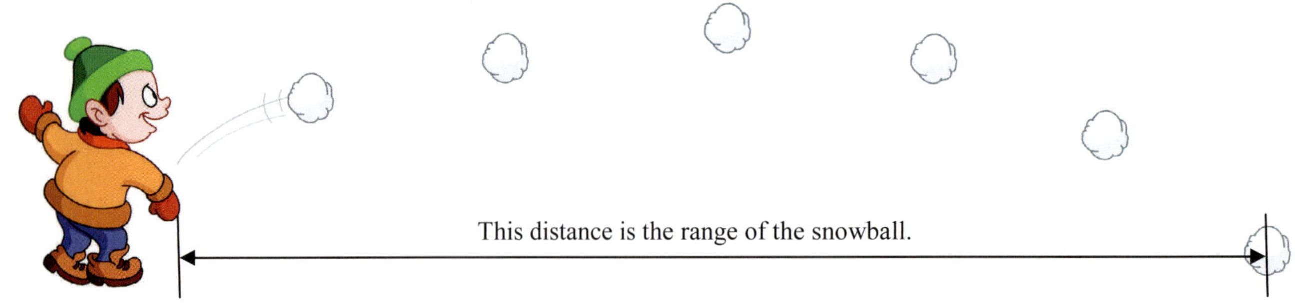

This child is throwing a snowball as far as he can. The different snowballs in the picture show where the snowball travels as it flies through the air. When it lands on the ground (the snowball farthest to the right), the distance from where the child threw it to where it lands is called the range of the snowball.

In your experiment, then, you looked at the range of three projectiles: a rock, a wadded-up sheet of paper, and some wadded-up toilet paper.

Now in the experiment, you threw each projectile as hard as you could, trying to make it go as far as possible. So you used pretty much the same force when you threw each projectile. The projectiles were all pretty much the same size as well. However, they had completely different ranges, didn't they? Why is that?

Jean Buridan said that Philoponus was right – the air actually slows down a projectile. It doesn't push it along. The reason a projectile moves through the air, then, is because the thrower gives it something that helps it to fight against the air so that it can continue to move. He called what the thrower gives the projectile **impetus** (im' puh tus). The longer the projectile is in the air, the more impetus the air takes from the projectile so that, eventually, the projectile cannot travel any farther. At that point, it stops.

According to Buridan, then, in your experiment, you gave each projectile an impetus by throwing it. The range of each projectile depended on how much impetus it had once it left your hand. The more impetus you gave it, the longer its range. This tells us that the rock had the most impetus, the wadded-up sheet of paper had less impetus, and the wadded-up toilet paper had the least impetus.

So what determines a projectile's impetus? Well, in your experiment, each projectile had roughly the same size and the same shape and was thrown with the same effort. However, there was one important thing that was different. Can you think of what that was? The weight. The rock was very heavy, the wadded-up paper was lighter, and the wadded-up toilet paper was the lightest of all. Notice that the range of the projectiles increased with increasing weight. The heaviest projectile had the longest range, while the lightest projectile had the shortest range. The range of the medium-weight projectile was in between.

Buridan said that the weight of an object is one thing that determines its impetus. Heavy projectiles will tend to travel farther than light projectiles, because heavy projectiles have more impetus. However, that's not quite enough, is it? After all, if you try to throw a *very* heavy rock, you might not be strong enough to throw it even as far as you threw the wadded-up toilet paper in your experiment. So while weight is one thing that determines the impetus of an object, there must be something else. What is that something else?

Think about it. You didn't do this in the experiment, but imagine what would happen if you threw only the rock, but you threw it with different strengths. Suppose at first you hardly put any strength into throwing the rock at all. How far would it go? It wouldn't go far, would it? Then suppose you threw it hard, but not as hard as you could. It would go farther, wouldn't it? Finally, if you threw with all of your might, it would go even farther, wouldn't it?

The range of the arrow that this boy is shooting will depend on how he aims the arrow, as well as the speed at which it leaves the bow and the weight of the arrow. The last two things determine its impetus, which will determine how well the arrow can fight the air that is trying to slow it down.

So what's the other factor that determines the impetus of a projectile? You might be tempted to say "the strength by which it is thrown," but that's not right. Remember, impetus is a property of the *projectile*, not the thrower. What property of the projectile is affected by the strength at which you throw it? The *speed* of the projectile. The harder you throw something, the faster it leaves your hand.

So according to Jean Buridan, a projectile stays in motion as it flies through the air because of its weight and the speed that it is given by whatever launched it. The projectile's weight and speed gives it impetus, and it uses the impetus to fight the air, which is trying to slow it down. The more impetus it has, the farther it can travel. It turns out that this is pretty much correct. In about 350 years, another scientist would come along and state this just a bit more clearly, but I am getting ahead of myself again!

LESSON REVIEW

Youngest students: Answer these questions:

1. Fill in the blank: The range of a projectile depends on the ________ that the thrower gives it.

2. Fill in the blanks: Impetus is determined by a projectile's ________ and _________.

Older students: In your notebook, define the range of a projectile in your own words. Then use your own words to explain what impetus is, how it is determined, and how it affects a projectile's range.

Oldest students: Do what the older students are doing. In addition, think about a bow. If you shoot an arrow by pulling back on the bow with just a little strength, the arrow will not go as far as when you pull back on the bow with all your strength. In your notebook, explain why.

Lesson 59: Albert of Saxony (c. 1316 – 1390)

Albert of Saxony was born in Germany about 1316 and died in 1390. Notice that he was born 16 years after we think Jean Buridan was born, so it is not surprising that he was Buridan's pupil. Like Buridan, he was a priest, and he was very interested in learning more about the creation that God made for us. He carried on Buridan's studies about projectiles, and he worked on other issues in natural philosophy as well. For example, he took a concept that was discussed by Buridan, the **center of gravity**, and refined it significantly. To understand the concept, perform the following experiment.

Tilting the Can

What you will need:
- An empty 12-ounce can like the ones containing soda
- A funnel
- A measuring cup that can measure ⅓ of a cup
- Water
- A flat surface that can stand to get a bit wet
- Freezer (You need this only if one or more of the students doing this experiment uses the "oldest student" review exercises.)

What you should do:
1. Look at the bottom of the empty can. You should see that it has a bottom rim, but then because of the way the bottom is shaped, there is a second rim a bit above it. Put the can on the flat surface and tilt it so that it rests on both the bottom rim and the second rim that is above it, as shown in the picture on the left. Can you tilt it so that it stays tilted once you let go of it? You shouldn't be able to.

2. Use the funnel and the measuring cup to add ⅓ of a cup of water to the can.
3. Now repeat step 1. Can you get the can to stay tilted once you let go of it? If you are careful enough, you should be able to.
4. Once the can is tilted and stable, gently push it on one side. You should see that as long as you don't push it too hard, the can will move, but it will continue to stay tilted. In the end, it will only fall if it is pushed hard.
5. Clean up your mess. If one of the students doing this experiment does the "oldest student" review exercise, put the can in the freezer with the water still inside it. Make sure the can is standing upright, not tilted. If not, dump the water out and throw the can away, or better yet, recycle it.

When the can was empty, it wasn't able to remain tilted on its own. Instead, it fell over. However, when you put the water in it and tilted it carefully enough, it remained tilted on its own. What explains this? It's Albert of Saxony's concept of a center of gravity. We know that when dropped, most objects fall to the ground. That's because they are attracted to the earth by gravity. According to Albert of Saxony, however, they aren't just attracted to the ground. They are actually attracted to a point near the center of the earth called the earth's center of gravity. In addition, the object itself has a center of gravity, which is attracted to the earth's center of gravity. In the end, any object acts like all of its weight is concentrated at its center of gravity.

So the center of gravity is a representation of how the weight in an object is distributed. For example, think about the soda can you used in your experiment. It has a center of gravity, and it behaves as if all of its weight is concentrated there. In addition, its center of gravity wants to fall to the earth's center of gravity. Well, when you tilted it, the can's center of gravity was off to the side of where the can was resting on its two rims. As a result, when you released the can, it fell, because its center of gravity wasn't supported by anything. The drawing on the right illustrates this. The red dot is the can's center of gravity. Notice that it is off to one side of where the can is resting on its two rims. Since there is nothing supporting the center of gravity, it falls. As a result, an empty can cannot remain tilted on its own. Its center of gravity just doesn't allow for that.

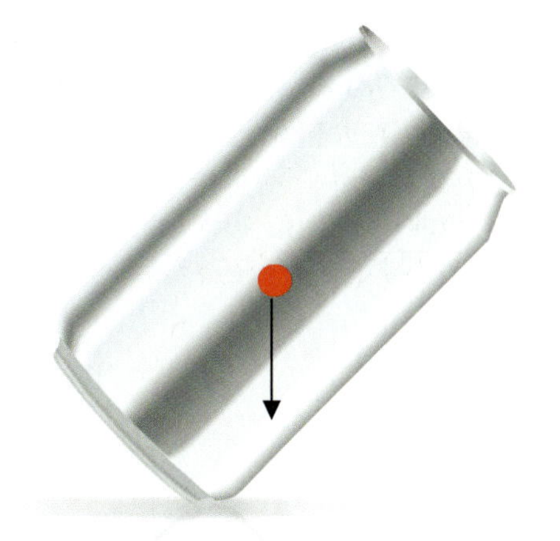

In your experiment, the empty can's center of gravity (the red circle) fell straight down, which means the can couldn't stay tilted.

Now in the second part of your experiment, you added some water to the can. You didn't fill the can up with water. Instead, you added just a little water. This meant you had a can that was only partially filled with water. When you tilted that can, it could remain tilted, because the water changed the can's center of gravity.

Before you added the water, the empty can was pretty much uniform. It had air filling its inside, and the aluminum on the outside was evenly distributed around the can. As a result, its weight was evenly spread throughout the can, and its center of gravity was the center of the can. When you added water, however, that changed. Once the water was added, there was a lot more weight near the bottom of the can. That shifted the can's center of gravity.

Remember, the center of gravity is a reflection of how the weight in the object is distributed. When the weight is distributed evenly, the center of gravity is at the center of the object. However, when the weight is not distributed evenly, the center of gravity will be closer to the highest concentration of the weight. Since there was a lot of weight in the water, and since the water was near the bottom of the can, the water caused the can's center of gravity to move down and over a bit.

In your experiment, the water shifted the center of gravity (red circle) so that it was above where the can rested. Since the center of gravity was supported by where the can rested on its rims, it did not fall.

Why does that matter? Well, when you added the right amount of water, it shifted the center of gravity so that it was *right above where the can was tilted.* Because of that, the center of gravity was supported. Since the center of gravity was supported, the can couldn't fall to the ground. As a result, the can remained tilted, even after you let go of it. In the end, then, how an object is affected by gravity is determined by its center of gravity. If the center of gravity is supported, the object will not fall. If it is not supported, it will fall.

Now think about what this tells us: *the center of gravity is not always at the center of an object.* This is what Albert of Saxony discovered. Since the center of gravity is a reflection of how the weight of an object is distributed, the center of gravity will only be at the center of the object if the weight is distributed evenly throughout the object. When that happens, the object behaves as if all its weight is concentrated at its center.

However, if the weight is not distributed evenly over the object, the center of gravity will be closer to where the majority of the weight is. Once again, the object will behave like all its weight is concentrated at that point, but the point will no longer be the very center of the object. It will be nearer to where the majority of the weight is found.

A tightrope walker can use a long pole to continually adjust her center of gravity so that it is always right above the rope. That keeps her from falling.

Have you ever seen a person walk on a tightrope? The person won't fall off the rope as long as his or her center of gravity stays right above the rope. That way, the rope supports the center of gravity, and the person doesn't fall. Some tightrope walkers use a long pole like the one shown in the drawing on the left. As she moves, she adjusts how she holds the pole. Since there is a lot of weight in the pole, small changes in how she holds the pole cause changes in the position of her center of gravity. She continually adjusts how she holds the pole so that her center of gravity stays above the rope. That keeps her from falling off the rope.

LESSON REVIEW

Youngest students: Answer these questions:

1. Fill in the blank: An object behaves like all its weight is concentrated at its ______ __ ________.

2. (Is this statement True or False?) An object's center of gravity is always at the center of the object.

Older students: In your notebook, draw pictures such as the ones on the previous page and use them to explain why the water allowed the can to remain tilted after you let it go.

Oldest students: Do what the older students are doing. In addition, I want you to try tilting the can again tomorrow, once the water inside it has frozen. Right now, predict in your notebook whether or not the can will remain tilted once you let it go. Try it out tomorrow and see if you were right.

NOTE: The next lesson requires something to sit in the freezer overnight. You should perform steps 1 and 2 of the experiment the day before you do the lesson.

Lesson 60: Guy de Chauliac (c. 1300 – 1368)

Do you remember who Galen was? He was a famous physician who died around AD 200. His works were considered the "final word" on medicine, at least until **Guy** (gee) **de Chauliac** (show' lak) came on the scene. He was born in France and studied medicine there as well. He quickly became known as an excellent physician, so it wasn't long before he was offered a position as Pope Clement VI's personal physician. He stayed on as the physician for the next two popes (Innocent VI and Urban V).

This 14th-century drawing is thought to be an actual portrait of Guy de Chauliac. The artist is unknown.

While he was performing his duties as these popes' personal physician, he started compiling all the information he could about anything related to medicine: anatomy, surgery, dentistry, and drugs. He eventually wrote a book that came in seven volumes and started replacing Galen's work in the minds of many physicians. While he borrowed a lot from Galen, he also added a lot of new information. Some of it was from his own work, and some was from the works of others.

While his book still contains many errors and recommends several procedures that would harm a sick person rather than help, it was still a welcome improvement over Galen. For example, Guy's book was much more accurate when it came to anatomy than was Galen's work. Why? Because Guy was allowed to observe the dissection of dead people. Remember, Galen could only dissect animals, because he had to obey the Roman government, and Rome outlawed the dissection of people.

However, Guy obeyed the church's rules, and the church had no rules against human dissection. In fact, about 100 years before Guy was born, Pope Innocent III actually ordered a human dissection to be performed as part of a murder investigation. Because Guy could actually see the insides of people's bodies, he had a much better understanding of human anatomy than did Galen.

His better understanding of human anatomy made him a better doctor, but what made him really special was his attitude. Have you ever heard of the **Black Death**? It was a terrible sickness that swept through Europe in the middle of the 1300s. Some historians estimate that it killed more than *100 million people*! Well, while Guy was Pope Clement VI's personal physician, the Black Death hit the town where he and the pope were living. Most physicians fled the city, but Guy stayed because he wanted to help as many people as possible and learn more about the sickness.

In the course of treating the Black Death, he noticed that it came in two different forms. One form caused large bumps to appear in certain regions of a person's body, followed by a blackening of the person's skin. That's why it was called the "Black Death." Modern doctors call this form of the disease the **Bubonic** (byoo bon' ik) **plague**. The other form of the disease caused problems with breathing, and the infected person would often die from not being able to breathe properly. Modern scientists call this the **Pneumonic** (new mon' ik) **plague**. We now know that both versions of the disease are caused by the same germ. The difference between them is how the person gets exposed to the germ. If the person breathes in the germ, he gets Pneumonic plague. If he is bitten by fleas that are carrying the germ, he usually gets Bubonic plague.

Guy says that he was infected with the Black Death but was able to survive it, and that's good, because his book wasn't finished until 15 years after the Black Death came to the city where he lived. The book not only contained a lot of knowledge about anatomy, surgery, and drugs – it also discussed how to care for a person's teeth. It recommended that people should swirl wine around in their mouth. This helped prevent tooth decay, because as you know from studying Hippocrates (Lesson 11), the alcohol in wine tends to kill germs.

Guy's book also mentions that people should avoid drinking hot liquids followed by cold liquids or cold foods. Do you know why that's good advice? Perform the following experiment to find out.

"Killing" a Life Saver

What you will need:
- Life Savers mints or candies (The mints work best.)
- A freezer
- A small plate
- A pot for boiling water
- Water
- A stove
- Kitchen tongs

What you should do:
1. Put six Life Savers on the small plate.
2. Put the plate (along with the Life Savers) in the freezer and let them sit there overnight.
3. The next day, put some water into the pot and get it boiling. There doesn't need to be a lot of water – just enough to completely submerge a Life Saver.
4. When the water starts boiling, take the pot off the heat and set it on a burner that isn't turned on.
5. Take the plate with Life Savers out of the freezer.
6. Drop one of the Life Savers into the hot water.
7. Listen carefully and watch the Life Saver. What do you hear and see?
8. Repeat steps 6 and 7 until all the Life Savers are in the water.
9. Use the kitchen tongs to pick the Life Savers out of the water and put them back on the plate. If the Life Saver is still whole once it is on the plate, it has "survived" the experiment.
10. Did any of your Life Savers survive the experiment?

When you dropped the freezing-cold Life Savers into the boiling water, what did you hear? You should have heard the sounds of the Life Savers cracking. If you were using candies, you probably saw them form lots of tiny cracks on their surface. When you tried to get the Life Savers out of the water, you probably found that most (if not all of them) broke apart as you grabbed them with the tongs, especially if you were using the mints. Why did this happen?

Well, as you probably already know, things get bigger when they get hot and they get smaller when they get cold. Scientifically, we say that most things expand when they are heated, and they contract when they are cooled. In your experiment, then, the Life Savers contracted when you put them in the freezer. Then, when you dropped them into the hot water, they started expanding. Since the Life Saver went from very cold to very hot in a short amount of time, it had to expand really quickly. That quick expansion actually caused the Life Saver to break.

The reason the Life Savers broke or cracked was because they were hard, and when things are hard, it is difficult for them to expand and contract. If you make them expand or contract too quickly, they end up cracking and breaking. Well, your teeth are also hard. They need to be in order to crush the food that you are eating. In fact, your teeth are covered by the hardest substance in your body. It's called **enamel**. If you drink or eat something hot, the enamel has to expand as it warms. If you drink or eat something cold, the enamel has to contract as it cools. If you change the temperature too quickly, you will make your enamel expand or contract so quickly that it will crack and break. Now it won't break as obviously as the Life Savers did, but it will develop tiny cracks.

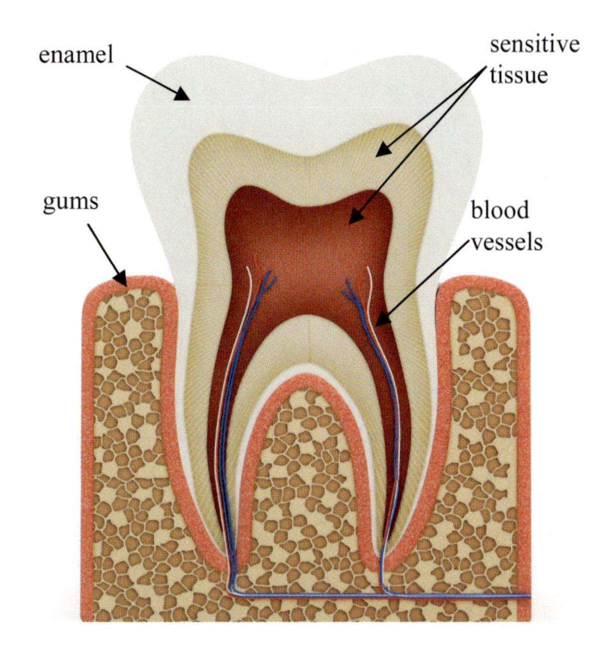

This is a drawing of a tooth. The enamel protects the sensitive tissue beneath it.

So what? Who cares if the enamel in your teeth develops cracks? Well, below the enamel is tissue, and germs would really like to infect the tissue. They can get to the tissue through those cracks. Also, the tissue has nerves in them. Do you know what nerves do? They allow you to sense things. One of the things the nerves in your teeth sense is pain. When you crack your enamel, you expose those nerves to what you are eating, and it can result in a *lot* of pain.

So definitely take the advice of Guy de Chauliac, and avoid changing the temperature of what you eat or drink quickly. If you drink something hot, for example, don't follow it right away with something cold. Let your mouth cool down first.

LESSON REVIEW

Youngest students: Answer these questions:

1. Why did Guy de Chauliac have better anatomy knowledge than Galen?

2. When a hard substance changes temperature quickly, what can happen?

Older students: In your notebook, explain why Guy de Chauliac had better anatomy knowledge than Galen. Also, explain why his advice regarding the temperatures of food and drink is good to follow. Be sure to use the concepts of expansion and contraction in your explanation.

Oldest students: Do what the older students are doing. In addition, can you think of a common household item that might break if the temperature is changed too quickly? Write it down in your notebook and explain how you came to that conclusion.

Science in the Late Middle Ages

This old printing press gives you an idea of what the first moveable type printing press was like.

Lessons 61 - 75: Science in the Late Middle Ages

Lesson 61: Nicole Oresme (c. 1320 – 1382)

Not long after Guy de Chauliac started writing about medicine, **Nicole** (nik' uhl) **Oresme** (or' em) began writing about several subjects related to science. He was particularly fascinated with the sun, moon, planets, and stars, as were many natural philosophers in his day. As you know from Lesson 21, Aristotle believed that the earth was at the center of everything, and the sun, moon, planets, and stars all moved around the earth. While there were a few philosophers who thought otherwise (such as Aristarchus from Lesson 22), most who lived in Oresme's time agreed with Aristotle that the earth was at the center of everything. Nevertheless, there were some who disagreed.

In Aristotle's view, not only was the earth at the center of everything, but it also stood still. It never moved in any way. Those who disagreed with Aristotle thought that the sun was at the center of everything and did not move. Not only did the earth orbit around the sun – it also rotated around, like a spinning top. After all, if the sun didn't move, the only explanation for how the sun travels across the sky every day would be for the earth to be rotating, continually exposing different parts of the earth to the sun.

Oresme actually agreed with Aristotle (who we now know was wrong), but he didn't like many of the arguments that people used to support Aristotle's views. For example, Jean Buridan (Lesson 58) argued that we know the earth doesn't rotate like those who disagree with Aristotle say, because if you shoot an arrow straight up into the sky, it falls to the ground at the same spot where you shot it. If the earth were rotating, the ground would move while the arrow was in the sky, and the arrow would land in a different spot. Oresme showed that this argument is quite wrong. Perform the following experiment to see the reasoning he used.

Those who thought that the sun was at the center of the universe also thought that the earth rotates while it orbits the sun. This rotation turns day into night.

Projectiles and Their Initial Motion

What you will need:
- A hand-sized ball like a baseball or tennis ball
- A wagon or something else that you can ride on as it is pulled by another person
- A short sidewalk or driveway along which the wagon can be pulled for a little while
- Two people to help you

What you should do:
1. Put the wagon at one end of the sidewalk or driveway.
2. Sit in the wagon.
3. Ask one of your helpers to grab the wagon handle so he can pull it when the time comes.
4. Ask your other helper to take the ball and stand off to one side of the sidewalk or driveway so that as the wagon is pulled down the sidewalk, it will pass her.

5. Tell the helper with the ball to throw the ball straight up in the air as soon as you pass her in the wagon. The moment you are right next to your helper is when she should throw the ball. She should then catch the ball when it falls back down.
6. When both helpers know what they are supposed to do and are ready, have the one helper pull you in the wagon so that you pass by the helper with the ball. The wagon shouldn't be moving really quickly or really slowly. A nice, easy speed will do.
7. Watch the ball from the time that the helper throws it until the time that she catches it. When the helper catches the ball, tell your other helper to stop pulling.
8. Note where you are compared to the ball and the helper who threw the ball.
9. Have your helper give you the ball as you ride in the wagon, and reset the wagon to where it was before.
10. Once again, have your helper pull the wagon at that same nice, easy speed.
11. This time, when you are right next to the helper who used to have the ball, throw the ball straight up in the air and catch it when it falls back down. Have that helper observe the ball while it's in the air.
12. When you catch the ball, tell your other helper to stop pulling the wagon.
13. Once again, notice where you are compared to the ball and the helper that you passed.
14. Put everything away.

What did you see in the experiment? In the first part, the ball was behind you by the time you stopped. That's because the ball took time to rise up into the air and then fall back down. During that time, you continued to move along the sidewalk, so you were well past your helper by the time she caught the ball. In the second part of the experiment, however, the ball fell right back into your hands.

What's the difference between the two parts of the experiment? The difference is who threw the ball. When your helper threw the ball, she was standing still. The ball went straight up and down, and it did not follow you. When you threw the ball, however, you were moving. The ball still went up and down, *but it also followed the motion that you had when you threw it*! This tells you something very important:

When a projectile is fired, it moves according to both how it is fired and whatever motion it had at the instant it was fired.

Think about it this way: When your helper threw the ball, it moved straight up and straight down, because your helper wasn't moving. The only motion the ball had, then, was the motion given to it by your helper's throw. When you threw the ball, you were moving. So the ball had two types of motion. First, it had the motion that you were moving with: along the sidewalk. It also had the motion you gave it when you threw it: up and then down. *So the ball moved both up and down and along the sidewalk*! That's why it ended up back in your hands – it moved up and down, but it also followed you, because it had that motion before you threw it.

Do you see what this means to the argument that Jean Buridan made? He said that an arrow shot straight up in the air would not land at the place where it was shot if the earth were rotating. However, as your experiment showed, the arrow would fall back to the same place even if the earth were rotating. After all, if the earth is rotating, everything on the earth is rotating with it. This includes the arrow that is being shot from a bow. So the arrow shot straight up would move up and down because of the bow, but it would also continue to rotate with the earth, because it was rotating with the earth as it was shot from the bow. Just like the ball in the second part of your experiment, then, it would move right along with the ground beneath it. That would make it fall back to the same place from which it was fired.

Now remember, Nicole Oresme did not believe that the earth rotates. Even though we now know that he was wrong, he was able to show how one argument that tried to show the earth wasn't rotating was wrong. This allowed philosophers who came along later to look at the other arguments used to support Aristotle's ideas and see how they were wrong as well. Eventually, this led to natural philosophers understanding that Aristotle was wrong. So even though Oresme agreed with Aristotle, his serious analysis of arguments that supported Aristotle helped to demonstrate, in the long run, that Aristotle was wrong. This is one of the great things about science. Even scientists who are wrong can help us eventually learn what is right!

Oresme argued against other ideas about the sun, moon, planets, and stars that were very popular in his day. For example, do you know what **astrology** (uh strah' luh jee) is? It is the odd belief that the movements of the stars and planets in the sky affect how we live our lives. As a result, people who believe in it think that they can predict a person's future by closely studying the patterns certain stars make. These patterns are called the constellations of the zodiac, and astrologers think that the position of each constellation affects each person in very specific ways. While this sounds like a silly idea, it was very popular in Oresme's time, and he argued forcefully against it.

This famous clock in Venice was built about 100 years after Oresme died. It tracks the constellations of the zodiac, which are very important in astrology.

Now don't get the word astrology mixed up with the word **astronomy** (uh strah' nuh mee). Astrology is a silly idea that we know isn't correct. Astronomy, on the other hand, is the science of studying the objects in the sky and the universe as a whole. Nicole Oresme did a lot of astronomy, but he argued against astrology, because he was a devout Christian, and he thought that astrology went against the teachings of the Bible.

LESSON REVIEW

Youngest students: Answer these questions:

1. Did Nicole Oresme believe that the earth rotates?

2. What is the difference between astronomy and astrology?

Older students: In your notebook, explain why an arrow that is shot straight up in the air falls down to where it was fired even though the ground underneath it is moving with the earth's rotation. Also, explain the difference between astronomy and astrology.

Oldest students: Do what the older students are doing. In addition, see if you can find any verses in the Bible that condemn astrology. Ask your pastor or other Christians if you can't find any.

NOTE: The next lesson requires a grid for graphing as well as either M&M or Skittles candies. If you don't have your own graph paper, you will need to photocopy the grid found on page A3 of your parent/teacher's *Helps and Hints* book. Also, start looking at the moon in the night sky. When the moon is a thin crescent in the night sky, do the experiment in Lesson 68 (page 206).

Lesson 62: Nicole Oresme and Graphing

While Oresme's work on astronomy and his denunciation of astrology were both very valuable to the progress of science, perhaps his most important contribution to natural philosophy was mathematical. He was the first to introduce the concept of illustrating the results of experiments using **graphs**. Do you know what a graph is? It is a way in which you can illustrate numbers using pictures. Perform the following experiment/activity to see what I mean.

Graphing Candies

What you will need:

- A bag of M&M or Skittles candies
- A bowl
- A graphing grid (You can get your own graph paper or use the grid that is on page A3 of your parent/teacher's *Helps and Hints* book.)
- Colored pencils or crayons (Optional – If you use them, you will need pencils of roughly the same colors as the candies you are using.)
- Your notebook (a blank sheet of paper if you are doing the review exercises for the youngest students)

What you should do:

1. Open the bag of candies and spill the contents into the bowl.
2. Put the bowl within arm's reach of where you are, but far enough away so that it is hard to see the candies in the bowl.
3. Reach into the bowl without looking at the candies, and pull out one of them.
4. Put the candy on the table.
5. Repeat step 3 many times, sorting the candies by color as you go. Continue the process until you have 10 of one color or until the bowl is empty.

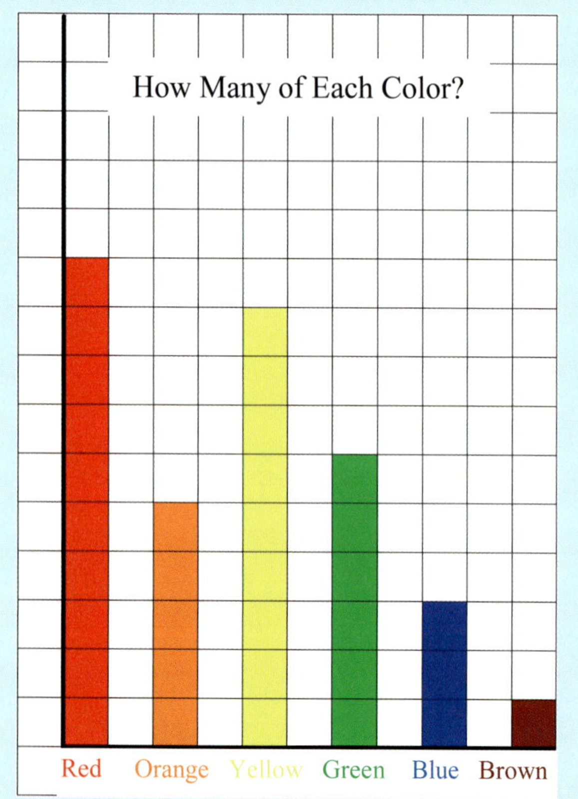

6. In your notebook, start a page that is titled, "Graphing Activity."
7. Count how many candies you have of each color. As you count them, record the results in your notebook. So your notebook should say something like, "Red: 10 candies, Blue: 3 candies, Yellow: 9 candies, etc."
8. Once you have counted the candies, take a piece of paper that has a grid on it, such as the one shown on the left. Write "How Many of Each Color?" across the top.
9. Starting on the far left, draw a thick black line down the page on the right side of the first set of boxes on the paper.
10. At the bottom, draw another thick black line across the page at the top of the bottom set of boxes.
11. Write all the colors you found at the bottom, putting a blank box between each color.
12. For every candy of that color you found, fill in a box above the label. So if you found 10 red candies, fill in ten boxes above the word "Red." When you are done, your grid should look like the drawing on the left.

13. Look at the grid that you have made. What does it tell you? Well, it tells you in a visual way how many of each candy you found when you put your hand in the bowl. For example, look at the grid on the previous page. Which color was the most common? You don't have to look at the numbers you wrote down in your notebook. You just have to look at the height of each bar on the graph. In the graph on the previous page, there were more red candies than any other candies. The next most common color was yellow, and the least common color was brown. Now, of course, you could have determined the same thing from the numbers you wrote in your notebook, but do you see that the drawing makes it easier to compare the colors? That's the beauty of graphing things. It gives you a visual way to compare numbers.
14. Now look at your own grid again. Paste or tape it into your notebook and below it write which color is most common and which color is least common.
15. Put everything away, and ask your parents if and when you can eat the candies!

Nicole Oresme was the first to come up with this idea of how to visually display numbers so that they are easily compared. Nowadays, we call it **graphing**. It turns out that there are a lot of different kinds of graphs, and not surprisingly, the one you just made is a very simple graph called a **bar graph**. In a bar graph, the height of the bar represents each number being displayed. In the bar graph on the previous page, for example, there were more red candies than any other color, so the red bar is the tallest one.

Nicole Oresme's first graph was also a bar graph, but it was more complicated than the one you drew. It is best for you to learn the simplest type of bar graph first, however. Let's make sure you understand everything you can about this kind of bar graph. First, the dark black lines you drew are called the **axes** (ak seez') of the graph, and the singular of the word is **axis**. The dark black line that goes up and down is called the **vertical axis**, while the dark black line that goes left and right is called the **horizontal axis**.

Look at the graph on the right. It represents the results of a poll in which I asked people which food they like the best: breads, fruits, vegetables, or meats. The different kinds of foods are listed on the horizontal axis, and the number of people who chose that kind of food is represented by the height of the bar, which follows the vertical axis. Now, looking at that graph, which type of food was liked by the most people? If you said "meats," you are correct, because the bar above "meats" is the tallest. Which type of food was liked by the fewest number of people? If you said "vegetables," you are correct, because the bar above "vegetables" is the smallest.

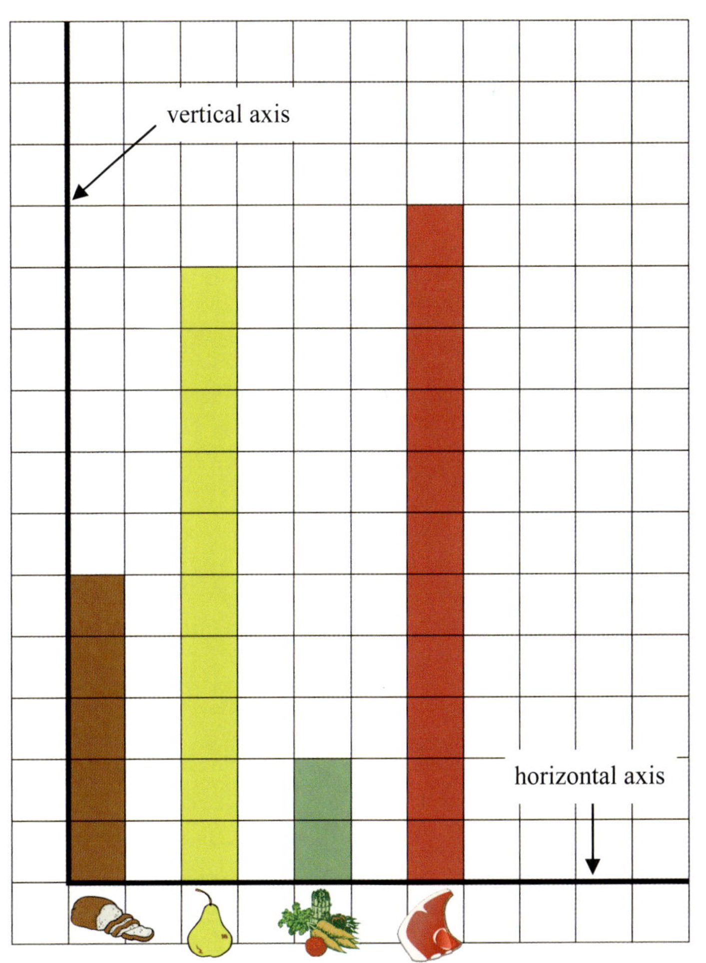

This bar graph represents the results of a poll regarding what kinds of foods people like the most.

It's important to understand that you can also read numbers from a graph, if you want to do that. For example, in the bar graph on the previous page, each square that is colored in represents one person choosing that type of food as his or her favorite. So, go back to the previous page, look at the graph, and tell me how many people chose meats as their favorite type of food. There are 11 squares filled in, so that means 11 people chose meats as their favorite food. Go back to the graph again and tell me how many people chose the least favorite food category. The answer is 2, because the bar above vegetables is 2 boxes high.

I hope you can see that a graph holds a lot of information. It not only tells you the actual numbers that are collected in an experiment, but it also shows them to you in a nice picture that allows you to quickly compare them. Because of this, graphs are used a lot in science to help scientists understand the results of the experiments they do. Nicole Oresme came up with the idea, and it has been used by scientists ever since.

LESSON REVIEW

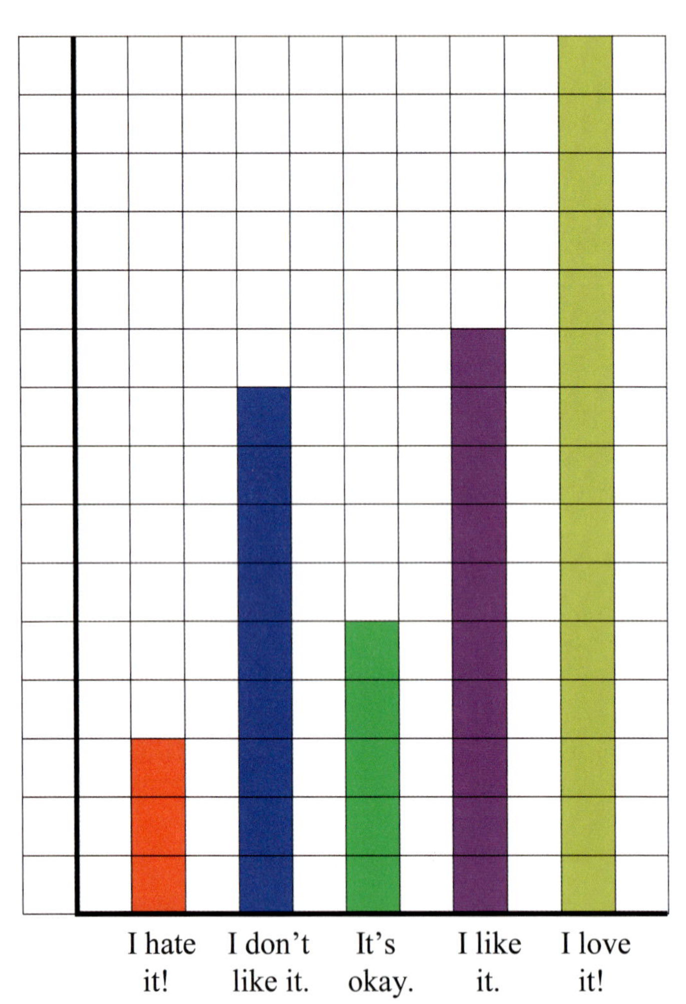

The graph on the left shows the answers I got when I asked several people the following question:

How much do you like science?

a. I hate it!
b. I don't like it.
c. It's okay.
d. I like it.
e. I love it!

Youngest students: Answer these questions:

1. Which answer was given by the most students?

2. Which answer was given by the fewest students?

Older students: Copy this graph onto a grid and paste it in your notebook. Then explain what is on the horizontal axis and what the vertical axis (or the height of the bars) represents. Write down which was the most common answer and how many students gave it as well as the least common answer and how many students gave it.

Oldest students: Do what the older students are doing. In addition, write down how many students answered the question. (**HINT**: Think about what each filled-in box represents.)

NOTE: The next lesson has an experiment that needs to sit for a while. You should start the experiment the day before you do the lesson.

Lesson 63: Nicholas of Cusa (1401 – 1464)

About 20 years after Oresme died, **Nicholas of Cusa** was born in Germany. He studied law at college and then started working for the church leaders, dealing with legal matters and trying to settle various disputes that arose. Eventually, he ended up representing the pope on such matters, and Pope Nicholas V was so impressed with his knowledge of the Bible that he made Nicholas a theologian, which is someone who studies the Bible as his career. Shortly after that, he was made Bishop of Brixen, which means he was in charge of all church matters in a specific region of northern Italy.

This portrait of Nicholas of Cusa was made by a 15th-century artist known as "Master of the Life of the Virgin."

While church matters took up a large amount of his time, he still was able to study the Bible, philosophy (including natural philosophy), and mathematics. This, of course, makes perfect sense, because many Christians find that they learn something about God while they study His creation. Nicholas was no exception. Much of his writings dealt with the nature of God and how it can be seen in both natural philosophy and mathematics.

Because Nicholas was drawn to mathematics, he thought it was very important to measure things if a person wants to learn about them. As a result, many of his scientific accomplishments involved developing a device that would measure something in nature. One of the things he invented was a **bathometer** (buh thom' ih tur). Do you know what that measures? It measures how deep water is.

Now you might think it is really easy to measure how deep water is. After all, you just stick a ruler in the water until one end hits the bottom of the lake, river, or container. Then, you can just read the depth of the water from the ruler. Well, if you are measuring the depth of a shallow lake, river, or container, that might be the best way to do it. What if you want to measure the depth of a lake that is very deep? Do you have to find a really long ruler?

Interestingly enough, you don't. All you have to do is understand the fact that water is heavy. As a result, it pushes down on anything below it, even if it's just more water! This leads to a very interesting effect, which is demonstrated in the following experiment.

The Depth of Water

What you will need:

- A plastic 2-liter bottle, like the kind that soda comes in (A plastic milk carton will also work.)
- A nail (One that is thick enough to make a hole that water can easily flow through.)
- An adult to help you
- A sink that has a water faucet

What you should do:
1. Have the adult help you use the nail to put three holes in the bottle. The first should be about 2 inches from the top. The second should be below it about 1 inch from the bottom. The third should be in between them. That way, you have a bottle with three holes in it: one hole near the top, one hole near the middle, and one hole near the bottom.
2. Have the adult cover the holes with his or her fingers so that water doesn't leak out of them.
3. Use the faucet to fill the bottle. You need to fill it very full, so that the water is well above the top hole.
4. Set the bottle on the edge of the sink so that the holes are pointing towards the sink.
5. Have the adult pull his or her fingers away from the holes so water starts flowing out each hole and into the sink.
6. Notice the difference between how the water flows through each hole.
7. Once the water has drained so that it no longer flows through the top hole, repeat steps 2-5 again.
8. Clean up your mess.

What did you see in the experiment? You should have seen that the water from the bottom hole squirted out so quickly that it traveled a long way from the bottle before it hit the bottom of the sink. The water coming from the hole in the top didn't squirt very far at all, and the water from the middle hole was able to travel farther than the water from the top hole but not as far as the water from the bottom hole.

Why did the water squirt out of the holes differently? Think about the bottom of the bottle for a minute. Any water near the bottom of the bottle had a lot of water *above* it. When the water near the bottom of the bottle was able to escape through the hole, all the water above it pushed it through the hole really hard. As a result, the water coming from the bottom hole was moving very quickly, and that made it squirt out far from the bottle.

The water near the top of the bottle didn't have nearly as much water above it, so there wasn't a lot of weight pushing it out of its hole. Because of that, it didn't squirt very far. The water leaving the middle hole had more water pushing on it than the water leaving the top hole, so it squirted a little farther than the water coming from the top hole. However, it didn't have nearly as much water pushing on it as the water leaving the bottom hole, so it didn't squirt as far as the water coming from the bottom hole.

This fireman's hose squirts a long way because the water is being pushed out with a lot of pressure.

The push that water feels is called pressure, and the more pressure there is behind water, the faster it squirts and the farther it travels. Think about a fireman's hose compared to your garden hose. The fireman's hose shoots water a lot farther than your garden hose, because the water has a lot more pressure behind it. In the end, then, the distance that water squirts depends on the pressure it is experiencing.

So what did your experiment show you? It showed you that the deeper the hole, the more pressure there was behind the water as it squirted out the hole. Now unlike your garden hose or a fireman's hose, there wasn't a machine making the pressure. Instead, there was just a bottle of water. The only thing responsible for the pressure was the depth of the water. The deeper the water, the higher the pressure.

When you are swimming deep in a pool, your ears might feel funny because of the pressure that the water above you is producing.

You might have experienced this when you went swimming. Have your ears ever felt funny when you dove deep into a pool? Your ears are very sensitive to pressure, and they felt funny because all the water above you was pushing on them with a lot of pressure. When you put your head right under water, your ears don't feel funny, because there isn't a lot of water above you, so your ears don't feel much pressure. The deeper you go, however, the more water pressure there is. If you go deep enough, you will eventually experience enough pressure to make your ears feel funny.

To measure the depth of water, then, all you have to do is measure the pressure. The more pressure that exists, the deeper you are. This is how a bathometer measures the depth of water, and we have Nicholas of Cusa to thank for it!

LESSON REVIEW

Youngest students: Answer these questions:

1. What does a bathometer measure?

2. Fill in the blank: The deeper the water, the higher the __________.

Older students: In your notebook, draw a picture of your experiment while water was coming out of all three holes. Explain why the water squirted out differently from each hole, and explain how that shows the way a bathometer measures the depth of water.

Oldest students: Do what the older students are doing. In addition, use what you have learned to explain why it is dangerous for a diver with no protection to go deeper than about 300 meters (1,000 feet) in the ocean, even if he has plenty of air in his scuba tanks.

NOTE: The experiment in Lesson 66 (on page 200) needs to sit for several days. It is best to complete steps 1-10 in that experiment today or tomorrow.

Lesson 64: Nicholas of Cusa and Humidity

The bathometer wasn't the only water-related thing that Nicholas of Cusa invented. He also made the first **hygrometer** (hi grah' muh tur). Do you know what a hygrometer measures? Perform the following experiment to find out.

Water in the Air

What you will need:

- A small container (like a juice glass or a small jar) made of clear glass
- A Ziploc bag large enough to put the glass inside and still zip shut
- A freezer

What you should do:

1. Make sure the glass is completely dry. The Ziploc bag needs to be dry as well.
2. Put the glass in the Ziploc bag and zip it shut.
3. Put the bag and glass in the freezer and wait for at least three hours; overnight is best.
4. After the bag and glass have been in the freezer for at least three hours, take them out.
5. Open the bag and pull the glass out.
6. Watch the glass as you pull it from the bag.
7. Let the glass sit on a counter for a minute and watch. What happens?
8. Now hold the glass up to the light and look at the light through the glass. How much of the light can you see?
9. Put the glass on the counter and throw the Ziploc bag away (or recycle it). Don't put the glass away yet, because you are going to look at it again later.

What did you see in the experiment? If things worked well, you should have noticed that the glass was a bit cloudy when you took it out of the Ziploc bag, but it should have gotten very cloudy very quickly once it was out of the bag. It probably got so cloudy that you could hardly see the light through it. Why did that happen?

If you don't already know, there is water vapor in the air. Remember, all substances have one of three phases that they can be in: solid, liquid, or gas. When you freeze water, it becomes ice, which is the solid phase of water. The water that you drink is water in its liquid phase. When you boil water, it evaporates and becomes a gas. We call that gas "water vapor."

Even if you don't boil any water, there is still water vapor in the air. When you put the glass in the Ziploc bag, there was some air in the bag, so there was some water vapor in the bag as well. When you put the bag and glass into the freezer, the water vapor cooled and eventually became a liquid. It then cooled more and became a solid, forming a thin layer of ice on the glass that is often called **frost**.

However, since there wasn't a lot of air in the Ziploc bag, there also wasn't a lot of water vapor, so not much frost formed. As a result, while the glass was a bit cloudy from the frost, it wasn't too cloudy. When you removed the glass from the Ziploc bag, however, you brought it into contact with a lot more air. Since the glass was still very cold, any water vapor in the air and near the glass turned first into a liquid and then into a solid. As a result, a lot more frost covered the glass, and it got very cloudy, making it hard to see through.

The water drops on this glass come from water vapor in the air condensing on the outside of the cold glass.

So in the end, your cold glass served as a way to see the water vapor that is in the air. You can't see it while it is a gas, but by exposing the gas to a cold glass, you made it turn into a liquid, which then turned into a solid, and that allowed you to see it.

Now you don't have to turn the water vapor in the air into frost to see it. Have you ever left a glass of ice-cold liquid out on a table or counter for a while, especially when the air temperature is warm? What happens? The glass starts to "sweat," doesn't it? Droplets of water form on the *outside* of the glass. What's happening? Is the glass leaking? Of course not!

Because of the liquid inside, the glass is cold enough to make the water vapor in the air become a liquid. You might already know that when a gas becomes a liquid, we say that it has "condensed." So the cold glass causes the water vapor to condense, forming the droplets that you see on the outside. That's another way you can detect water vapor in the air.

Scientists have a measure for the amount of water vapor that is in the air. We call it **humidity**. When there is a lot of water vapor in the air, the humidity is high. When there isn't much water vapor in the air, the humidity is low. Guess what a hygrometer measures? It measures humidity. So Nicholas of Cusa was able to figure out a way to measure how much water vapor is in the air.

How did he do it? He put a ball of wool on a balance that measured the weight of the wool. Wool tends to absorb water, and it will even absorb water that is in its gas phase. So guess what happened to the weight of the ball of wool when there was a lot of water vapor in the air? It got heavier! When there wasn't much water vapor in the air, the ball of wool got lighter. In the end, the weight change in the wool told Nicholas how much water was in the air, and so the weight change in the wool was a way to measure humidity.

Why did Nicholas want to measure humidity? Well, part of the reason was curiosity. Natural philosophers wanted to understand why there is water vapor in the air and how it affects the rest of creation. Nicholas believed strongly that to understand anything, you had to measure it. By figuring out how to measure humidity, he helped us understand it better.

However, there was a very practical reason for measuring humidity as well. In this time period, most things were sold by weight. So if someone wanted to buy wool, he bought it by the pound. Well, since wool tends to absorb water from the air, it was heavier on very humid days. Therefore, if you wanted to get the most for your money, you would need to buy wool on a day when there was very little water vapor in the air. In other words, you would want to buy it on a day when the humidity was low. Nicholas of Cusa knew his invention would help people get their money's worth of anything that tended to absorb water from the air!

Nowadays, we find measuring humidity to be important because it helps us to predict the weather and learn how to prepare for our day. Have you ever heard the phrase, "It's not the heat, it's the humidity"? People in very humid climates say this, because a high humidity makes a hot day seem even hotter. However, we haven't gotten to the point where I can tell you why this is the case, because natural philosophers hadn't figured that out in this time period.

When the humidity is high, a hot day feels even more uncomfortable.

Because a high humidity makes a hot day even more uncomfortable, most places that report on the weather have a measurement called **heat index**. It tells you what the temperature feels like to the body once you take into account both the air temperature and the humidity. The higher the heat index, the hotter the day will feel to you. For example, the actual temperature on a given day might be 30 degrees Celsius (86 degrees Fahrenheit), but if the humidity is very high, it could feel like 42 degrees Celsius (108 degrees Fahrenheit). Without Nicholas of Cusa's first hygrometer, the weatherman might not be able to tell you about the heat index!

Before you do your lesson review, look at the glass again. The frost should have started to go away by now. What's there in its place? You should see little drops of water. How did they get there? Well, as the glass warmed up, the frost melted, and that turned the water from solid (frost) to liquid (little drops). At the same time, while the glass was cold, more water vapor in the air condensed on it, making even more water drops.

LESSON REVIEW

Youngest students: Answer these questions:

1. Where did the frost on the glass in your experiment come from?

2. What is humidity?

Older students: In your notebook, explain why a cold glass forms water drops on the outside. Define humidity, and explain what a hygrometer measures. Write how a high humidity affects you on a hot day.

Oldest students: Do what the older students are doing. In addition, think about a cold day. Do you think the humidity would affect how cold it feels on a cold day? If so, how? Explain your reasoning in your notebook, and correct it if it is wrong once you check the answer.

NOTE: The next lesson has an experiment that requires clear gelatin. You should make it the day before the lesson so that it is ready to go the next day.

Lesson 65: Nicholas of Cusa and Lenses

Do you remember Roger Bacon? In Lesson 52, you learned about his experiments with lenses. He determined that curved glass bends light in an interesting way. A circular or oval piece of glass, for example, tends to focus the light that hits it to a single point, called the focal point. He realized that this is why a round or oval piece of glass magnifies whatever you see when you look through it, and he reasoned that eventually, we would be able to shape glass in a way to make it easier for people with poor eyesight to read. Indeed, the first eyeglasses were made only 18 years after Bacon wrote his book on optics.

A farsighted person sees distant things clearly and nearby things as blurry (top). A nearsighted person sees nearby things clearly and distant things as blurry (bottom).

Interestingly enough, however, the eyeglasses that were made back in Bacon's time helped only *some* of the people who had poor eyesight. They helped people who could see things clearly when they were far away but had trouble seeing things clearly when they were up close. Nowadays, we call such people **farsighted**. Other people have the opposite problem. They can see things clearly when they are up close, but they cannot see things clearly when they are far away. We call these people **nearsighted**. The eyeglasses made based on Bacon's work helped farsighted people, but they did nothing to help the nearsighted.

Now remember, the reason a lens helps people see better is because the curve of the lens bends the light, focusing it to a point. Well, in addition to all the other things that Nicholas of Cusa studied, he also studied optics. He reasoned that if a round or oval piece of glass bent light so as to help farsighted people, there ought to be a way to make a piece of glass that is curved so that it would help nearsighted people. Perform the following experiment to see that he was right!

A Diverging Lens

What you will need:

- Unflavored gelatin (This kind of gelatin is clear when it sets.)
- A round cookie cutter or a small glass
- A butter knife
- A spoon
- A spatula

- A pie pan, preferably with a completely flat bottom
- A white paper plate
- A flashlight
- Black construction paper
- Tape
- Scissors
- A dark room

What you should do:
1. Make the gelatin almost according to the instructions, but use **½ cup less boiling water** than instructed, because you want the gelatin to be very firm.
2. Pour the gelatin into the pie pan and put it in the refrigerator so the gelatin can set.
3. Once the gelatin is set, you are going to make another lens, like you did in Lesson 52. This lens will be very different, however. Use the butter knife to cut a square that is larger than the cookie cutter or the opening of the small glass.
4. Use the spatula to put the square of gelatin in the middle of the paper plate.
5. Use the cookie cutter or small glass to cut a half circle out of one side of the square. To get rid of the gelatin inside the cookie cutter or glass, don't lift it up. Instead, pull it away from the square of gelatin so that the bottom of the cookie cutter or glass scrapes the gelatin away. Once the cookie cutter or small glass is far from the gelatin square, you can lift it up and get rid of the gelatin inside.

6. Repeat step 5 on the other side, so that the gelatin has the shape shown in the picture on the left.
7. Try to clean up the plate so that there is no stray gelatin on it. The only gelatin on the plate should be a part of the odd shape shown in the picture.
8. Use the black construction paper, scissors, and tape to make a covering for the flashlight like you did in Lessons 23 and 52.
9. Take the paper plate with gelatin on it and the flashlight into a dark room.
10. Shine the flashlight so that the beam of light comes out of the flashlight and hits one side of the gelatin shape straight on, as you did in Lesson 52. The beam needs to hit it between the two black lines in the picture.
11. Move the flashlight up and down between the two lines in the picture so that the beam hits the circle at different places. Don't change the angle, however. The beam needs to hit the side of the gelatin straight on, just at different locations on the shape's side. Notice how that changes the way the light comes out the other side of the circle.
12. Clean up your mess.

What did you see in the experiment? If things went well, you should have seen that the light beam coming out the other side of the gelatin was bent away from its original path, but unlike what happened with the circle of gelatin in Lesson 52, the light wasn't focused towards a single point. Instead, it was spread out. That's because the shape of the gelatin in this experiment formed a **diverging lens** – a lens that spreads light out rather than focusing it to a single point.

Remember, it's the curve of the gelatin (or glass) that makes a lens work. A circle is curved outward. In science, we say that it has a **convex** (kahn' vex), or "curving outward," shape. The shape you made in this experiment is called a **concave** (kahn' kayv), or "curving inward" shape. You can remember the difference between convex and concave by thinking about something "caving in." The shape of gelatin you made in this experiment looks like it caved in on both sides.

As shown in the drawing on the right, a concave lens takes light that is hitting it straight and bends it so that it spreads out, away from the center. In other words, the light beams all *diverge* (move away) from each other, which is why it is called a diverging lens. Go back to the drawing in Lesson 52 (page 158) and see the difference between the diverging lens you made in this experiment and the circular lens you made back then. That lens focused the light beams to a single point, so it is often called a **converging lens**.

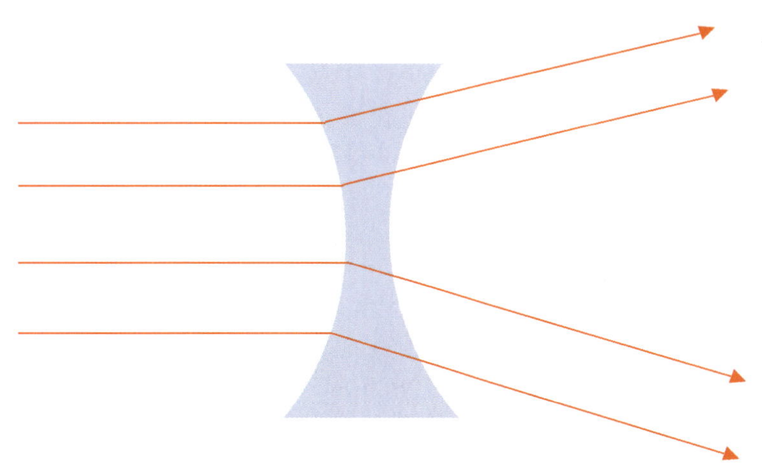

A concave lens works the opposite of a convex lens. It makes light beams diverge from each other, so it is a diverging lens.

Now don't get lost in all the terms. Here's the main point. A convex lens, like the circular one you made in Lesson 52, focuses light to a single point. That's why we call it a converging lens. It behaves the exact opposite of the concave lens you made today. A concave lens spreads light beams out, away from the other light beams, which is why we call it a diverging lens. Now, a convex lens helps farsighted people see better. Nearsightedness is the opposite of farsightedness, right? Well then, what kind of lens should help nearsighted people? A concave lens like you made today. Nicholas of Cusa was the first to point this out. Because of him and Roger Bacon, we can now help anyone with poor eyesight, be they nearsighted or farsighted!

LESSON REVIEW

Youngest students: Answer these questions:

1. Fill in the blanks: Nearsighted people have a hard time seeing things that are ______ _______.

2. Fill in the blanks: The lens you made in the experiment today was a con____ lens.

Older students: In your notebook, draw a convex lens like you made in Lesson 52, and show what it does to light beams that hit it straight on. Label it as a convex lens. Next to that drawing, draw a concave lens like you made today, and show what it does to light beams that hit it straight on. Label it as a concave lens. Explain why convex lenses are also called converging lenses and concave lenses are also called diverging lenses.

Oldest students: Do what the older students are doing. In addition, explain the difference between farsightedness and nearsightedness. Convex lenses help farsighted people. Write that in your notebook, and then explain why, once you know that, you know concave lenses help nearsighted people.

Lesson 66: Nicholas of Cusa and Plants

Nicholas of Cusa studied a great many things, which was common for natural philosophers in this time period. In addition to studying humidity, water depth, and optics, he also studied the nature of plants. Back in his time, most natural philosophers thought that plants absorbed most of what they needed from the soil. After all, they observed that when a tiny seed is planted in the soil, it can grow into a tall plant. There is a lot more "stuff" in the tall plant than there was in the tiny seed. Where did that "stuff" come from?

The oak tree in this picture grows from a tiny acorn, which the inset shows isn't even as large as one of its leaves. Where does the material that makes up the tree come from?

Natural philosophers knew that all plants had roots, and those roots were in the soil. Roots seem to be designed to absorb things from the soil, so most natural philosophers thought that plants must get all the "stuff" that makes them grow from the soil itself. However, Nicholas knew better. He described an experiment that clearly demonstrates this. His experiment would be hard for you to do in a reasonable amount of time, but this experiment demonstrates the same thing, and it takes a lot less time.

Growing Plants without Soil

What you will need:

- A one-liter plastic bottle
- Dried beans from the supermarket
- Scissors
- A paper towel
- Water
- Legos (marbles or beads that are made of glass or plastic will work as well)
- An adult to help you

What you should do:

1. Have the adult help you cut the top off the bottle, as shown in the picture on the right.
2. Roll up the paper towel and stick it through the opening of the bottle's top so that most of it is out of the bottle's top but a small part of it is inside.

3. Turn the top of the bottle upside down and put it on top of the bottom part of the bottle. If the paper towel is too long (it probably will be), cut it down until the upside-down bottle top rests comfortably on the rest of the bottle and the paper towel almost touches the very bottom, as shown in the picture on the top right.
4. Pour some water into the upside-down top of the bottle to get the paper towel wet.
5. Lift up the upside-down top of the bottle and put enough water in the bottom part of the bottle so that it nearly reaches where the opening of the upside-down top of the bottle will rest.
6. Replace the upside-down top of the bottle.
7. Put a few bean seeds in the part of the paper towel that is in the upside-down top of the bottle. Cover the seeds with the paper towel ends so that they are completely surrounded by wet paper towel.
8. Put the Legos (or marbles or beads) on top of the paper towel to hold it down around the seeds. Make sure the seeds are completely surrounded by paper towel, because they need to stay wet all over. Your experiment should now look like the contraption pictured on the bottom right.
9. Put the bottle contraption somewhere that gets sun but doesn't get too hot.
10. Let the contraption sit for several days. Check it once each day. If the water level in the bottom part of the contraption gets low, add some more to the bottom so that the part of the paper towel that sticks down is always submerged in water.
11. After a few days, you should see bean seedlings growing up out of the Legos.
12. The day you are actually working on this lesson, pull the Legos out of the contraption carefully.
13. Lift up the upside-down top of the bottle that is holding the seeds.
14. Look at the seedlings and the paper towel. You can see that the roots of the plants go into the paper towel, but does it look like any of the paper towel is actually gone?
15. If you are doing the challenge lessons, replace the Legos and put the plant by a window. Ideally, it should be by a window that gets a lot of sun, and it should be set 30 centimeters (a foot) from the window. Check the plants from time to time to make sure there is plenty of water in the bottom bottle. If you aren't doing the challenge lessons, you can dispose of the contraption and the plants.

In your experiment, the plants grew to be much larger than the seeds. However, the Legos weren't used up, and neither was the paper towel. The roots went into the paper towel to be sure, but there weren't big holes where the roots had absorbed the paper towel, were there? In the end, the plants got a lot bigger, but they didn't use up the paper towel or Legos in the process. What did they use up? The obvious answer to this question is the water. The water level got lower, partly because the water evaporated, but also because the plant used it.

Nicholas of Cusa was convinced that unlike many natural philosophers of his day thought, a plant doesn't use the soil in which it grows. He said that anyone who doubts this should weigh a plant and the soil in which he puts the plant. After a long while, the plant will grow, and the person should weigh the plant again. He should then weigh the soil after it is allowed to dry, and he will find that the soil's weight hasn't changed much at all. In the end, this demonstrates that the plant doesn't use the soil to grow.

Now it turns out that Nicholas of Cusa wasn't completely correct. First, a plant does use *some* things from the soil. There are certain chemicals in the soil that act a bit like vitamins for the plant. They make the plant healthier, but the plant doesn't need large amounts of them. Thus, a plant doesn't need to take *much* from the soil (except the water), but the better the soil, the healthier the plant will be. Second, we know that plants need something besides just water to grow. They need light. Nicholas of Cusa thought the sun's light played a role in plant growth, but he wasn't quite sure what it did. If you continue to study science, you will learn how plants use light to live and grow.

These plants are growing well despite the fact that they are not in soil. This is an example of hydroponics.

The fact that plants don't need soil to grow is used in a process called **hydroponics** (hi druh pahn' iks). In this process, plants are grown without soil at all. Instead, their roots are simply kept wet with a solution of water and the chemicals that a plant would normally get in small amounts from the soil. In the end, this allows the plant to get everything it needs to grow and stay healthy.

LESSON REVIEW

Youngest students: Answer these questions:

1. (Is this statement True or False?) Plants need soil in order to grow.

2. What do a plant's roots absorb from the soil?

Older students: In your notebook, explain why we know that plants must absorb something as they grow. Describe your experiment and how it demonstrates that plants don't actually absorb the soil in which they grow. Instead, indicate what they do absorb.

Oldest students: Do what the older students are doing. In addition, explain what hydroponics is. Note that in the picture of hydroponics above, the roots are not submerged in water. Can you explain why? If so, write your explanation down in your notebook and check to see if it is correct. Change it if it isn't correct.

NOTE: The next lesson requires a photocopy of the letters found with the answers to the review exercises for Lesson 67 in the *Helps and Hints* book.

Lesson 67: Johannes Gutenberg (c. 1395 – 1468)

This is a portrait of Johannes Gutenberg.

The next person I want to discuss was not a natural philosopher. In some ways, however, he contributed more to the progress of science than most natural philosophers. **Johannes** (yo hahn' ess) **Gutenberg** (goot' n burg) was a blacksmith. He also worked with metals, specializing in gold. However, in about 1439 or so, he ended up designing and building what has been called the most important invention in the second millennium (which means the entire period from AD 1001 to AD 2000). What was this amazing invention? A new scientific instrument? A new method of producing food to feed the hungry? A new means of transportation? No. It was the **movable type printing press**.

Prior to Gutenberg's invention, producing a book was an incredibly expensive, time-consuming process. A person who knew how to read and write had to copy (by hand) the book that was being produced. He had to copy slowly so that his copy would be the same as the original and so that his writing would be neat enough to be read easily. Often, another person was employed to read over his shoulder, checking the accuracy of the copy as it was being produced. Can you imagine the time it would take to do this? Perhaps after completing the following activity, you will have some idea.

Copying vs. Movable Type Printing

What you will need:

- A photocopy of the "Block Letters" that are found in the *Helps and Hints* book for this lesson.
- Your notebook (a blank sheet of paper if you are doing the review exercises for the youngest students)
- Tape
- Scissors
- A pencil
- A stopwatch or a watch with a seconds hand
- Someone to help you

What you should do:

1. Cut out each letter on the photocopied page so you have a bunch of squares of paper, each with a single letter on in.
2. Put the letters next to your notebook.
3. Start a new page in your notebook. Title it, "Copying Versus Movable Type Printing."
4. Have your helper get ready to time how long it takes you to complete step 5.
5. Copy the following phrase into your notebook:

   The quick red fox jumps over the lazy brown dog

   Be sure to copy the phrase neatly so that anyone can read it.

6. Once you are done, write down the time it took you to complete step 5 (in seconds) right under where you wrote the phrase.
7. Have your helper get ready to time how long it takes you to complete step 8.
8. Search through the letters you cut out. Use them to produce the same phrase. Each time you find the correct letter, tape it in your notebook. For example, search through the letters to find a "T," and then tape it in your notebook. Then find an "H" and tape it next to the "T." Finally, find an "E" and tape it next to the "H." You have now spelled "THE" with the letters. Continue the process until the phrase is complete.
9. Now you have two versions of the same phrase. You copied the first one, and you used the letters you cut out to build the second one.
10. Write down how long it took you to complete step 8 (in seconds).
11. Put away the tape.

Which step took less time? Obviously, step 5 took a lot less time, because you didn't have to sort through the letters and tape them into place. All you had to do was copy the phrase. However, even though step 5 was faster when it came to copying the phrase *once*, it would take the same amount of time if you wanted to copy the phrase again.

Metal letters and numbers like these were covered in ink and pressed onto paper in Gutenberg's movable type printing press. The lettering left behind on the page would be the reverse of the metal letters, making them oriented correctly for reading.

Now suppose the letters you cut out for your activity were made out of metal, and suppose you could cover them with ink. What would happen when you pressed those ink-covered metal letters onto a blank sheet of paper? The phrase would be printed on the paper, wouldn't it? That's almost how Gutenberg's movable type printing press worked. In actuality, the printing press had to use reversed letters, because what would come out when they were covered with ink and pressed on the page would be the reverse of the letters themselves. Look at the metal letters in the photograph. Imagine covering them in ink and pressing them on paper. In the end, whatever the letters spelled would be transferred to the paper, wouldn't it?

Now imagine that you want to copy the phrase "The quick red fox jumps over the lazy brown dogs" a few hundred times rather than just once. Even though finding individual metal letters and arranging them to make the phrase would take time, once you did that, all you would have to do is cover the letters in ink and then press them on paper. That would be very fast, wouldn't it? Let's do some math to see exactly how much time you would save using the movable type rather than copying by hand.

If you copied the phrase by hand 1,000 times, it would take 1,000 times as long as step 5 did in your experiment. Multiply the number of seconds it took for you complete step 5 by 1,000. Divide that number by 60, and that will be the number of minutes it would take you to copy the phrase 1,000 times by hand. Divide that number by 60, and you have the number of hours it would take.

Now let's say the letters you used in step 8 were really metal letters. How long would it take to cover them in ink? Probably about 3 seconds. How long would it take to press those ink-covered letters onto a piece of paper? Maybe 3 more seconds. That means it would take a total of 6 seconds to use the letters to copy the phrase on a piece of paper. Once you got the letters together, then, it would only take you 6,000 seconds to make 1,000 copies (6 seconds x 1,000 copies). So take the number of seconds it took you to complete step 8 and add 6,000 to it. That's the number of seconds it would take you to set up the letters and then use them to make 1,000 copies. Once again, divide by 60 to get the number of minutes, and divide again by 60 to get the number of hours.

Now compare the two times. How long would it have taken you to make 1,000 copies by hand? It took me 25 seconds to copy the phrase once. That means it would take 25,000 seconds (*almost 7 hours*) to copy the phrase 1,000 times. It took me 400 seconds to put the letters in order and tape them in my notebook. Adding 6,000 to that, it would take only 6,400 seconds (*less than 2 hours*) to use a movable type printing press to make 1,000 copies of the phrase.

Do you see the difference? A movable type printing press takes a long time to set up, but once it is set up, it can make lots of copies very quickly. In the end, this made it much easier to produce a large number of books. It has been estimated that there were about 3 million books produced in Europe from AD 1300 to AD 1400. From AD 1500 (after Gutenberg invented his printing press) to AD 1600, more than *200 million* books were produced! When you make a large number of the same thing, we say that what you have made has been **mass-produced**. When something is mass-produced, it is usually a *lot* more affordable. Gutenberg's printing press allowed for the mass production of books, which made them more affordable.

This changed the world in an incredible way. For example, prior to the printing press, books were so expensive that almost no one could own a copy of the Bible. Even after the invention of the printing press, only universities, monasteries, and very wealthy people could afford a Bible. Eventually, however, the cost of books came down as a result of mass production, and personal Bibles became affordable. Imagine not being able to read God's Word for yourself! Prior to the printing press, that was true for almost everyone. Even after the printing press was invented, Bibles were mostly printed in Latin, which few people could read. So in addition to the printing press, translations of the Bible had to be made so that people could finally read God's Word on their own.

The printing press also changed science forever. After all, when natural philosophers made discoveries, the way they told the world was to write about the discoveries. When books were expensive and hard to make, that limited how well their discoveries could be communicated. The printing press made it much easier for natural philosophers to communicate their discoveries to one another. No wonder many historians consider Gutenberg's invention to be the most important thing that happened in the past 1,000 years!

LESSON REVIEW

Youngest students: Answer these questions:

1. What does it mean when someone says that a product has been mass produced?

2. How did Gutenberg's printing press change the world?

Older and Oldest students: In your notebook, explain how a moveable-type printing press works. Explain how even though it takes a long time to set up, in the end, it makes the mass production of books take a lot less time. Explain how this changed the world.

Lesson 68: Leonardo da Vinci (1452 – 1519)

This is a portrait Leonardo da Vinci drew of himself using red chalk.

Have you heard of **Leonardo da Vinci** (dah vin' chee)? If so, you might be surprised to see him included in a book about science. After all, most people think of him as an artist. He is, after all, the man who painted *The Last Supper*, the best-known painting that illustrates the meal Jesus and His disciples ate before Christ was betrayed by Judas (Matthew 26:20-30, Luke 22:7-22, Mark 14:17-26). He had many other famous paintings that illustrated things that happened in the Bible. In addition, he painted the famous *Mona Lisa*, which has been called "the best known, the most visited, the most written about, the most sung about, the most parodied work of art in the world." [John Lichfield, "The Moving of the Mona Lisa," *The Independent*, 4/2/2005]

Why are you learning about an artist in a science book? Because while da Vinci was a very accomplished artist, he was also very interested in studying the world around him. In fact, it can be argued that his study of the natural world helped improve his art. For example, most artists agree that his portraits seem significantly more realistic than the portraits made by other artists in this part of human history. Studies of his paintings indicate that he was able to make them so realistic because he understood how light would interact with them.

That's where I want to start with da Vinci's natural philosophy. He was fascinated by light and spent a lot of time studying how it allows us to see things. Initially, he believed what most natural philosophers had taught through the centuries – that we see things because the eye transmits light, and that light interacts with the light in the world, producing images. However, the more he studied light, the more he realized that couldn't be true. Eventually, he came to the correct conclusion that we see things because light reflects off them and enters our eyes.

As a part of his studies about light, he was able to explain something that had puzzled natural philosophers for centuries before him. Perform the following experiment to see what I mean.

Earthshine

What you will need:
- A night in which the moon is out and is not full (Ideally, it should be a thin crescent.)
- A sheet of dark (preferably black) construction paper
- Scissors
- A pen or colored pencil whose markings you can read on the dark construction paper
- A nickel

What you should do:
1. Put the nickel in the middle of the construction paper but on one side so that half of the nickel is on the paper and the other half is off the paper.
2. Use the pencil or pen to draw the outline of the part of the nickel that is on the paper.

3. Use the scissors to cut the half circle you drew out of the paper. You should now have a sheet of dark construction paper that has a half circle cut out of one side.
4. Go outside and look at the moon.
5. Look through the half circle and move it around until the paper just barely covers up the part of the moon that is lit up. Your goal is to look just at the part of the moon that is not lit up. Can you see that even though the rest of the moon looks "dark," it is still lighter than the night sky behind it?
6. Now that you have the idea that the dark part of the moon isn't completely dark, look at the moon without using the paper. Can you now see that even though only part of the moon is really bright, the rest of the moon is still barely visible?
7. Go back inside.

You probably already know that the moon doesn't make its own light. Instead, it glows in the night sky because it reflects light that comes from the sun. The sun's light reflects off the side of the moon that faces the earth, and that makes it look like the moon is making its own light. You probably also already know that the reason the light coming from the moon makes different shapes, called the phases of the moon, depends on where the moon and earth are compared to the sun.

When the earth is between the sun and moon, for example, the sun shines full on the side of the moon facing earth. This means that whole side of the moon reflects light from the sun, and the moon appears full to us. When the moon is between the sun and the earth, however, none of the sun's light hits the side of the moon that faces the earth. As a result, none of the sun's light gets reflected to the earth by the moon, so the moon looks dark. In between, different amounts of the moon's side facing the earth gets hit by the sun's light, so we see only a part of the moon lit up.

However, as your experiment demonstrated, even when a portion of the moon is "dark," there is still a little bit of light coming from the dark part. After all, even though the moon was in a phase other than full, you could still see the rest of the moon, even though the rest of the moon wasn't reflecting the sun's light. Well, if the rest of the moon wasn't reflecting the sun's light, what light was it reflecting?

In this photograph, you can see the entire moon, even though only a small sliver is lit by the sun. The rest of the moon is lit by light reflecting off the earth. The small light above the moon is the planet Venus.

This was a mystery until Leonardo da Vinci figured it out. Because he was fascinated by light, he studied how it reflected off different things, and he realized that it could reflect off one object, then hit a second object, and then reflect off that second object and come back to the first object. True, the light would be a lot dimmer when it reflected the second time, but it would still be visible.

Da Vinci realized that this is why the rest of the moon is often visible, even when the moon is not full. Even though the sun's light isn't hitting the darker parts of the moon, the sun's light is hitting the earth, reflecting off the earth, and hitting the darker parts of the moon. Some of that light reflects off the moon, coming back to earth. As a result, we see the rest of the moon. It is a lot dimmer than the part of the moon that is reflecting the sun's light, but it is still bright enough to be visible against the night sky.

This picture was taken by astronaut Bill Anders in 1968. It shows the earth shining in the moon's sky. Note that there is only a "half earth," because the sun is on the left. The photo looks like it's sideways because Anders was not on the surface of the moon when he took the picture. He was orbiting the moon in the Apollo 8 spaceship.

When you are sitting outside on a bright, sunny day, you might say that you are enjoying the bright sunshine. Guess what scientists call the light that reflects off the earth and hits the moon? They call it **earthshine**. In other words, if you were on the moon and looked up in the sky at the right time, you would see the earth shining in the sky. It would not be as bright as the sun, by any means, but because the earth is much bigger than the moon, the earth shines in the moon's sky more brightly than the moon shines in the earth's night sky.

Now even though Leonardo da Vinci correctly explained why we can see the darker parts of the moon when it is not full, he did get some things wrong. For example, he thought that the sun's light reflected off the waters of the earth (like the oceans), and that's where earthshine came from. However, we now know that most of the earthshine that hits the moon comes from sunlight reflecting off the clouds in the sky. Some of it does come from light reflecting off the water, and some of it also comes from light reflecting off the land. However, most of it comes from light reflecting off the clouds.

LESSON REVIEW

Youngest students: Answer these questions:

1. How did Leonardo da Vinci's scientific studies help him with his painting?

2. What is earthshine?

Older students: In your notebook, explain what earthshine is and how it allows us to see the rest of the moon dimly, even when only a part of it is lit up by the sun.

Oldest students: Do what the older students are doing. In addition, imagine that you looked at a crescent moon for several days. The crescent would get thinner or thicker, depending on when you observed it. What about the rest of the moon? Would its brightness vary as well? If so, why? Answer these questions in your notebook and then check to see if you were right. If not, correct your answers.

Lesson 69: How Leonardo da Vinci Wrote Things

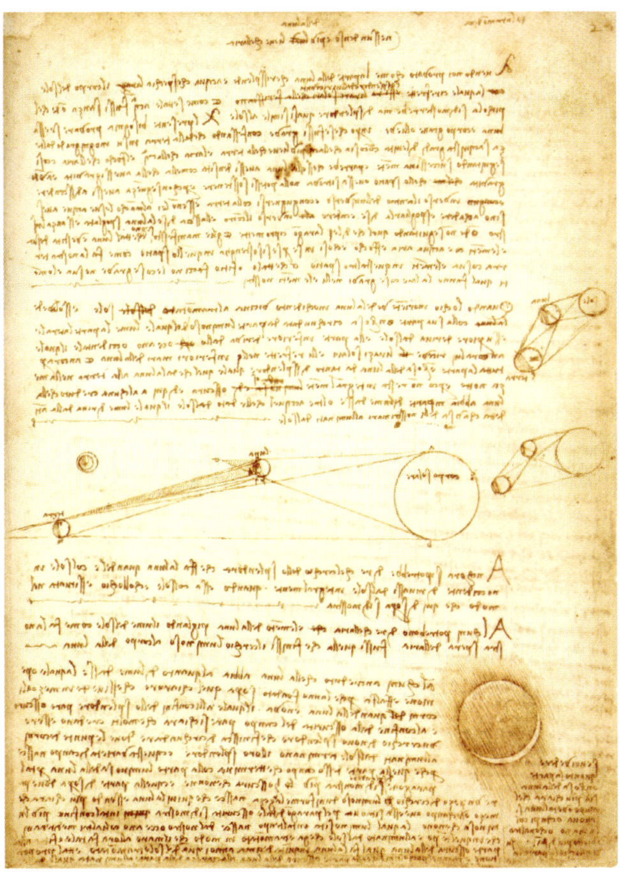
This is the page in Leonardo da Vinci's notebook that explains earthshine.

Before we discuss anything else about Leonardo da Vinci's contributions to science, I want to spend some time discussing something about his work that is rather unusual. In the previous lesson, you learned that da Vinci was the first to explain earthshine. On the left, you will see one of the pages from his notebook where he has written his explanation. The drawing in the middle of the notebook page, for example, shows the sun on the right, the moon in the middle, and the earth on the left. It then traces the rays of the sun, showing how they not only form the crescent moon that is seen from earth, but also reflect off the earth to dimly light up the rest of the moon. On the lower right, you can see a picture of the crescent moon with the rest of the moon dimly visible.

Now look at the words on the page. Do you notice anything strange about them? They are Italian words, but there are two things that make them different from most Italian words. First, they are written in a shorthand that da Vinci made up himself. Do you know what shorthand is? It's a way of abbreviating words so that you don't have to write every letter in the word. Some people do this today when they text each other on their cell phones. They write things like "Its gd 2 hr frm u" instead of typing out the entire phrase "It's good to hear from you." That way, they spend less time typing. That's shorthand. Leonardo used it so that he could write a lot of information in a short amount of time.

The other thing you might notice about the words is that they are written backwards! Even though you can't read Italian, can you see that the right sides of the sentences are all lined up with one another, but the left sides are not? Now look at the page you are reading. On this page, the left sides of the sentences line up, but the right sides do not. English and Italian are both written from left to right, and generally we line up our words where we start writing, but we don't worry about lining them up at the end of the line. Even though you probably don't read Italian, you should now be able to see that da Vinci's writing is from right to left. It's backwards!

All of da Vinci's personal notebooks are written this way. It's called **mirror writing**. Can you think of why it is called that? Try the following experiment to see.

Mirror Writing

What you will need:
- Two white sheets of paper
- A pencil
- A mirror

What you should do:

1. Look at yourself in the mirror. You look normal, don't you?
2. Use the pencil to write "Hello There" on the piece of paper. Write the letters large.
3. Go to the mirror and hold up the paper so that the words reflect in the mirror. Can you read them? You probably can, but you should see that they are backwards. That's one thing a mirror does. It reverses an image. Whatever is on the left side in the real world is on the right side of the mirror, and whatever is on the right side in the real world is on the left side of the mirror. As a result, even though you wrote the letters from left to right, in the mirror, they go from right to left.
4. Use the pencil to write the same phrase backwards. In other words, you should write something that looks like this:

   Hello there

   It might take some practice, but you should eventually be able to do it.
5. Go back to the mirror and hold this paper up to the mirror so that the backwards letters face the mirror. Notice now that they appear correct in the mirror.
6. Hold both pieces of paper up at the same time so that the letters face the mirror. Compare how the two phrases appear in the mirror.

Because a mirror reverses right and left, when someone writes backwards, it is called mirror writing. Essentially, if you do mirror writing, you are writing so that the words will look correct in a mirror. Do you remember the letters used in Gutenberg's movable type press? They were reversed, just like mirror writing. When Leonardo da Vinci was writing things he wanted other people to read, he did not use mirror writing. However, when he wrote in his personal notebooks, he always used it.

Now, of course, the obvious question is, "Why?" You probably found it very difficult to mirror write in the experiment, didn't you? Even if you found it relatively easy, it still took more time than writing the other way, didn't it? If da Vinci developed his own shorthand so that he could write things quickly, why would he slow himself down by mirror writing? The short answer is that we don't know. However, three ideas have been put forth over the years.

The first idea is that da Vinci didn't want people reading his personal notebooks. They represented his own thoughts, and they were private. If he wanted to communicate anything, he would write normally, so the fact that he used mirror writing in his notebooks suggests that he didn't want to communicate what was in those notebooks. While that makes some sense, it's not hard to read mirror writing. As you saw in your experiment, all you have to do is hold it up to a mirror! So if anyone really wanted to read da Vinci's personal notebooks, all he would have to do is hold them up to a mirror. At the same time, good mirrors were not all that common in da Vinci's day, so perhaps this was a good way for him to keep his notes secret.

Of course, the mirror writing isn't the only thing that would make his personal notebooks hard to read. Remember, he had his own shorthand as well. It would be hard for someone to read da Vinci's shorthand if he had not been told what all the abbreviations meant. Perhaps da Vinci thought that the combination of having to hold the pages up in a mirror and figure out his shorthand was enough to keep most people from reading.

Another suggestion is that da Vinci really wanted to think about the words he was writing. As you will see in the next few lessons, da Vinci was a very careful observer, and those careful observations led to some brand new thoughts about the natural world. It is possible that while he wanted to write quickly, he also wanted to force himself to think about each word so that he would be writing as carefully as he made his observations.

The last suggestion is practical. Leonardo da Vinci was left-handed. In those days, writing from left to right with your left hand was a messy business, because back in those days they didn't have nice, clean ballpoint pens. Instead, they dipped sharpened feathers in ink in order to write. The ink was wet, and it would take time to dry. If you were writing with your right hand, this wouldn't be a problem, because your right hand would never pass over the ink. If you were writing with your left hand, however, your hand would pass over the ink you just put on the paper. This would smudge the ink, making it hard to read what you just wrote. It would also make your left hand black with ink. If a left-handed person used mirror writing, however, his hand would never pass over wet ink. So it's possible that da Vinci used mirror writing so that he could write without worries of smudges or messes.

This person is holding the quill pen in his left hand. Once he starts writing, his hand will pass over the ink that he has just put on the page. This could lead to a very messy situation unless he waits for the ink to dry!

Whatever the reason, we know that all of da Vinci's personal notes were in mirror writing, and they contain a lot of incredible science, as you will learn in the upcoming lessons!

LESSON REVIEW

Youngest students: Answer these questions:

1. What is shorthand?

2. What is mirror writing?

Older students: In your notebook, write the phrase "Hello There" normally and then in mirror writing. Explain why the term "mirror writing" is used. In addition, explain what shorthand is. Then make a note that Leonardo da Vinci used both shorthand and mirror writing in his personal notebooks.

Oldest students: Do what the older students are doing. In addition, give the three reasons people have suggested for why da Vinci used mirror writing. Explain which you think is most likely and why.

NOTE: The experiment in the next lesson involves collecting some leaves. You will need to reserve a nice day for it. In addition, you might want to look at the activities in Lessons 71 and 72 as well. They involve observing tree branches, leaves, and a stump. You might want to do all those activities in one trip.

Lesson 70: Leonardo da Vinci and Observing Plants

Leonardo da Vinci not only wrote in an unusual way – he also observed nature in a way that was rather unique for his time. Once again, his work as an artist probably influenced the way he observed nature. Remember, as an artist, da Vinci strove to be realistic. He wanted to accurately depict what he was drawing, painting, or sculpting. Well, in order to accurately draw, paint, or sculpt something, you must first observe it very carefully. You must become very familiar with its details. The more familiar you become with something, the more accurately you can draw it.

Do you ever wonder why I ask you to draw things in your notebook? One reason I ask you to draw things (even if you aren't very good at drawing) is that some things are best explained using drawings. There is another reason, however. When you draw things, you look at them in a more detailed way. As a result, you observe them more closely and are more likely to notice important details that you wouldn't notice if you were just looking at them. Da Vinci understood this. He drew things specifically so he would observe them carefully. As a result, he learned things about nature that no one else knew, because no one else had taken the time to observe parts of nature as carefully as he had.

This is just one example of a plant drawing from Leonardo da Vinci's notebooks. The detail and realistic nature of the drawing show how committed he was to drawing an accurate representation of nature.

Look at the drawing on the left. It comes from one of da Vinci's personal notebooks. Notice how much detail is in the drawing. You can see the shape of the leaves and how they curl downward near their tips. You can see that when the leaves branch away from the main stem, they branch off in pairs – one leaf on one side of the stem, and the other leaf on the other side of the stem. You can see how the fruit of the plant forms in bunches. Da Vinci couldn't have drawn this plant in so much detail had he not observed it very carefully.

Now even though da Vinci was a great artist and observed things very carefully, he wanted to do more. He wanted to add to the accuracy of his drawings. One way he could do that, of course, would be to attach real examples of what he was studying to his notebooks. However, that would make his notebooks hard to handle, and more importantly, things like plants would just decay away rather quickly. He wanted to come up with a way to preserve something directly into his notebooks, and he ended up doing something that no one had ever done before. To see what he did, do the following activity.

Leaf Prints

What you will need:

- Several leaves from different plants (When you pull them off the plant, make sure the stalk that connects them to the branch comes off with each leaf.)
- Fingerpaint (Regular paint and brushes will work as well.)
- White paper with no lines on it
- Newspapers
- Some heavy books
- Someone to help you

What you should do:

1. Look at one of the leaves you collected. It should have one side that is a darker shade of green than the other side.
2. Put some newspaper down and place your leaf on the newspaper with the darker side facing up.
3. If you are using fingerpaint, dip a finger into the paint and use your finger to cover the darker side of the leaf with a thin layer of paint. Use your other hand to hold the stem while you are doing that. If you aren't using fingerpaint, run your finger across the darker side of the leaf to understand how it feels, then use the brushes to cover that side of leaf with a thin layer of paint. Once again, you can use your other hand to hold the stem while you are doing that.
4. Once the darker side of the leaf is covered in a thin layer of paint, move it to a part of the newspaper that has no paint or ink on it, keeping the painted side facing up.
5. Lay a white sheet of paper on top of the leaf so it is pressing against the leaf.
6. Put another sheet of newspaper on top of the white paper so that it covers both the leaf and the white paper.
7. Wipe your hands clean.
8. Put a couple of heavy books on top of the newspaper so that they are pushing down on the white paper and leaf that are underneath the newspaper.
9. Slowly count to 20.
10. Remove the books, and then remove the newspaper that is covering the white paper and the leaf.
11. Pull up on a corner of the white sheet of paper. Most likely, the leaf will start to come up with the paper. If it does not, you should see that the paper now has an image of the leaf on it.
12. If the leaf does come up with the paper, grab the leaf by the stalk and gently peel it from the paper. Now you should see an image of the leaf on the paper. That's what I wanted you to make.
13. Use this method to make an image of each leaf that you collected. It might take some practice to get it exactly right, but with time, you should be able to make a nice image of the leaf, such as the one you see on the right. One of the keys to making a good image is to make the layer of paint on the leaf very thin. You want enough paint to leave an image, but not so much paint that it smudges and clumps.
14. Once you have the practice down, make a couple of images that you will put in your notebook, if you are keeping one.
15. Feel free to be creative. If you use different colors of paint on different parts of the leaf, for example, you will have a multicolored image when you are done.
16. Clean up your mess.

This leaf image is from one of Leonardo da Vinci's notebooks. He made it using a process similar to the one you used in this experiment.

In your experiment, you made a "print" of an actual leaf. Look at one of the prints you made for your notebook. Do you think you could draw a picture of a leaf with so much detail? Probably not. Since Leonardo da Vinci strove to make realistic images of nature, he realized that drawing alone wasn't enough. As I mentioned before you did your experiment, he also knew that he couldn't keep a leaf in his notebook. As a result, he decided to use a leaf to make its own image in his notebook.

The glass on this lamp should be clear, but it is black with soot – the carbon from the oil that was not burned in the flame.

Now unlike you, da Vinci didn't use fingerpaint to make his leaf print! In fact, he actually used **soot** mixed with oil. Do you know what soot is? Have you ever seen a fireplace that has been used over and over again? You might see some black "stains" near the top of the fireplace. Those "stains" come from a black powder that is generally called soot. It's a chemical called **carbon**. Carbon is in every plant, so whenever you burn wood, you burn carbon. However, when wood burns quickly, not all of the carbon has a chance to burn. As a result, some of it is left behind as a black powder.

Back in da Vinci's time, if you wanted light when it was dark, you typically burned wood, candles, or oil. Most people who could afford to do so would use oil, because it burned evenly, and a little bit of oil would last a long time. The oil was put in a lamp, and there was usually a wick in the lamp that would draw up the oil, allowing it to burn. However, oil lamps produced a lot of soot. A cool object could be held over the lamp, and the soot would collect on that object. It could then be collected and mixed with oil (or egg white and honey) to make ink, which could be used in a quill pen or on a leaf to make a print.

LESSON REVIEW

Youngest students: Answer these questions:

1. Why did da Vinci make a print of a leaf in his notebook?

2. What is soot?

Older students: Paste or tape the prints that you want to keep in your notebook. Explain in your notebook how you made them and why someone studying nature would want to make such prints.

Oldest students: Do what the older students are doing. In addition, explain how da Vinci made the ink that he used to make his leaf print. Make sure you define the term "soot" in your explanation.

NOTE: The experiment in Lesson 73 (a challenge lesson) needs at least one bean plant. If your bean plants from Lesson 66 are doing fine, that's all you need. If not, you should grow some more and sit them 30 centimeters (a foot) or so from a window that gets a lot of sun. They can just be in a cup of soil if you want; they don't need to be hydroponic.

Lesson 71: Leonardo da Vinci and Branches and Leaves

Because Leonardo da Vinci was such a careful observer, he noticed things about plants that other people missed. Let's see if you can notice two of the things that he noticed about plants.

Examining Branches and Leaves

What you will need:
- Several different kinds of trees and bushes to investigate
- Your notebook
- A pencil

What you should do:
1. Find a tree (one that is not an evergreen) whose wooden branches are reasonably easy to see underneath all the leaves. Don't be afraid to crawl into the tree a bit in order to get a good look at them.
2. Look at the thick wooden branches near the trunk. There are probably a lot of stalks that lead to leaves, but that's not what you should focus on right now. Look at the branches coming from the trunk.
3. Find a branch that splits off into two new branches, like the one pictured on the upper right.
4. Look at the part of the branch before it splits into two new branches, and then look at the two branches that form after the split. Draw the branch in your notebook. Do you notice a relationship between the branches before and after the split?
5. Look at several other branches that split off into two new branches. See if you can determine a general pattern in how the branches look before and after the split. If you see a pattern, write it down in your notebook.
6. Now ignore the thick branches and find a thin branch that has several leaves attached to it, like the one pictured on the lower right. Examine how the leaves are arranged on the branch. Specifically, look at how the stalks that lead up to the leaf split off from the branch. Draw the branch and its leaves.
7. Go to another tree, or even a bush. It should have leaves that are shaped differently from the leaves of the tree you just examined. Once again, it is best to stay away from evergreens. Find a thin branch on it that has several leaves on it, and look at how the stalks split off from the branch. Draw this new branch in your notebook.
8. Repeat step 7 for two more trees or bushes. In the end, you should have a total of five drawings: One drawing of a thick branch that splits into two branches, and four drawings of different leaves that split off from a thin branch.

Let's start with the thin branches and the leaves. Did you see any difference in the way the leaves split off from the thin branches in the different trees or bushes that you examined and drew? You probably should have. You see, each type of tree or bush makes its own kind of leaves. By looking at four trees and bushes with differently shaped leaves, you were looking at four different types of trees and bushes.

Well, in his studies, da Vinci noticed that there are three basic ways that leaves split off from the branches of a plant, and they depend on the type of plant. A given type of plant will always use the same arrangement for its leaves. The three basic arrangements that da Vinci saw are now called **opposite**, **alternate**, and **whorled**.

Opposite **Alternate** **Whorled**

A plant that uses the opposite arrangement has its leaves split off from the branch at the same point. Notice that in the leaf arrangement labeled "opposite" above, when you see one leaf split off from the branch, there is another leaf right on the opposite side. However, a plant that uses an alternate arrangement has a leaf split off from one side of the branch and then, farther up the branch, a leaf will split off from the other side. In other words, the branch alternates which side a leaf will split off from it. In a whorled arrangement, multiple leaves all split off at once, making a circular pattern.

Did you notice this about the leaves that you drew? Remember, I asked you to look at trees or bushes that had differently-shaped leaves. That's because each type of tree or bush has its own leaf shape. By looking at four plants with four different kinds of leaves, you were looking at four different types of plants. Since each type of plant always has the same leaf arrangement, it is likely that among the four types you looked at, you saw at least two of the arrangements shown above. Had you looked at many more plants, you might have found all three!

It really doesn't matter whether or not you noticed that there were different ways a plant arranges its leaves on a branch. However, now that you know it, go back and look at the four drawings you made. Can you tell which of the three basic arrangements exists in each of the plants that you drew? Identify each drawing as having an arrangement that is opposite, alternate, or whorled.

Now let's move on to the branches. What did you notice about the thick wooden branches that split off into two new branches? You should have noticed that the two new branches are always thinner than the branch from which they split. In other words, a tree branch starts out thick and round, but each time it splits into two new branches, each of the new branches is thinner than the original. Da Vinci noticed this in his observations, and it interested him. As a result, he made careful measurements of tree branches before and after they split, and he noticed a precise, mathematical relationship between the original branch and the two branches that split off from it.

To understand this mathematical relationship, you need to know what the term **area** means in math. Do you know what that means? It refers to the amount of space inside a specific place. For example, look at the picture on the right. It shows the flat surfaces of a bunch of tree trunks that have been cut. Which flat surface has more space, the one labeled "Log 1" or the one labeled "Log 2"? There is a lot more space on the flat surface of Log 1, because it's bigger than Log 2. So Log 1 has more area than Log 2.

The area of Log 1's flat surface is bigger than the area of Log 2's flat surface.

Now it turns out that mathematically, you can measure the area of something, like the flat surface of a log. I don't want to teach that to you here. Just know that there is a way to measure area. What da Vinci found was that there is a relationship between the area of the branch before and after the split. Specifically, suppose you cut down a branch that splits in two and then measure the area of the flat surface of the branch before it splits. That's the area of the big part of the branch. Then, suppose you measure the area of the two branches that are found after the split. What da Vinci found was that when you add the areas of the two smaller branches, you will get the area of the original branch! So a branch splits according to its area. The total area of the new branches has to equal the area of the original branch.

This rule is called **da Vinci's rule**, and it has interested scientists since da Vinci's time. Until very recently, no one knew why it was true. However, in 2012, some researchers found a possible reason. It turns out that if a tree follows Da Vinci's Rule, it will stand up to wind better than a tree that doesn't follow Da Vinci's Rule. So God probably designed trees in the manner da Vinci observed so that they wouldn't be broken down by any but the strongest of winds.

LESSON REVIEW

Youngest students: Answer these questions:

1. You see two trees. One arranges its leaves in an opposite way, and the other in an alternate way. Are they the same type of tree?

2. Look at the picture above. Can you find the log that has the smallest area?

Older students: If the branches you drew in your experiment don't have all of the possible leaf arrangements, draw a picture to represent the missing leaf arrangement and label it so that you have examples of all three. Write that a given type of plant will always use the same leaf arrangement.

Oldest students: Do what the older students are doing. In addition, write down Da Vinci's Rule and indicate why trees might have been made according to that rule.

Lesson 72: Leonardo da Vinci and Tree Rings

Have you ever seen the stump that is left over when a tree is cut down? Have you ever looked at it really closely? In this lesson, you will learn about the information Leonardo da Vinci found in a tree stump (or a thick branch of a tree).

Tree Rings

What you will need:

- A tree stump sitting outside (It would actually be a bit better to find a large tree branch that has fallen on the ground and have an adult use a saw to cut it in the middle. The idea is that you need a nice, flat surface that you can sand to bring out the tree rings. It is also best to use a stump or branch from a tree that loses its leaves in the winter.)
- Sandpaper
- Your notebook (a blank sheet of paper if you are doing the review exercises for the youngest students)

What you should do:

1. Look at the surface of the stump or the cut surface of the tree branch. Do you see any patterns?
2. Use the sandpaper to smooth the surface of the stump or the cut tree branch. If the stump is really large, don't feel like you have to sand the entire surface. Sand a circle near the center of the stump.
3. Once the surface is smooth to the touch, brush it off and look at the surface of the wood again. Do you see any patterns now? You should see that the wood is actually composed of rings.
4. Look at the way the rings circle each other on the surface of the wood. Do you see that they all seem to circle around the same point? Are the rings circular?
5. Look at several of the rings more closely. Are they all the same thickness? Are they all the same color?
6. Draw a picture of the surface of the wood, showing the rings. Be as accurate as you can be.

How clear were the rings on the branch or stump that you examined? I bet they were easier to see when you sanded the wood, weren't they? Wood that has been sitting out for a long time (like a stump) gets all sorts of dirt on it, and the wood changes color, too. As a result, it makes it hard to see the tree rings. Sanding the wood helps get rid of the dirt as well as the wood that has changed colors, making the tree rings easier to see. Even if you used a cut tree branch, the saw probably scratched the wood, so sanding it helped reveal the rings.

A tree forms rings as it grows. Note that the rings are not the same thickness.

What are those rings? Da Vinci realized that they are a record of the tree's growth. A tree forms a new ring every year, so by counting the rings on a tree, you can figure out how old it is. A tree that has 100 rings is 100 years old, and a tree that has 500 rings is 500 years old. Now, of course, if you were looking at a branch in your experiment, the number of rings didn't tell you how old the tree was. Instead, it told you how old the branch was. If a branch has 20 rings, that means the branch started growing 20 years ago. The tree, of course, could be much older, since a tree adds branches as it grows.

Of course, tree rings tell you a whole lot more than just how old a tree (or a branch) is. To understand the information in tree rings, you need to know how they form. When spring arrives, a tree grows a lot. As a result, it starts adding wood quickly, and that produces one color of wood. As summer gives way to fall, the tree doesn't grow as much, and the wood gets added more slowly. This produces another color of wood. So as you look at the rings on a tree, you see the difference between the wood added in the spring and the wood added in the late summer or fall.

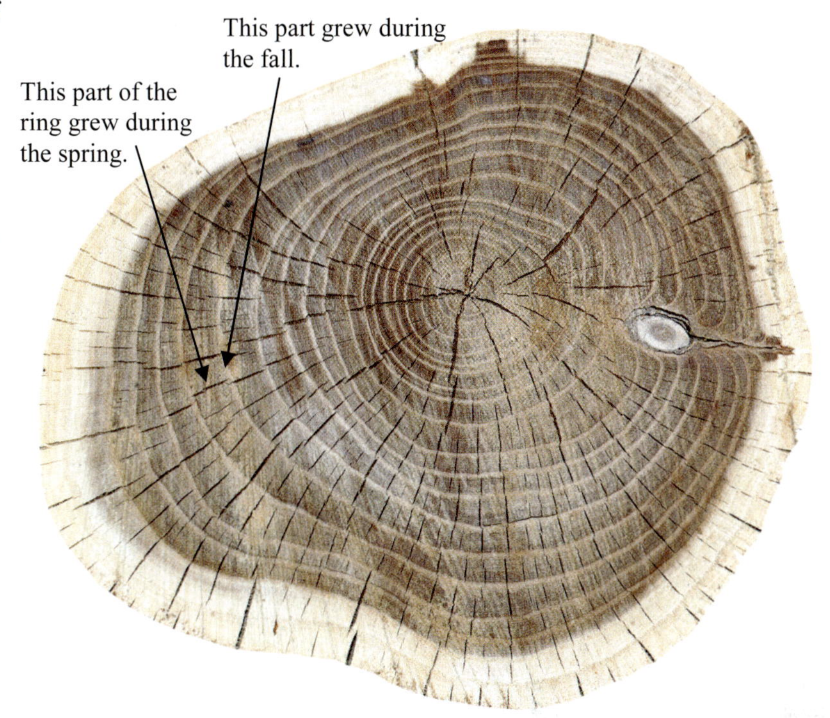

The alternating colors in the rings of this fruit tree tell you the difference in the way the tree grew during different parts of the year.

Think about what that means for a minute. The thick part of a tree ring tells us about how well the tree grew during the spring. Well, imagine a spring where it doesn't rain much. You know that plants need water to grow, so if there isn't a lot of water, how much wood do you think will be added to the tree? Not much, right? So what would you expect when it comes to the thickness of the tree ring? It wouldn't be very thick, would it? Now imagine the opposite. What about an ideal spring with lots of rainfall. You would expect a thick ring, wouldn't you?

That means the rings on a tree tell you something about the weather during the year that the ring formed. A thin ring means a spring that wasn't all that great for growing, while a thick ring means a spring that was very good for growing. Look at the picture of the tree trunk above. Can you see the rings that were formed during springs that weren't great for growth (the thin rings)? Can you see the rings that were formed during springs that were great for tree growth (the thick rings)?

I want you to notice something else about the tree trunk in the picture above. Do you see that the rings aren't perfect circles? Instead, the rings are generally thicker on the left side of the picture and thinner on the right side. This has to do with how much light the tree got on each side. The side that got more sunlight would grow better than the side that didn't get much sunlight, so the rings will be thicker on the side that gets more sunlight.

So not only do the tree rings tell you how old a tree is and what the weather was like when the ring was formed, they can also tell you which side of the tree got more light! Can you see why I said that there is a lot of information in a tree ring?

Now before I leave this subject, I do have to make a very important point. Some trees will form clearer rings than others because of the nature of how they grow or the place in which they grow. For example, there are two different kinds of trees: **deciduous** (dih sih' juh wus) trees and **evergreen** trees. Deciduous trees lose all their leaves during the winter, while the leaves (needles) of an evergreen never completely go away at the same time. In general, deciduous trees make clearer rings. They don't grow when their leaves are gone, so the addition of new wood in the spring usually makes a much easier-to-see boundary against the wood that was added in the late summer or fall.

Most of the trees in this picture are evergreens, but the silver birch in the picture is deciduous, because it has lost its leaves for the winter.

Also, there are some trees that don't form any noticeable rings. Can you imagine where such trees live? They live in a place where the weather is always great for growing! As a result, they are always adding wood quickly. Trees like palm trees, that live in places that are warm, sunny, and get plenty of rain all year will often have no rings at all, because the speed at which they grow doesn't vary with the season.

I want to add one thing before you complete this lesson. We now know that another natural philosopher, Theophrastus (thee' uh fras' tus) of Eresus (c. 372 BC – c. 287 BC), was actually the first to notice that a tree produces rings when it grows. However, da Vinci is still considered the father of this method for analyzing tree growth, even though the honor may actually belong to Theophrastus of Eresus.

LESSON REVIEW

Youngest students: Answer these questions:

1. If a tree has 139 rings, how old is it?

2. What is the difference between a deciduous tree and an evergreen tree?

Older students: In your notebook, right under the drawing you made, explain why trees form rings. Also, explain how the rings tell us about the weather during the year it formed and which side of the tree got more light. Finally, explain the difference between deciduous and evergreen trees.

Oldest students: Do what the older students are doing. Explain why deciduous trees generally have easier-to-see rings than evergreen trees.

NOTE: The next lesson (a challenge lesson) has an experiment that needs to sit for 2 hours but not more than 6 hours. You should start the lesson at the beginning of the day and come back to it later on.

Lesson 73: Leonardo da Vinci and Plant Movements

You might already know that plants tend to grow towards the light. For example, look at the bean plant that you have been growing. If it was sitting about 30 centimeters (a foot) from a sunny window, the stem of the plant is bent towards the window, and the leaves are turned so that their tops are facing the window. This is called **phototropism** (foh' toh troh' piz uhm). The "photo" part of the word refers to light, and the "trop" part comes from a Greek word that means "to turn." So phototropism is the process by which plants turn towards the light.

These seedlings are all bent to the right because that's where the light is coming from. This is an example of phototropism, where plants grow towards light.

While the name is modern, the fact that phototropism exists was known among natural philosophers long before the birth of Christ. Aristotle thought that plants could not move, but it was well known even back then that they grew towards the light. Theophrastus of Eresus (the man I mentioned in the previous lesson) argued that this wasn't anything the plant did. Instead, the sun removed water from the nearest side of the plant, which caused the plant to "tilt" towards the sun.

Da Vinci noticed something that clearly showed such reasoning was incorrect. While da Vinci didn't show that phototropism was caused by plants themselves, he did show that plants could move, which later on helped scientists to understand that phototropism is, indeed, the result of plants themselves moving. What did da Vinci discover? Perform the following experiment to find out.

Phototropism and Heliotropism

What you will need:

- The bean plant that has been sitting near a sunny window
- A room that can be made completely dark for a few hours

What you should do:

1. Take the bean plant from the place it has been growing and put it in the room that you are going to make dark. The room should be lit up right now.
2. The plant should already be bent in one direction, because it was growing towards the window. In addition, if you look at the leaves, you should see that they are turned so that their tops face the window. Move the plant around so that the leaves are facing one of the walls. It doesn't matter which wall, but be sure you remember which wall it is.
3. Make the room completely dark and leave it alone for at least 2 hours but no more than 6 hours.
4. After the plant has been sitting in the dark room for at least 2 hours, turn on the lights and look at the plant. First, look at the leaves. Are they pointing towards the wall that they were pointing towards when you left the room? What about the stem? Is it still bending towards the wall that it was bending towards before you left the room?

5. You are officially done with the plant. You can let it grow if you want, or dispose of it and the contraption in which it is growing.

What did you see in the experiment? You should have seen that while the stems were still bending towards the wall that they were originally bending towards, the leaves were no longer pointing towards the wall. Instead, they were probably hanging down from their stems, pointed towards either side of the plant.

What do these results tell us? First, they tell us that plants do, indeed, move on their own. If Theophrastus of Eresus was right, the absence of light shouldn't have caused the leaves to move while the stem remained bent. If the sun was tilting the stem and leaves by getting rid of water on one side of the plant, then when the plant was left to sit in the dark, the stems and the leaves should have reacted in the same way. However, that's not what happened. The stems didn't change noticeably at all, but the leaves did.

Second, it tells us that whatever caused the bean plant's stem to bend towards the window as it grew is different from what caused the leaves to move so that their tops were facing the window. Both of these effects are due to light, of course, but obviously whatever causes the leaves to turn towards the light acts much more quickly than whatever causes the stem to grow towards the light.

Da Vinci didn't know what caused either process to happen, but he recognized that the two processes are different, and that's the key. He knew that plants grow towards the light, but he also knew that was a slow process. If you take a plant that is bent one way and you expose it to light from a new direction, it will eventually bend in that new direction, but the change generally takes at least a day, if not longer. In addition, he knew that *all* plants grow towards the sun.

This is a field of cotton that is just starting to produce. The leaves of the plants change position throughout the day to keep their tops pointed in the direction of the sun.

In addition to this, he noticed that *some* plants change the position of their leaves as the day continues. In the morning, they point the tops of their leaves to the east, but by sunset, the leaves have completely reversed direction, and their tops are pointed to the west. In other words, the plants track the sun as it moves in the sky. They continually change the direction their leaves are pointing so that the tops of their leaves are always pointed towards the sun! That way, they get as much sunlight as possible all day. Now once again, this is not true for all plants, but it is true for some, including alfalfa, cotton, and bean plants, for example.

Now remember, the process by which some plants track the sun is different from phototropism. Phototropism is done by all plants (to one extent or another) and generally takes a while, at least on the order of a day or so. The process by which a plant tracks the sun must be much faster. As a result, we call it something different. We call it **heliotropism** (he' lee uh troh' piz um). Once again, you can see the "trop" part of the word, which refers to "turning." The "helio" part comes from a Greek word that means "sun." So heliotropism is the process by which a plant turns towards the sun. As the sun's position changes, so does the plant's position.

In addition to the fact that not all plants perform heliotropism, there is something very important that I must add: Some plants only perform heliotropism during part of their lives. For example, when a sunflower plant is first growing, it performs heliotropism. Even the buds of the sunflowers track the sun as it moves across the sky. However, once the flowers are fully formed, the stem stiffens and the flowers themselves no longer perform heliotropism. As a result, don't expect to go to a field of sunflowers and watch the blooms turn as the sun travels across the sky. It won't happen. Even once the flowers have formed, however, the leaves continue to perform heliotropism to some extent. They do not follow the sun as well as they did before the flower formed, but they still move a little in response to the sun's motion in the sky.

Notice how the sunflowers in this field are pointed in different directions. That's because once they bloom, the flowers themselves do not perform heliotropism.

I haven't told you what causes the difference between phototropism and heliotropism, because natural philosophers hadn't figured that out at this point in human history. Da Vinci just noticed that they are two different processes.

LESSON REVIEW

Youngest students: Answer these questions:

1. What is phototropism?

2. What is heliotropism?

Older students: In your notebook, describe your bean plant before and after it sat in a dark room. Explain how the stem shows phototropism, while the leaves show heliotropism.

Oldest students: Do what the older students are doing. In addition, discuss the differences between phototropism and heliotropism, including how quickly they happen and whether or not all plants perform them.

Lesson 74: Leonardo da Vinci and Air

Leonardo da Vinci studied a lot more than just plants. In fact, there were other things that he studied significantly more, and you will learn a lot about that in the final section of this course. For right now, however, I want to discuss something that da Vinci learned about fire and air. Do you remember learning about Roger Bacon? One thing he noticed was that fire needs something in air. If you cover a candle with a jar, for example, the candle goes out once whatever the fire needs is gone from the air. Well, da Vinci improved our understanding of the effect, and in the end, he learned something new about air.

Fire Consumes Only a Portion of the Air

What you will need:

- Some modeling clay, like Play-Doh
- Wooden matches
- A birthday candle
- Three quarters
- A tall glass
- A glass pan (or at least a pan with low edges)
- Food coloring
- A spoon for stirring
- Water
- An adult to help you

What you should do:

1. Put a small amount of water in the pan. It shouldn't be much more than 1 centimeter (under half an inch) high.
2. Add some food coloring to the water and stir so the water is colored.
3. Put a lump of clay in the middle of the pan. The lump should be tall enough to be just barely covered by the water.
4. Put a birthday candle in the lump of clay so it stands upright. If this changed the shaped of the clay so that it is now above the water, use your fingers to mold the clay so it is just below the water again.
5. Put the three quarters in a triangle around the birthday candle. They should be spaced far enough so that the opening of the tall glass will rest on them.
6. Break the head off one of the matches and throw it away.

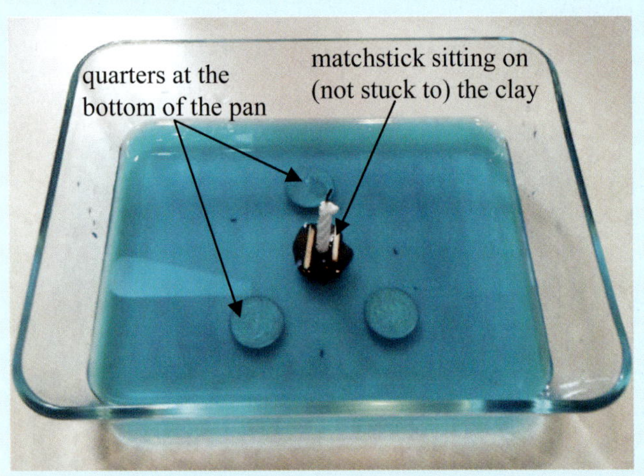

7. Take the match stick that remains and break it in half.
8. Put one half of the matchstick on one side of the candle so it rests on the clay.
9. Put the other half of the matchstick on the other side of the candle so it rests on the clay. Right now, your setup should look like the picture on the left.
10. Have an adult light the candle so that it is burning nicely.
11. Turn the tall glass over and use it to cover the candle so that its mouth rests on the three quarters.
12. Watch the water inside the glass as the candle burns.

13. Once the candle is completely out, note what the matchsticks are doing. If they aren't floating, knock the pan gently a couple of times, and they should float.
14. Clean up your mess.

What happened in your experiment? You should have seen that when you first put the glass over the candle, there was very little (if any) water in the glass. After all, the glass was full of air. That air pushed the water (or at least most of it) out of the glass, making the inside of the glass almost dry. However, as the candle burned, water should have come in, especially when the candle went out. The water level in the glass should have risen enough to allow the matchsticks to float, even though the matchsticks weren't originally floating at all.

What causes the water to rise in the glass? Well, as the candle burned, it used up oxygen in the air. That caused less air to be in the glass. Do you remember learning about air pressure in Lesson 33? Air presses on anything that it contacts. So air was pressing down on the water, both outside the glass and inside the glass. Well, as the candle used up some of the air inside the glass, there was less air, so there was less air pressure inside the gas. However, there was still the same amount of pressure pushing on the water outside the glass. That imbalance in pressure caused water to be pushed into the glass, making the water level inside the glass rise.

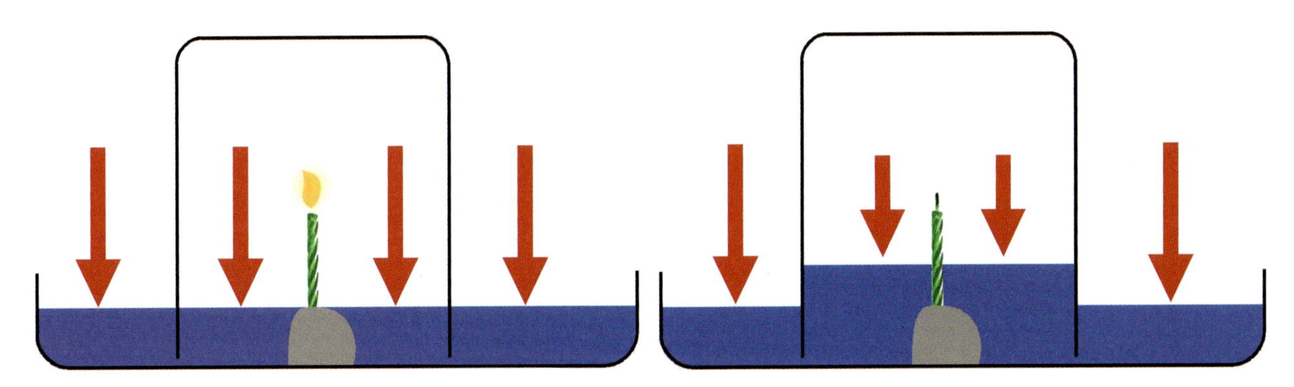

In your experiment, the air pressure (represented by the red arrows) was the same pushing down on the water inside and outside the glass. However, as the flame used up something in the air, the pressure inside the glass reduced, which meant the larger pressure outside the glass pushed water inside the glass.

None of this is surprising based on what you already know. After all, as Bacon figured out, fire uses up something in air. Given what you already know about air pressure, it makes sense that as the fire inside the glass used up something in the air, there would be less pressure in the glass. As a result, the water in the glass would rise. However, here was the new concept that da Vinci added: *Fire doesn't use up all the air.*

What would have happened in your experiment had the fire used up all the air in the glass? The water level would have risen a lot higher. After all, the water was being pushed into the glass due to the fact that the air outside the glass was pressing down on the water harder than the air inside the glass. The imbalance in the pressure is what caused the water to rise in the glass. If there had been no air left in the glass, there would have been no pressure, and that would have been a *huge* imbalance, causing a lot of water to be pushed into the glass. Since only some water was pushed into the glass, we know that the flame of the candle didn't use all of the air inside the glass.

Why is that so important? Well, natural philosophers had for a long time thought of air as an element. In case you have forgotten, you learned about elements in Lesson 7. An element cannot be broken down into two simpler substances. Iron, for example, is an element. If I have a lump of iron, I can split it into smaller lumps, but no matter how small the lumps, they will always be iron. Likewise, sulfur is an element. It can be split into smaller amounts of sulfur, but it cannot be broken down into anything else. Iron sulfide, however, is not an element. I can split it down into two simpler things: iron and sulfur.

Iron sulfide (left) can be broken down into iron and sulfur (right). However, iron and sulfur cannot be broken down further. That means iron and sulfur are elements, but iron sulfide is not.

Elements, then, are chemicals that cannot be broken down into simpler chemicals. If a chemical, like iron sulfide, can be broken down into simpler chemicals (iron and sulfur), then it is not an element. Now you should see the importance of what da Vinci figured out. He realized that air cannot be an element. After all, if it were an element, it couldn't be broken down into other things, so the flame would use up all of the air around it. Since the flame used only some of the air around it, there must be at least two chemicals in air: the chemical the flame used, and the chemical it didn't use. Because it contains at least two separate things (like iron sulfide contains iron and sulfur), it is not an element.

Now it turns out that air contains a lot more than two different things. It is an amazing mixture of gases that is perfectly designed to support life on earth. However, I am getting ahead of myself again, because natural philosophers won't figure that out until much later!

LESSON REVIEW

Youngest students: Answer these questions:

1. What is an element?

2. How did Leonardo da Vinci figure out that air is not an element?

Older students: In your notebook, make a drawing of what happened in your experiment, like the one on the previous page. Use it to explain what happened in your experiment and why it tells us that fire uses only a portion of what is in air. Also explain how this tells us that air is not an element.

Oldest students: Do what the older students are doing. In addition, natural philosophers thought that water, earth, and fire were elements as well. Based on what you know, explain why earth is not an element. Fire and water aren't elements, either, but you might not know enough right now to explain that. If you do, go ahead and explain why they aren't elements.

NOTE: The next lesson (a challenge lesson) has an experiment that needs to sit for 24 hours. Don't read any of the lesson; just start the experiment at least 24 hours before you do the lesson.

Lesson 75: Leonardo da Vinci and Plastic

How many of the things you use every day are made from plastic? Plastic is a very popular substance when it comes to making things, because it is light and easy to mold into lots of difference shapes. In addition, it is easy to make thin or thick, and depending on the type of plastic you use, it can be very flexible or very hard. When discussing the history of plastics, most people start with a man named Alexander Parkes. In 1862 at the Great International Exhibition in London, he revealed to the world a plastic he had made. Of course, the term "plastic" wasn't used until sometime later. In fact, people called what Parkes made "Parkesine" in honor of its inventor.

Plastics are so popular today that they are used to make all sorts of things. Nearly everything in this picture (even the man-made sponge) has at least some plastic in it.

However, the history of plastic starts much earlier. In fact, it was made in the Middle Ages, long before Leonardo da Vinci. However, as far as history can tell, da Vinci wrote down the first detailed recipe for making it. It's not exactly clear what he used the plastic for. Some historians think he used it to preserve interestingly-shaped objects in nature that would otherwise decay over time. Others think he used it to strengthen some of the mechanical objects that he designed (you'll learn more about them later).

His plastic started out as a liquid that he could apply to a surface like paint. He could cover an object with a thin layer of the "paint" and allow it to dry. He could then add more layers, allowing each one to dry before adding the next one. In the end, once he had added several layers, he would have a hard mold around whatever he painted. This mold would certainly preserve the shape of what he painted, and it would probably make the object stronger as well. To give you an idea of how da Vinci made his plastic, perform the following experiment.

Making Plastic Out of Milk

What you will need:

- Skim milk (Even 1% fat does not work nearly as well.)
- Vinegar
- A tablespoon
- A 1-cup measuring cup and a ½-cup measuring cup
- A saucepan
- A stove
- A spoon (wooden or plastic is best) for stirring

- A funnel
- A coffee filter
- A sink
- A lamp that can be pointed or tilted so that its light bulb (not a compact fluorescent or LED light) can be put close to the surface of a counter (If it's a warm, sunny day, don't worry about the lamp.)
- A plate (It can't be made of paper, foam, or plastic.)
- A spatula
- A cookie cutter (The more interesting the shape, the better.)

What you should do:

1. Pour 2½ cups of milk into the saucepan.
2. Put the saucepan on a burner that is set on medium heat.
3. Stir the milk as it heats. Test its temperature with your finger from time to time.
4. When the milk is warm but not hot, add two tablespoons of vinegar.
5. Continue stirring. Fairly quickly, the milk will begin to curdle.
6. When the curds form big clumps and the rest of the liquid is fairly clear, take the saucepan off the stove.
7. Put a coffee filter into the funnel so that you can pour the contents of the saucepan into the filter and it will drain through the funnel.
8. Hold the funnel over the sink and slowly pour the contents of the saucepan into the filter. Allow the liquid to run out of the funnel and down the drain of the sink until only the curds remain.
9. If there are any leftover curds in the pan, use the spoon to push them into the coffee filter.
10. Carefully pull the coffee filter out of the funnel and wrap it around the curds.
11. Squeeze the paper so that water comes out of the curds.
12. Once you can't get any more water out of the coffee filter, unwrap the curds and remove them.
13. Squeeze the clump of curds in your hand to get out even more water. Dry your hands and squeeze again, to get even more water out.
14. You should now have a clump of moist curds in your hand. Put them on the plate and press them into a thin, flat sheet.
15. Use the cookie cutter as if the flat sheet of curds is a sheet of cookie dough, cutting out the shape that is in the cookie cutter.
16. Roll the excess of the curds into a ball.
17. Put the ball on the plate near the shape that was made by the cookie cutter.
18. Put the plate on a counter somewhere and tilt or aim the lamp so that the curds are warmed by the lamp. **Don't put the lamp too close to the plate, as the curds could catch fire**, depending on the power of the bulb. The goal is simply to use the light bulb to heat the curds. If it is a warm, sunny day outside, you could also set the plate out in the sun.
19. Wait for about 24 hours. If you are setting the curds out in the sun, take them in overnight in case it rains. The next morning, put them back out in the sun.
20. If you are using the lamp, remove the lamp after 24 hours.
21. Touch the curds that are in the shape made by the cookie cutter. What do they feel like?
22. Pull the shape off the plate. Use a spatula if it is stuck to the plate.
23. Feel the shape. What does it feel like?
24. Try bending the shape. What happens?
25. The ball might take a few more days to dry out, depending on its size. Once it is dry, however, pick it up and roll it around in your hand. What does it feel like?
26. Try to bounce the ball. How does it bounce?
27. Clean up your mess.

What did that mess of curds turn into once it dried? It turned into something that was a lot like plastic, didn't it? That's the kind of chemical that Leonardo da Vinci made more than 300 years before the "first plastic" was made by Alexander Parkes. Now it turns out that da Vinci didn't use milk to make his plastic. However, he used the same basic kind of chemical.

You see, all living things have chemicals in them called **proteins**. You have probably heard about proteins, because you eat them. When you have a steak or a hamburger, you are eating proteins. Well, it turns out that proteins are very important chemicals in life, and anything that is living (plants included) has proteins in it. Proteins are long molecules that are folded up into particular shapes in order to do the things they need to do in a living creature.

However, under certain conditions you can force those proteins to unfold. They then will link together, making even longer molecules. It turns out that long molecules, which are called **polymers** (pah' luh merz), allow plastics to have their useful properties. When you added vinegar to milk, you caused the proteins in milk to unfold, and they started linking together. As the curds dried, the linking continued, making really long molecules. In the end, then, they made something that behaves like plastic.

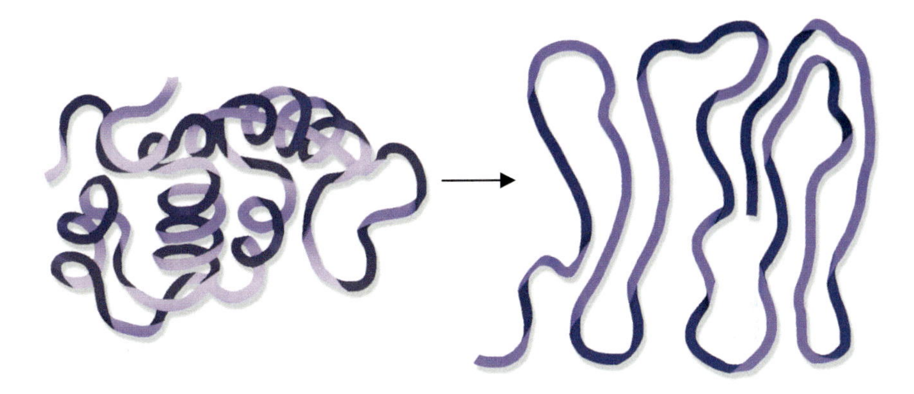

The proteins in milk (and other things that come from living creatures) are folded into very interesting shapes, like what you see on the left. When you added vinegar and heat, you caused the milk proteins to unfold, making them look more like what you see on the right. When they are unfolded, they can link with other proteins to make long molecules that behave like plastic.

As I said, da Vinci didn't use milk, but he used plant and animal products that had a lot of proteins. He didn't know about proteins, of course, but he experimented around with the plant products and animal products until he was able to come up with something very much like the "first plastic" that was produced more than 300 years later.

LESSON REVIEW

Youngest students: Answer these questions:

1. What is it about the molecules in plastic that gives it such interesting properties?

2. What kind of chemicals can you find folded into interesting shapes in all living creatures?

Older students: In your notebook, write down the "recipe" you used to make plastic. Say what kind of chemical formed the basis of your plastic.

Oldest students: Do what the older students are doing. In addition, explain what the vinegar and heat did to the chemicals in the milk that allowed them to form a plastic.

Science in the Early Renaissance

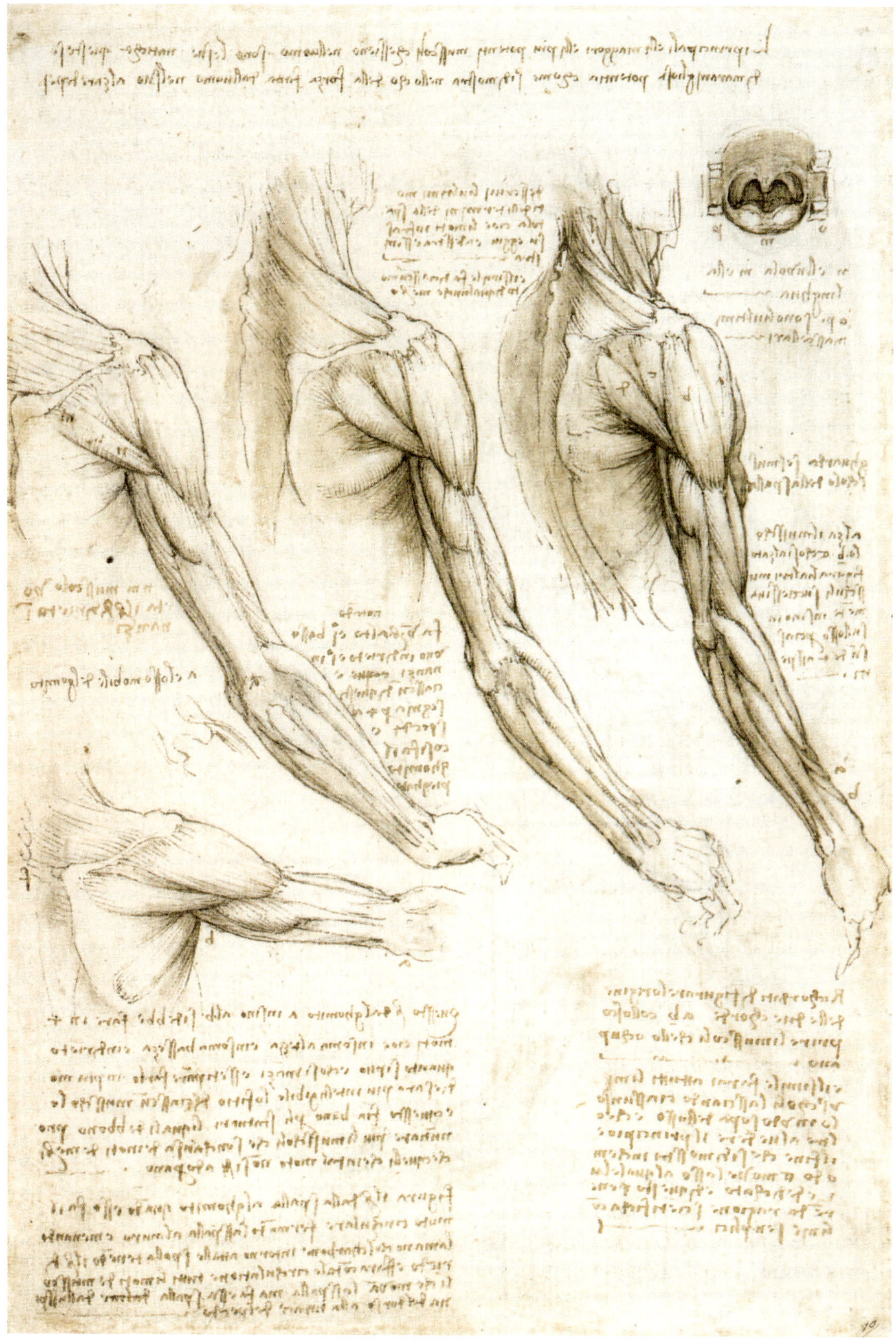

This page from one of Leonardo da Vinci's personal notebooks is dedicated to a study of arm muscles.

Lessons 76 - 90: Science in the Early Renaissance

Lesson 76: Leonardo da Vinci and Density

Have you ever noticed that when an object is made out of a substance like metal it is heavier than when that same object is made out of plastic? For example, think about pipes that carry water in your home. Some are made of metal, while others are made of plastic. If you have ever been in a situation where you had to lift such pipes, you would find that the pipes made out of plastic are a lot lighter than the ones made out of metal, even when the pipes are exactly the same size. If you have ever played with a toy car made out of metal, you probably noticed it is heavier than a similar toy car made out of plastic.

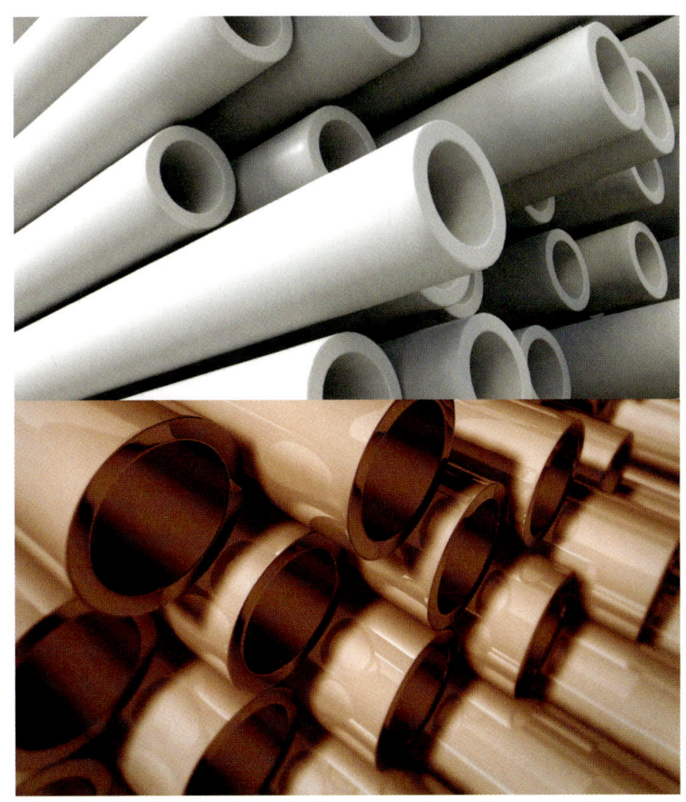

The plastic pipes in the top picture are about the same size as the metal pipes in the bottom picture, but they weigh a lot less.

Why is that? To understand the answer to this question, you first have to learn a new term: **matter**. You should remember that nearly everything you see around you is made up of atoms, and those atoms often join together to make molecules. Because there are lots of different atoms and lots of different ways they can join together, there are all sorts of different molecules in creation. That means, of course, that there are all sorts of different substances in nature. Well, even though there are a lot of different atoms and molecules, we can refer to them all as "matter."

If something is made up of atoms or molecules, it is made up of matter. Most everything you see around you is made up of matter, but there is one thing you see right now that has no matter in it at all. Can you guess what it is? It's light. Light is pure radiant energy. It is not made up of atoms or molecules, so it has no matter in it. However, nearly everything else in creation is made up of matter.

You also need to remember a term that you probably already know: **volume**. Volume is a measure of how much space something takes up. Something that is really big (like a house) takes up a lot of space and therefore has a large volume. Something that is really small (like a pencil) takes up only a little space and therefore has a small volume.

Now that you know these terms, you can understand the answer to the question of why metal pipes weigh more than plastic pipes of the same size. It's because the matter in metal is packed very tightly together, but the matter in plastic is packed much more loosely. So think about two objects that are the same size and shape but one is made of plastic, and the other is made of metal. The one made of plastic won't have as much matter in it, because the matter in plastic is packed loosely. If the other object is made of metal, it will have more matter in it, because its matter is packed more tightly.

If two objects are the same size and shape, they have the same volume. However, if they are made of different substances, they will pack their matter into that volume in different ways. The one that packs its matter more tightly will be the one that has more matter in that volume and therefore will weigh more. So plastic pipes (or plastic toys) are lighter than metal pipes of the same volume (or metal toys of the same volume) because the matter in metal is packed more tightly than the matter in plastic.

Now it turns out that each substance packs its matter a bit differently, so it is nice to come up with a way of measuring how tightly matter is packed in a substance. As a result, scientists use the term **density**, which is a measure of how much matter exists in a given volume of a substance. If a substance packs its matter loosely, it has a small amount of matter in a given volume, so it has a low density. If a substance packs its matter tightly, it has a large amount of matter in a given volume, so it has a high density. Using this term, then, plastic has a lower density, while metal has a higher density.

Well, it turns out that measuring density is pretty important if you want to work with a substance. If you are building something that carries a large amount of water, for example, it would be nice to know how much weight that amount of water will have. Leonardo da Vinci faced just such a problem. He worked with systems that would move or hold large amounts of water so it could be used to transport goods or provide water for crops.

As a result, he wanted to know how much a large amount of water would weigh, so he needed to measure water's density. As a result, he came up with the first **hydrometer** (hi drah' muh tur). Now don't get this confused with a hygrometer, which measures how much humidity is in the air. You learned about that in Lesson 64. A hydrometer measures the density of a liquid. Perform the following experiment where you make your own hydrometer.

Making a Hydrometer

What you will need:

- Two glasses of the same size
- A straw
- Some modeling clay, like Play-Doh
- Water
- Salt
- A spoon for stirring

What you should do:

1. Fill the two glasses with equal amounts of water.
2. Dissolve as much salt as you can in one of the glasses. Don't worry if there is leftover salt at the bottom of the glass. That won't matter.
3. Roll the Play-Doh into a small ball and stick one end of the straw into it. You do not want to stick the straw all the way through it. Make it so the straw goes about halfway into the ball. You want the Play-Doh to form a watertight seal with the bottom of the straw.
4. Put the straw in the glass without salt in it so that the ball of Play-Doh is on the bottom. Watch to see if water gets in the straw. If it does, redo the Play-Doh ball so that it forms a watertight seal with the bottom of the straw.
5. See how far the straw/ball contraption sinks. If it sinks so deep that the ball rests on the bottom of the glass, take it out and remove some Play-Doh from the ball. If it floats really high in the glass,

add more Play-Doh. Your goal is to make the Play-Doh ball big enough so that the ball floats about an inch (3 centimeters) from the bottom of the glass, as shown in the picture on the previous page.

6. Once you have the contraption floating in the right place, make a careful observation of how high it is floating in the glass.
7. Remove the contraption and put it in the glass that has saltwater in it.
8. Once again, make a careful observation of how high it is floating in the glass. How is it different from the height at which it floated in the glass with water but no salt? You can take it out and put it in the other glass again to compare if you want.
9. Clean up your mess.

What did you see in the experiment? You should have seen that the contraption floated higher in the saltwater than it did in the freshwater. Can you explain why? Think about it. You probably already know that because of the salt in saltwater, a given volume of saltwater weighs more than the same volume of freshwater. Using the term you just learned, then, saltwater has a higher *density* than freshwater.

You probably also know that in order to float, an object must weigh less than the water that it has to move out of its way. In fact, it will float at the exact point where the weight of the water it moves out of the way is equal to its own weight. A ship that is full of cargo, for example, will float lower in the water than the same ship when it is empty, because when it is full, it has to move more water out of the way to get to the point where the amount of water moved is equal to its own weight. Well, the contraption you built has to move more water out of the way the deeper it goes, because more of the straw sinks into the water. Since freshwater has a lower density than saltwater, the contraption had to move more freshwater out of the way to reach its own weight in water. As a result, it had to sink deeper in the freshwater.

So what you just made is something that tells you the density of the liquid in which it is floating. If the liquid has a low density, the contraption will sink lower in the liquid before it starts floating. If it has a high density, the contraption won't sink as far before it starts floating. In other words, you made a hydrometer! While Leonardo da Vinci didn't use a straw and modeling clay to make his hydrometer, it worked on the same principle.

LESSON REVIEW

Youngest students: Answer these questions:

1. Two objects have exactly the same volume, but the first one is heavier. Which has the lowest density?

2. What is a hydrometer?

Older students: In your notebook, draw the results of your experiment, showing the difference in how your hydrometer floated in freshwater and saltwater. Explain what density is and why the thing you made is a hydrometer.

Oldest students: Do what the older students are doing. In addition, vegetable oil floats in freshwater. In your notebook, compare how high your hydrometer would float in vegetable oil to how high it floated in freshwater. See if you were right, and correct your answer if it is wrong.

Lesson 77: Leonardo da Vinci and Water Flow

NOTE: This lesson has an outdoor experiment. Feel free to skip it and come back to it on a nicer day if the weather isn't suited for going outdoors.

As you learned in the previous lesson, da Vinci worked with systems that moved a lot of water. For example, in 1482, he was hired by the Duke of Milan (Ludovico Sforza). The Duke had da Vinci do all sorts of things, including making sculptures, designing weapons of war, designing new buildings, and working on canals. Back then (and even today), Milan had a network of canals that connected the city to rivers and lakes in that part of Italy. Since a lot of trade goods were transported by ship back then, a city that was connected to a lot of waterways became an important center for trade.

While most people think of Venice as the Italian city with canals, Milan has had a network of canals since the Middle Ages. This is a section of the canal network called a "Naviglio," which da Vinci supervised and improved.

The Duke of Milan wanted his city to remain an important trading center, so he wanted the canal system to be the best in the world. Therefore, he had da Vinci oversee their upkeep and design improvements to the system. As a result, da Vinci spent a lot of time studying how water moves from one place to another. He even applied what he learned from his experience with canals in order to design a system that would move water from lakes and rivers to places where crops grew. When you bring water from one place to water crops in another place, it is called **irrigation** (ihr uh gay' shun), and da Vinci designed ways for it to be done.

The more he worked with water, the more he learned about it. One thing he learned was a general law that is now known as the **Law of Continuity**. You are probably already familiar with this law, but I want you to do an experiment so you understand it fully.

The Law of Continuity

What you will need:

- A garden hose attached to a spigot
- Clothes you can get wet in
- An open space outdoors
- A rock that is small enough for you to completely close your hand around it

What you should do:

1. If the hose has a spray nozzle on it, unscrew the nozzle so the hose ends in just its normal opening.
2. Turn the water to the hose on full, so that water streams out of the hose.
3. Hold the hose in front of you, pointing away from you.
4. Watch the water as it streams from the hose. How far does it go? How big is the stream of water coming out of the hose?
5. Use your thumb to cover a *small part* of the opening at the end of the hose. The part of your thumb that is covering the opening should block the water, so now water has to flow out of the part of the opening that is not covered by your thumb.
6. Holding the hose the same way you did in steps 3 and 4, watch the water as it streams from the partly-blocked end of the hose. How far does it go? How big is the stream of water?
7. Now use your thumb to cover even more of the opening at the end of the hose.
8. Repeat step 6.
9. Use your thumb to cover as much of the opening as you can while still allowing a stream of water to spray from the hose.
10. Repeat step 6.
11. Put the rock on the ground.
12. Using only the stream of water coming from the hose, try to move the rock. What do you have to do to make the stream of water powerful enough to move the rock?
13. Turn off the water, and put the hose back the way you found it.

What did you see in your experiment? If things went well, you should have seen that the more your thumb covered the opening of the hose, the farther the stream of water sprayed. Of course, the stream of water was smaller the more your thumb covered the hole. Also, to get the water to spray strongly enough to move the rock, you had to use your thumb to cover a lot of the opening of the hose. Otherwise, the water wouldn't hit the rock hard enough to move it.

What explains these results? Do you remember studying how you can measure the depth of water by measuring its pressure (Lesson 63)? One thing you learned there was that a stream of water travels farther in the air the faster the water shoots out of the opening. So one thing you can conclude from your experiment is that the more your thumb covered the opening of the hose, the faster the water came streaming out of it. This is one way a hose nozzle changes the kind of spray coming from the hose. If you want to get the water to shoot farther, you adjust the nozzle so that it blocks part of the opening of the hose. That way, the water shoots out of the hose faster, and therefore it travels farther.

You can adjust the nozzle on a hose to block more of the opening. That makes the water come out faster, which means it will spray farther.

This also explains why you had to cover most of the hose's opening to get the water to hit the rock hard enough to make it move. After all, if the water was going to make the rock move, it had to hit the rock hard. That means it had to be traveling quickly. So the more your thumb covered the opening of the hose, the faster the water came streaming out of it. The stream of water was not as wide, but the speed at which it moved was a lot faster.

Da Vinci was the first to describe why this works. He said that if water is flowing through an enclosed container that is full (like a hose, for example), the amount of water that comes out has to equal the amount of water coming in. If that didn't happen, water would start "backing up" in the container, and the container would eventually burst. In other words, water has to leave the hose to make room for the new water that is coming in. So in the end, if water is flowing through a hose or a pipe, the amount of water coming out has to equal the amount coming in.

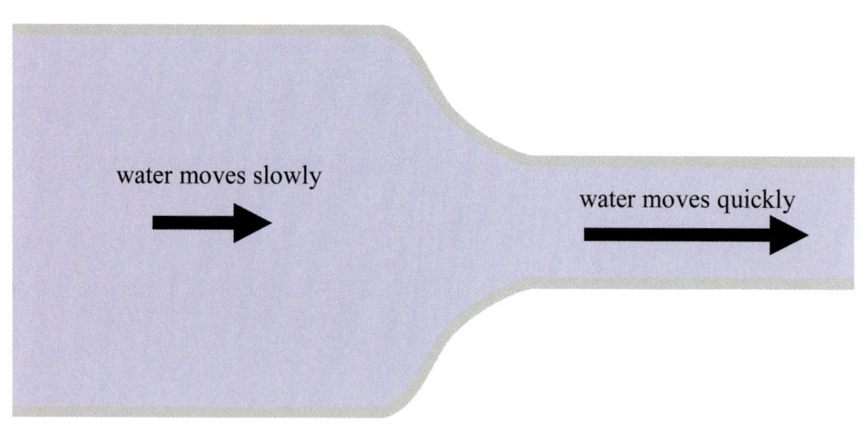

The Law of Continuity tells us that in this system, the water moves more quickly in the smaller pipe.

Well, if my hose has a big opening, there is a lot of room for water to come out, so the water doesn't have to move really quickly to make room for the new water that is being put into the hose. However, if I cover part of the opening, then there is less room for the water to get out. The same amount of water has to get out every second, however, because it has to make room for the new water that is coming in. As a result, the water has to travel faster for the same amount to get out of the hose every second. As the opening gets smaller, the water has to travel even faster.

So in a system where water is being fed in at a constant rate, the smaller the opening through which the water leaves, the faster the water must move through that opening. As I mentioned previously, this is called the Law of Continuity, and da Vinci was the first to understand it.

LESSON REVIEW

Youngest students: Answer these questions:

1. What is irrigation?

2. Fill in the blanks: When water flows through pipes, the amount of water leaving the pipe has to be ________ __ the amount entering the pipe.

Older students: In your notebook, define the word irrigation. In addition, state the Law of Continuity in your own words and explain why it works.

Oldest students: Do what the older students are doing. In addition, look back at the experiment you did in Lesson 63. Would the Law of Continuity work there? Why or why not? Answer that question in your notebook and check to see if you are right. Correct your answer if you are not.

Lesson 78: Leonardo da Vinci and Erosion

Because da Vinci worked with the movement of large amounts of water, he was interested in how water is affected by its surroundings as well as how it affects its surroundings. He drew very detailed pictures, such as the ones shown on the right, of how water flows in different situations. Of course, as I have mentioned before, by drawing something, a person studies it in a very detailed way. As a result, da Vinci was able to learn things about how water flows that no one else had ever documented before.

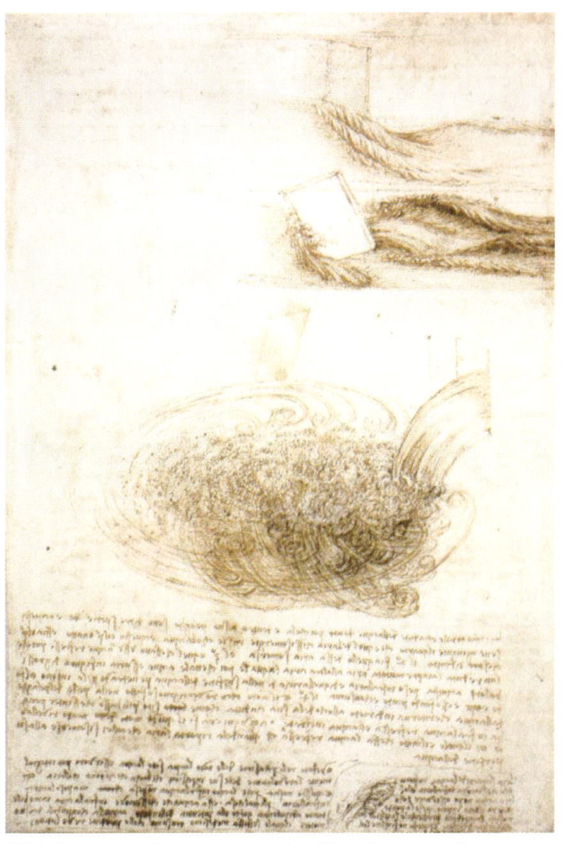

This is a page from one of da Vinci's journals, showing how he studied the way water flows in different situations.

One thing he noticed was that there is a lot of power in the flow of water. As you noticed in the previous lesson, if you get water flowing fast enough, it can move rocks. The faster the water is flowing, and the more water that flows at that speed, the heavier the rocks that it can move. Well, da Vinci noticed in his studies of water that when water moves, it not only has the ability to move the things with which it comes into contact, it also has the ability to change the area through which it is flowing. Da Vinci studied such changes carefully, and he noticed that given enough time (or enough water), the flow of water through an area can change the very appearance of that area. See for yourself a bit of what Leonardo saw in his investigations by doing the following experiment.

Erosion

What you will need:

- A pie pan or other cooking pan that has a raised edge all the way around it
- Some dirt
- A sink that can get dirty and has a water tap
- Depending on how your sink's tap is positioned, you might need a pitcher.

What you should do:

1. Use the dirt to make a "mountain" in the center of the pan, as shown in the picture on the right. Take a good look at the "mountain" so you know what it looks like.
2. See if you can put the pan in the sink so that the "mountain" is right under the tap. If your sink's tap just isn't in the right position to do that, fill a pitcher with water instead.
3. Turn the tap on really low so that only a small trickle of water comes from it and hits the "mountain." If you can't use the tap, pour water on the "mountain" from the pitcher, but pour it as

slowly as possible. Observe what happens as the water hits the "mountain" and flows over it. Notice how the "mountain" changes, and notice what the water that is filling the pan looks like.
4. After the water has been flowing gently over the "mountain" for a while, turn the tap off or stop pouring the water.
5. Gently tip the pan to drain most of the water out of the pan and into the sink. Don't tip it enough to move the "mountain," however. Look at the water as it drains. How dark is it?
6. Observe the "mountain." How has it changed since you observed it in step 1?
7. Now turn the tap on so that it is about half open. That way, the water should be flowing out quickly, but not as quickly as is possible. If you are using a pitcher, just pour the water so that it flows a lot more quickly than before.
8. After the water has been flowing for a little while, repeat steps 4 – 6.
9. Now turn on the tap or pour from the pitcher so that water flows as quickly as possible over the "mountain" (without splashing everyone around the sink).
10. After the water has been flowing for a little while, repeat steps 4 – 6.
11. Clean up your mess.

What did you see in your experiment? You should have noticed that when the water ran over the "mountain" slowly, not much changed. A small hole might have formed in the top of the "mountain," indicating that the water was cutting into the "mountain" a bit. In addition, the water that accumulated in the pan was dirty, indicating that it picked up some of the "mountain" as it flowed over it. As you increased the flow of water, however, you should have seen that the "mountain" was affected a lot more. A large hole probably developed where the water fell on the "mountain," and the water got dirtier the quicker the water flowed.

In your experiment, you saw the effects of **erosion** (ih roh' zhun), the process by which rocks and soil are broken up and washed away. Water has a lot of power when it comes to erosion, so it can completely change the shape of the land over which it is flowing. In your experiment, erosion probably cut holes and valleys into your "mountain." Eventually, the shape of the mountain changed due to the erosion caused by the water.

The banks of this river are high because over time, the water has cut down through the land over which it is flowing. That's the power of erosion.

The picture on the left shows another way erosion can affect the land over which water flows. Notice how high the banks of the river are. At one time, they weren't nearly that high. However, as water continued to flow in the river, it eroded the land, chipping off bits of it and carrying it away. As time went on, this process cut down through the rock, forming the steep banks that you see now.

Can water really cut through rock? Yes, it most certainly can! As your experiment showed, the power of erosion depends on two things: how much water is flowing and how quickly it flows. The faster the water flows, the stronger it is. The more water that flows, the more material it can carry away. As a result, water can cause a lot of erosion. If water is flowing slowly, it takes a lot of time for a lot of erosion to occur. However, if water is flowing quickly, a lot of erosion can happen in a short amount of time.

Let me give you two examples of the power of erosion when water is flowing quickly. Look at the machine pictured on the right. It is cutting a special tool out of metal. Do you see the tool's outline in the metal? It looks a lot like a wrench, doesn't it? Do you know what the machine uses to cut through the metal? *It uses water mixed with some sand.* That mixture is shot at the metal in a very thin stream that moves incredibly quickly. This erodes the metal, cutting right through it!

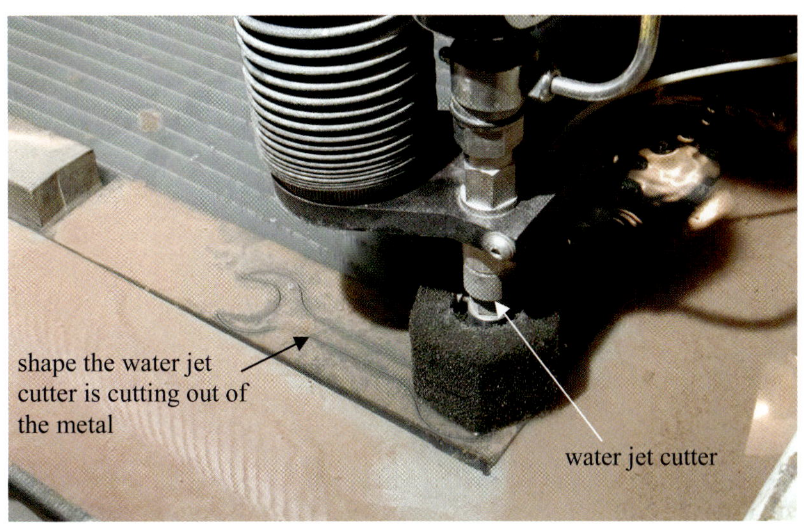

Water is so powerful that a thin stream of water and sand moving quickly enough can cut through metal! That's how this water jet cutter works.

Here's my second example: In 1980, a volcano in the state of Washington (Mt. St. Helens) erupted violently. Several smaller eruptions followed over the next few years. In 1982, a small eruption ended up melting a large amount of snow that had built up on the volcano over the winter. As the large amount of melted snow flowed quickly down the mountain and across the land, it carved a system of canyons that were up to 40 meters (140 feet) deep. All this happened in a single day! Thus, while erosion *can* take a long time to produce big changes in the surrounding land, such changes can also happen in a very short time, as long as there is a lot of water flowing quickly.

As da Vinci noted, flowing water is very powerful. Over a long time, even a gentle flow of water can cause a lot of erosion. If a lot of water flows quickly, it can cause an enormous amount of erosion in a very short time, changing the shape of the land in amazing ways!

LESSON REVIEW

Youngest students: Answer these questions:

1. What is erosion?

2. (True or False?) Water is strong enough to cut through metal and rock.

Older students: In your notebook, define the word erosion. In addition, explain what two things determine how much erosion takes place as water flows over land.

Oldest students: Do what the older students are doing. In addition, consider a deep canyon that has a river flowing gently at the bottom. One scientist tells you that it took the river millions of years to erode the land and form the canyon. Another scientist tells you the canyon was formed in a few months by rapid erosion. In your notebook, explain each scientist's assumption about how water flowed to carve the canyon. Check your explanation and correct it if it is wrong.

Lesson 79: Leonardo da Vinci and the Human Skeleton

I have told you all sorts of scientific facts that da Vinci uncovered during his studies, but I haven't even touched on the scientific achievement for which he is best known today. I want to start discussing that now. It turns out that da Vinci's artistic and scientific abilities combined to produce the best understanding of human anatomy of his time.

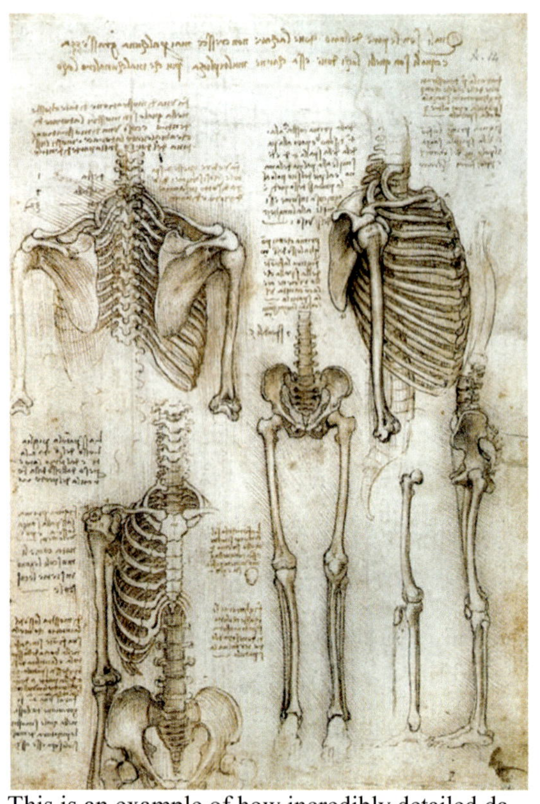

This is an example of how incredibly detailed da Vinci's drawings of human anatomy were.

Why was da Vinci's work so important in the area of human anatomy? Because he was the first to produce very detailed and accurate drawings of human anatomy based on the dissection of human bodies. Do you remember what "dissection" means? You learned about it in Lesson 20. It means cutting something open that was once alive so you can study it more closely. Also, do you remember learning about Guy de Chauliac back in Lesson 60? He was a great physician for his time specifically because he was able to watch human dissections, unlike many of the previous medical authorities such as Galen.

Well, if people like Guy de Chauliac observed dissections, what made da Vinci's work so special? The drawings he made of what he saw were not only detailed, but they were amazingly realistic. This made his drawings the first really accurate diagrams illustrating human anatomy. So it was the special blend of da Vinci's scientific and artistic talents that made his work on human anatomy so special.

Nowadays, we appreciate how important the blending of such talents is, so there are people who are called **scientific illustrators**. They combine their knowledge of science and their artistic abilities like da Vinci did, producing illustrations that are used in textbooks and other scientific works. If you enjoy science and art, you might want to investigate being a scientific illustrator one day.

As shown in the illustration above, Leonardo da Vinci was very interested in the human **skeleton**. That's where we are going to start learning about human anatomy. The human skeleton is a network of hard structures called **bones**. One of the skeleton's main jobs is to support the body. Without your skeleton, you couldn't sit up, stand up, or do much of anything. You would be just a blob of flesh. To give you some experience with the human skeleton, do the following activity.

The Human Skeleton

What you will need:

- A photocopy of page A4 from the Appendix of your parent/teacher's *Helps and Hints* book
- Scissors
- Glue (A glue stick works best, but any glue will do.)
- Your notebook (a blank sheet of paper if you are doing the review exercises for the youngest students)
- Something you can use to cover the illustration on the next page

What you should do:
1. Cover the diagram on this page so you won't look at it as you do this activity.
2. Use the scissors to cut out the human body outline that you find on page A4 of the Appendix.
3. Start a new page in your notebook and label it "The Human Skeleton." If you are using a blank sheet of paper, label the paper that way.
4. Paste the human outline on that page of your notebook (or the blank sheet of paper).
5. Use the scissors to cut out each set of bones on page A4 of the Appendix. You don't have to cut them out perfectly. If there is white space around the bones, don't worry about it. If you are having trouble, get an adult to help you.
6. Without using any books or resources, lay each set of bones in the outline of the body where you think they are in a human body. Don't glue them onto the outline yet. Just lay them there. Some parts will lie on top of other parts. That's fine. Just put them where you think they should go.
7. When you are done, check your work against the drawing that is found in your parent/teacher's *Helps and Hints* book on the page that has the answers for this lesson's review questions.
8. Fix any mistakes you made, and then glue the bones in the right place.
9. Clean up your mess.

How well did you do in terms of putting the bones in the right places? What matters more than the answer to that question is how well do you know where those bones go now? The illustration on the right shows you the human skeleton in a slightly different pose than the one that you used in your experiment. Can you still recognize the bones?

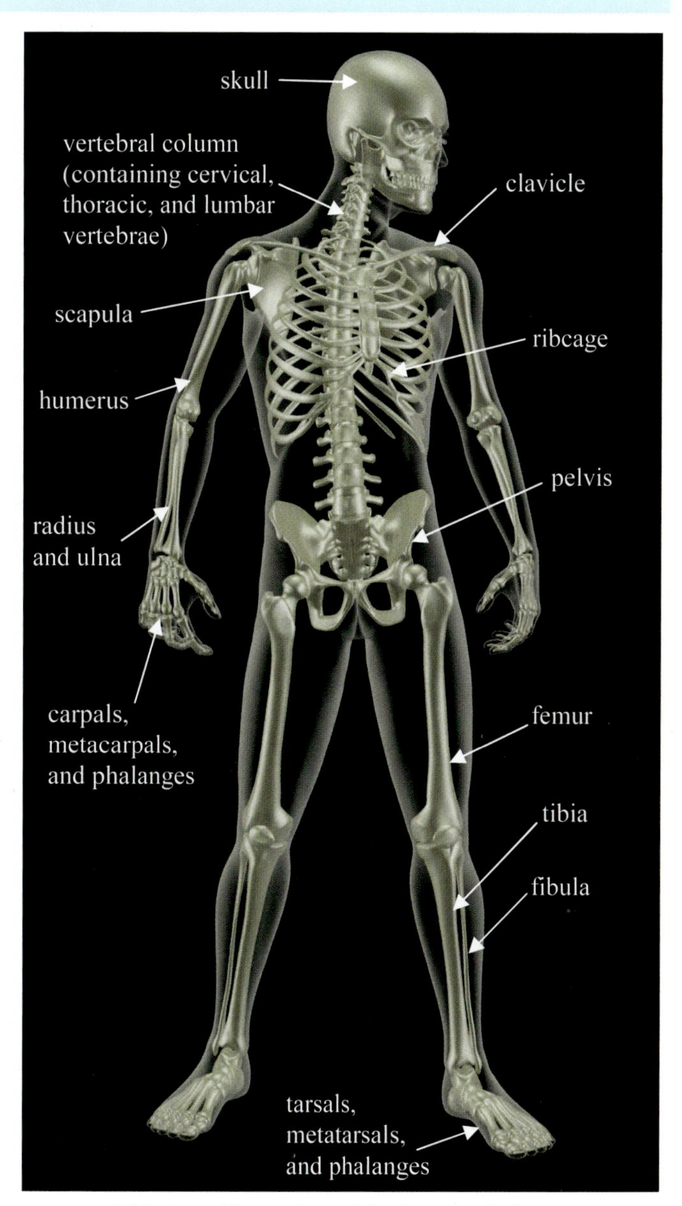

This is an illustration of the human skeleton.

As I told you before, one of the main jobs of the skeleton is to support the body. However, parts of the skeleton have a second important job as well – to protect some of the more fragile parts of the body. Looking at the skeleton on the right, can you guess which parts protect as well as support? What's inside your **skull**? Your brain. It's very fragile. The skull protects the brain so that even if something hits your head hard, your brain isn't hurt too badly. Also, look at the **ribcage**. It looks like a little prison, doesn't it? Your stomach, heart, kidneys, liver, and some other organs are in there. The ribcage protects those organs so that even if something hits your chest hard, they aren't hurt.

There's at least one other part of the skeleton that protects something. Can you guess what it is? It's the **vertebral** (vur' tuh brul) **column**. You have a long organ called the **spinal cord** that carries messages from the brain to the rest of the body and back again. It's very sensitive, so it is inside the vertebral column.

That way, it is well protected in case you fall on your back or something hits your back. You'll learn more about your vertebral column in an upcoming lesson.

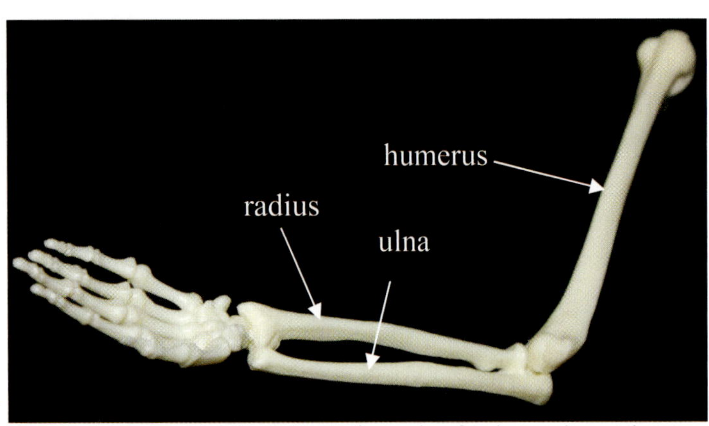

This model shows that your arm is made of only one bone above your elbow, but two bones below your elbow.

You can feel some of these bones in your own body to get an idea of what they are like. For example, wrap one hand around your arm below your wrist, toward your elbow. Twist your arm back and forth. You should feel two different bones moving underneath your hand. Now do the same thing with your arm above the elbow, toward your shoulder. You should feel only one bone twisting around. That's because above your elbow, your arm is made of only one bone, the **humerus** (hyoo' muh rus). However, below your elbow, your arm has two bones, the **radius** (ray' dee us) and **ulna** (uhl' nuh).

Another set of bones you can feel easily can be found at your sides. Push your thumbs into each side of your body just under your armpits. Now slide them down your sides. Do you feel bumps as your thumbs slide down your sides? Those are your **ribs**, which make up your ribcage. Put your hands on your legs right below your knees and push in on your legs. Now slide your hands down. Do you feel a hard ridge running down your legs? That's your **tibia** (tih' bee uh), which is the bigger of the two bones that run from your knee to your foot.

Based on what you felt, you should recognize that your bones are hard. They have to be, if they are going to support and protect you. However, they are also surprisingly flexible. If they weren't, they would break easily. The skeleton is an incredible feat of engineering, and da Vinci was excellent at showing us this fact. Like da Vinci, you should give thanks to your Creator for the marvelous skeleton He gave you!

LESSON REVIEW

Youngest students: Answer these questions:

1. What is a scientific illustrator?

2. What are the two main jobs of the skeleton?

Older students: In your notebook, label the bones that are glued into the outline of the person. You can use the picture on the previous page to learn which bones you should label. Below that, list the two main jobs of the skeleton.

Oldest students: Do what the older students are doing. In addition, do a little research to find out how many bones are in the human skeleton. Write in your notebook what you find.

Lesson 80: Joints

The bones in the human skeleton are somewhat flexible, but they are mostly hard. That allows them to support the body and protect the organs that are inside. Because the bones are hard, they don't bend very well. Nevertheless, your body can bend in all sorts of wonderful ways. That's because your skeleton isn't made up of just one big bone. Instead, it is made up of a lot of bones (as many as 350, depending on how old you are), and those bones are connected to one another by **joints**. The body's joints help it to bend in some amazing ways.

Your body can bend because your bones are connected to one another with joints, which allow the bones to move in relation to one another.

Consider your arm, for example. Stretch it out in front of you so it is completely straight. Use your index finger to point at something that is in front of you. Now, without moving your shoulder or wrist at all, change the position of your hand so your index finger is pointing straight up. How did you do that? You did it by bending your arm at the elbow. Do you remember what the bones of the arm look like? Look at the previous page or your notebook to remind yourself.

Near your shoulder, your arm is made up of one bone – the humerus. The lower part of your arm, however, is made up of two separate bones – the radius and ulna. The humerus is connected to the radius and ulna at your elbow. The humerus can't bend, and neither can the radius or ulna. However, your arm can bend because your elbow is a **joint** that allows the radius and ulna to change position relative to the humerus.

Perform the following experiment to learn more about the elbow joint and how it works.

A Hinge Joint

What you will need:
- Two craft sticks (or popsicle sticks)
- An index card or other piece of cardboard that is sturdy but is thin enough to fold easily
- Tape
- Scissors

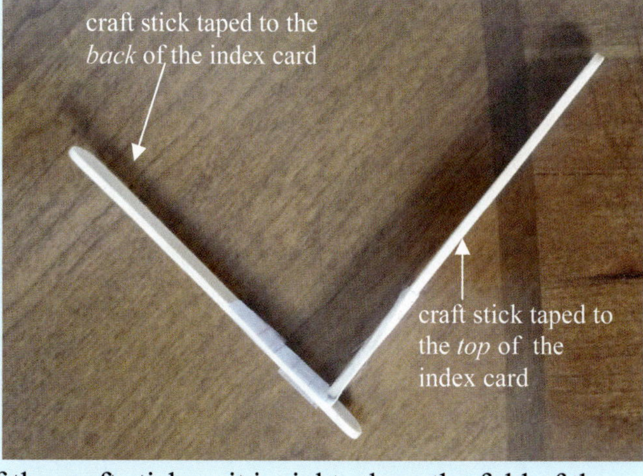

What you should do:
1. Cut a strip of index card that is only a bit wider than the craft sticks.
2. Fold the strip of index card so that it is bent in the shape of an "L."
3. Use tape to attach one craft stick to the top side of the bottom of the "L." Line up the end of the craft stick so it is right where the fold of the "L" is.

4. Use tape to attach the other craft stick to the back of the part of the "L" that is sticking straight up. The end of this craft stick needs to go a bit below the fold. Your contraption should look something like the picture on the previous page.
5. Hold one craft stick in each hand and gently push them together so that they touch one another. You can do this because the "L" that you made out of the index card folds down.
6. Slowly pull the two sticks apart until the whole contraption is completely flat. Once again, you can do that because the "L" that you made out of the index card unfolds. However, because of the way the sticks are taped to the "L," there is a limit to how far the index card will allow the sticks to bend. They can get close enough to touch each other, but the farthest they can go the other way is for them both to lay flat. In other words, your popsicle-stick contraption can bend one way, but it can't bend the other way.
7. Hold your arm in front of you so it is "folded," with the lower part of your arm touching the upper part of your arm.
8. Straighten your arm out. Once your arm is straightened out, can you continue to move the lower part of your arm farther in the same direction? No, you can't. Like your craft-stick contraption, your arm can bend one way, but it can't bend the other. It can only straighten out.
9. Clean up your mess.

In your experiment, you made a model of a **hinge joint**, and your elbow is an example of a hinge joint. I want you to examine the elbow a bit more carefully, but you need to know a couple of terms. Scientists who study human anatomy don't use the terms "upper part of your arm" and "lower part of your arm." Instead, when these scientists use the term "arm," they are actually talking about the upper part of your arm, which has the humerus bone. When scientists refer to your lower arm (which has the radius and ulna), they use the term **forearm**. According to scientists, then, your elbow is between your arm and your forearm.

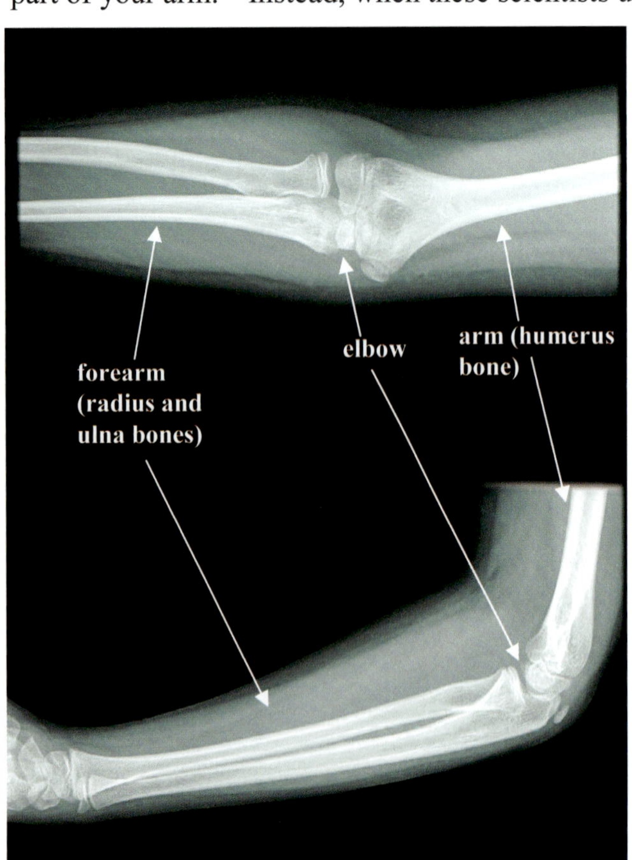

These X-rays show a person's arm and forearm. In the top picture, the elbow is extended. In the bottom picture, it is flexed.

Do you know what an X-ray is? An X-ray is an image taken of things inside your body. By using a special kind of radiation and something that detects that radiation, a doctor can actually make an image of something that is inside you, like your bones. The picture on the left is made up of two X-rays of someone's arm, forearm, and elbow. In the top picture, the forearm and arm are stretched out. In the bottom picture, the forearm has been brought about halfway towards the arm.

When your forearm and arm are straightened out as shown in the top X-ray, we say that your elbow joint has been **extended**. When your forearm is brought closer to your arm, as is shown in the bottom X-ray, we say that your elbow has been **flexed**. Now think about what happens when you flex your elbow. Which bones actually move when you do that? Your humerus doesn't move, does it? You humerus *can* move, but it doesn't move when you flex your elbow. Instead, when you flex your elbow, only your radius and ulna move. In other words, when you flex your elbow, you are moving your *forearm*, not your arm.

You might think that this is an unimportant fact, but as you will learn in an upcoming lesson, it tells you a lot about how your muscles work. The elbow joint doesn't allow both the arm and the forearm to move. Think about the contraption you made in your experiment. The index card allowed both of the craft sticks to move. However, in reality, when you flex your elbow, only your forearm moves. You will learn why in an upcoming lesson. For now, I just wanted you to notice this fact.

Now why do we call the elbow a hinge joint? Well, look at the drawing on the right. A hinge is formed when a rod rotates in a groove, as shown on the left side of the drawing. If you look at the elbow, it looks a lot like a hinge. In addition, think about the motion of a door. The door can open and close, but it can't move any other way. You can't tilt the door or turn it around. The only thing you can do is open and close it. That's a lot like the elbow – it flexes and extends like a door opening and closing. So like the hinges on a door, your body's hinge joints don't allow for a lot of different types of motion. However, even though they don't allow for a lot of different types of motion, they are very stable joints. It is hard to pull a hinge joint in the body apart. As you will learn in the next lesson (if you are doing the challenge lessons), you can't say that about every joint in the body!

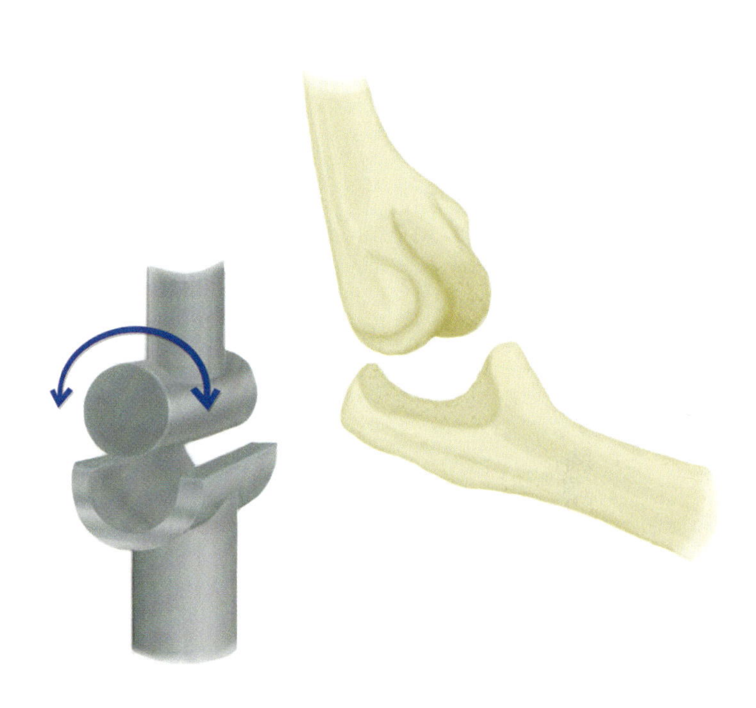

A hinge joint, like the elbow, is similar to the hinge on a door.

LESSON REVIEW

Youngest students: Answer these questions:

1. What parts of your skeleton allow the bones to move in relation to one another?

2. What kind of joint is the elbow?

Older students: In your notebook, draw a picture like the X-rays on the previous page. Label the top one as "extended" and the bottom one as "flexed." Label the elbow, forearm, and arm. Below the drawings, write that the elbow allows the forearm to flex and extend, but that's the only motion allowed, which is why it is called a hinge joint.

Oldest students: Do what the older students are doing. In addition, write down at least one other joint in your body that is a hinge joint. Check your answer and correct it if it is wrong.

Lesson 81: More on Joints

Since your body can move in all sorts of wonderful ways, you shouldn't be surprised to learn that there are many different kinds of joints in your body. Think about how you can move your arm at the shoulder compared to how you can move your forearm at the elbow. Hold your arm at your side and then raise it up so your hand is pointed toward the ceiling. Now twirl your arm in a circle. Your arm can really move a lot at your shoulder, can't it? Can you twirl your forearm around at your elbow? Of course not! Your forearm can move toward or away from your arm, but that's about it.

So your shoulder joint allows you to move your arm a lot more than your elbow joint allows you to move your forearm. Why? Do the following activity to find out.

A Ball-And-Socket Joint

What you will need:
- A small Styrofoam cup
- A lot of modeling clay, like Play-Doh
- A pencil

What you should do:
1. Roll the Play-Doh up into a ball that is big enough so that when the ball is put in the cup, it rests in the opening without sinking into the cup, as shown in the picture on the left.
2. Push the pencil into the center of the ball of clay. The pencil should go deep into the ball of clay. It can even push through the other side a bit.
3. Hold the bottom of the cup in one hand. Be sure not to hold the cup near the top, because that will squeeze the cup and mess up the activity.
4. With your other hand, grab onto the pencil and move it around in all sorts of different directions. Do you see how the ball rolls in the cup as you move the pencil?
5. Think about how much motion the pencil has compared to the motion you saw in the previous experiment, where you made a hinge joint.
6. Pull the ball of clay out of the cup.
7. Remove the pencil from the ball.
8. Pull about one-third of the clay out of the ball and put it aside.
9. With the clay that remains, make a new ball.
10. Put the ball in the cup. This time, it will sink down into the cup.
11. Push the pencil into the center of the ball of clay again.
12. Once again, hold the cup by the bottom and use your other hand to hold the pencil.
13. Move the pencil around, like you did in step 4. Now that the ball is deeper in the cup, do you notice anything different about how freely the pencil can move?
14. Clean up your mess.

What did you see in the activity? Because the ball of clay was able to roll around in the cup, the pencil could move nearly any way you wanted it to move until it hit the edges of the cup. That's a lot of freedom of motion! When you made the ball smaller, however, what happened? It went deeper into the cup, and as a result, the pencil's movement was not as free. It hit the edge of the cup before it was tilted nearly as much as when the ball was big.

What you made in your experiment is a model of a **ball-and-socket joint**. In this kind of joint, one bone ends in a round "ball," and the bone it joins with has a round "socket" into which the ball fits. Just like what happened in your experiment, the ball can roll around in its socket, allowing the bone that has the round ball on it to move quite a bit. That's the kind of joint you have in your shoulder. Look at the drawing on the right. The red and grey thing is a muscle, but I want you to look at the bones. The upper part of your humerus ends in a round ball that fits into the round socket of your scapula.

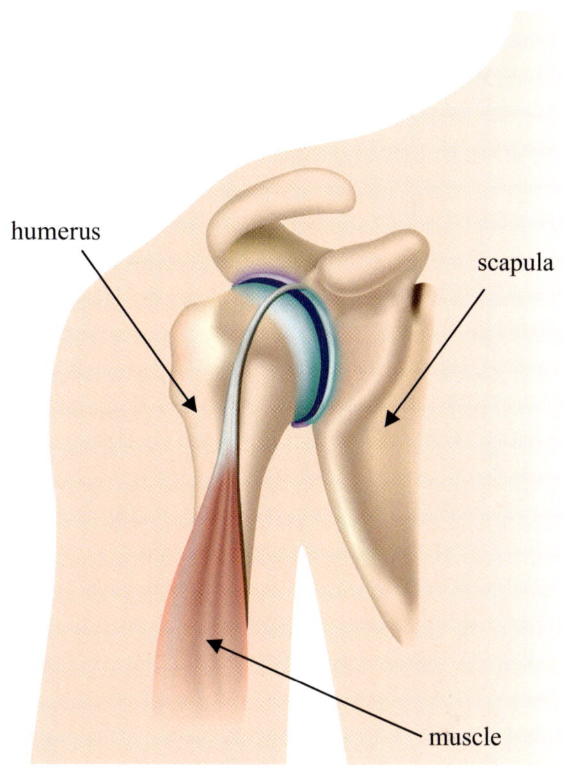

Your shoulder joint is a ball-and-socket joint, which allows your arm to move in a lot of different ways.

Because it is a ball-and-socket joint, your shoulder joint allows your arm to move in a lot of different ways. As you move your arm, the ball of your humerus moves smoothly in its socket, a bit like the ball of clay moved in your experiment. This is why you can move your arm at your shoulder much more than you can move your forearm at your elbow. While a hinge joint (like your elbow) limits the amount of movement, a ball-and-socket joint allows for a lot of movement.

Why does your body have different kinds of joints? Why aren't they all hinge joints or ball-and-socket joints? Well, there are two things you have to think about when you are studying joints. The first is **range of motion**. How much can the bones move in the joint? The range of motion of your elbow joint is fairly small. On the other hand, the range of motion of your shoulder joint is large. You might think that it would be great for all of your joints to have a large range of motion, but that's because you aren't considering the other issue: **stability**.

Joints are wonderful things because they allow your bones to move, but if you move the bones too much or two violently, the joint can become **dislocated**, which means the bones don't line up properly in the joint. Have you ever known someone with a dislocated joint? It's very painful, because the joint is designed for the bones to fit together in a certain way. When the joint is dislocated, the bones aren't fitting together the way they are supposed to fit together. That not only causes pain, but it limits the amount of motion the bones have.

Well, it turns out that the more motion a joint has, the less stable it is. In other words, a joint that can move a lot is more likely to become dislocated than a joint that can move only a little. So while a hinge joint doesn't allow for a large range of motion, it is rarely dislocated, because it is very stable. On the other hand, a ball-and-socket joint allows for a large range of motion. As a result, it is not nearly as stable as the elbow, and it can become dislocated much more easily. So when you think about the joints in the body, you have to think about *both* range of motion and stability. A large range of motion is good, but some joints need to be very stable, so they cannot allow for a large range of motion.

Now there are ways to take a ball-and-socket joint and make it more stable. You explored that in the second part of your activity. When you used less clay to make the ball, it fell deeper into the cup. So in your model "joint," your "ball" was deeper in your "socket" in the second part of the activity. What did you notice about the range of motion when that happened? It was less, wasn't it? The pencil could still move around a lot, but the edges of the cup limited how far it could go.

It turns out that there are ball-and-socket joints in your body that are like the "joint" in the second part of your activity. The ball fits more deeply into the socket. While that limits the range of motion a bit, it allows the joint to become more stable. It's still not as stable as a hinge joint, but it is more stable than it would be otherwise. However, the "cost" of that stability is a more limited range of motion. Can you think about a joint like that in your body? It's your hip joint.

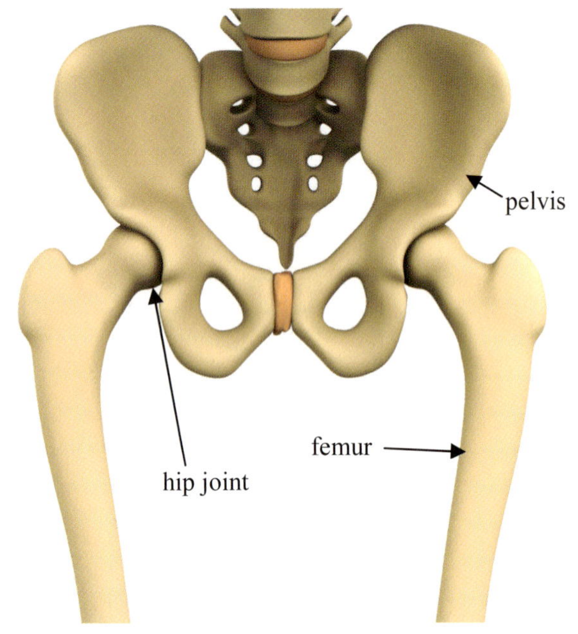

The hip joint is a ball-and-socket joint, but the ball fits deeply in the socket. This limits the range of motion a bit, but it adds stability.

You can move your leg at your hip a lot like you can move your arm at your shoulder, but the range of your leg's motion is a bit more limited. That's because the ball structure at the end of your femur rests deeply in the socket of the pelvis. While you can't move your leg quite as freely as your arm, the hip is not as easily dislocated as the shoulder.

There are other types of joints in the body as well. Gliding joints, for example, allow bones to slide across one another. In each case, however, there is a balance between range of motion and stability. This shows us how carefully God designed our bodies. He gave us many different joints so that we could move our bodies as freely and safely as possible!

LESSON REVIEW

Youngest students: Answer these questions:

1. What does it mean for a joint to become dislocated?

2. What kind of joint is the shoulder joint?

Older students: In your notebook, draw either a shoulder or hip joint. Under the drawing, explain what a ball-and-socket joint is. Compare the range of motion of a ball-and-socket joint to that of a hinge joint. Indicate what it means when a joint is dislocated.

Oldest students: Do what the older students are doing. In addition, explain the relationship between the range of a joint's motion and its stability.

Lesson 82: Leonardo da Vinci and the Vertebral Column

Because Leonardo da Vinci was a great artist, he was able to portray human anatomy more realistically than anyone before him. Even though he lived 500 years ago, some of his human anatomy drawings are considered as accurate as modern ones. This is particularly true for his drawings of the **vertebral column**, which is sometimes called the backbone or the spine. When you think of your vertebral column, how do you imagine its shape? It changes when you bend, but what do you think of as its "normal" shape? Most people think of the vertebral column as straight. Indeed, that's how all natural philosophers who drew the vertebral column before da Vinci depicted it. However, as you can see from da Vinci's notebook page on the right, he correctly showed that the vertebral column is not straight. Instead, it is shaped a bit like an "S." It curves one way near your head, another way in the middle of your chest, and another way near your rump. Da Vinci was the first to show this correct shape in a drawing of the vertebral column.

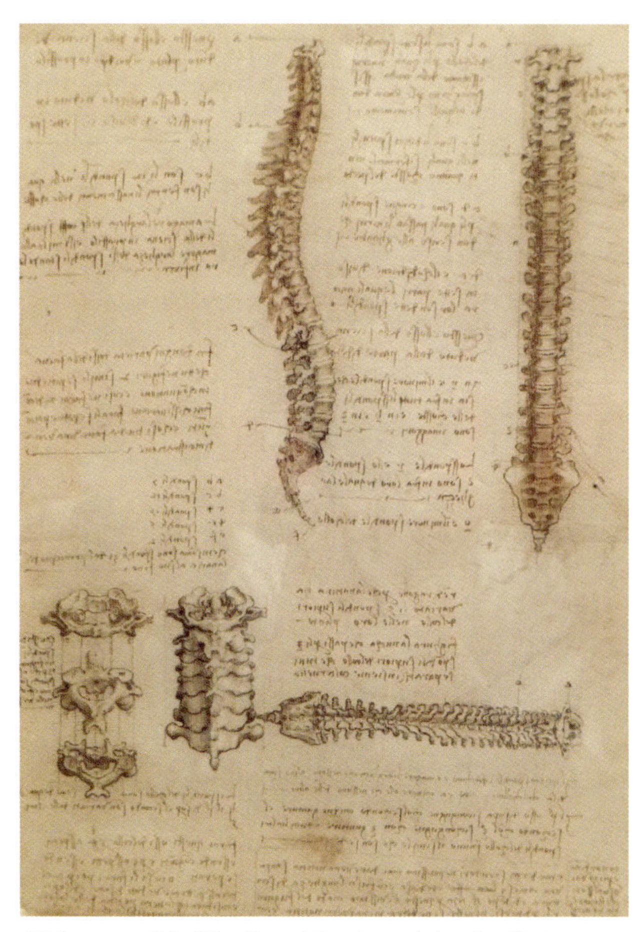

This page of da Vinci's notebook contains the first accurate drawing of the vertebral column.

It turns out that this "S" shape is vitally important for you when you are walking or running, because it allows the vertebral column to absorb some of the shock that your body experiences each time your foot hits the ground. You might not notice this shock so much when you are walking, but when you are running, you can feel each time your foot hits the sidewalk or floor, can't you? The "S" shape of the vertebral column allows it to act as a tight "spring," absorbing that shock so your body feels it less. Without an S-shaped vertebral column, walking and running would be much harder on your body!

To get an idea of how the vertebral column is put together and how it protects the spinal cord, do the following activity.

The Vertebral Column

What you will need:

- Four spools of thread (If the spools are empty, that's ideal, but spools with thread on them will work as well.)
- Two long rubber bands or several smaller rubber bands
- Scissors
- Tape
- String (It needs to be strong string. Yarn will do if you don't have any strong string.)

What you should do:

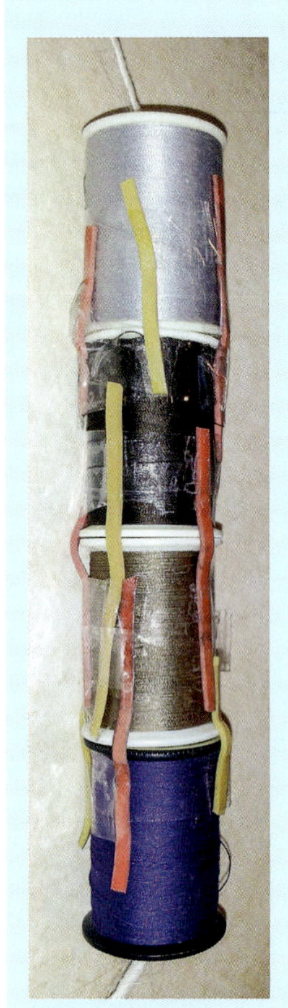

1. Thread the string through the center hole of each spool one at a time until you have all four spools on the string.
2. Cut the rubber bands into 12 pieces, each of which is about as long as one of the spools.
3. Push two of the spools together and lay a piece of rubber band so that half of it lies on one spool and half of it lies on the other spool.
4. Tape the rubber band down so it holds the two spools together tightly. It is best to wrap the tape all the way around the spool so the rubber band is really stuck to the spool.
5. Repeat steps 3 and 4 three more times so that you have four equally-spaced rubber bands holding the two spools together.
6. Repeat steps 3-5 with the next spool in line so that now you have three spools of thread held together with rubber bands. In the end, the first spool should be connected to the second spool with rubber bands, and the second spool should be connected to the third spool with rubber bands.
7. Repeat steps 3-5 again with the last spool. In the end, your contraption should look something like the picture on the left.
8. Hold your contraption by the bottom spool so it stands straight up. The bottom of the string will be trapped between the surface you are standing your contraption on and the spool you are holding. As you hold the bottom spool, push down on it so that the bottom of the string can't move.
9. Gently pull the top of the string towards you. What happens to the spools?
10. Gently pull the top of the string away from you. What happens to the spools?
11. Gently pull the top of the string to one side and then the other. What happens to the spools?
12. Once you are done with the lesson, you need to peel the tape off so the spools can be used again. You can throw away the rubber band pieces and used tape.

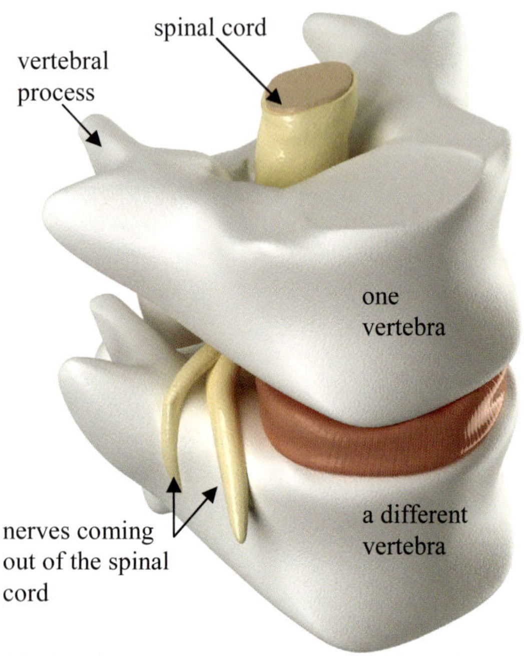

This drawing shows two vertebrae and how they come together to make the vertebral column, which protects the spinal cord.

The contraption you made is a model of your vertebral column. Like the contraption, your vertebral column is made of a bunch of individual bones stacked on top of one another. These individual bones are called **vertebrae** (vur' tuh bray). Like the spools, each vertebra (vur' tuh bruh – the singular of "vertebrae") has a hole in its center. Your **spinal cord** runs through those holes, just like the string did in your contraption. You'll learn more about your spinal cord later. For right now, just realize it is a very sensitive organ that is vitally important to your body. Your vertebral column protects it by forming a tunnel of bone for it to run through, just like the spools formed a tunnel for the string in your contraption.

Your vertebral column is easy to feel. Just press your fingers into the center of your back right above your rump. Then slide your fingers straight up your back. You should feel bumps all the way up the center of your back. Those bumps are called **vertebral processes**, and they are on each vertebra.

In your contraption, then, the spools represented the vertebrae, and the string represented the spinal cord. What did the rubber bands represent? They represented **ligaments**, which are strips of elastic material that hold the vertebrae together. After all, how stable would the vertebral column be if it was just made of individual bones stacked together? It would fall apart really easily, right? The ligaments hold the vertebrae together so that doesn't happen. It turns out that every joint in your body has ligaments that hold the bones together in the joint.

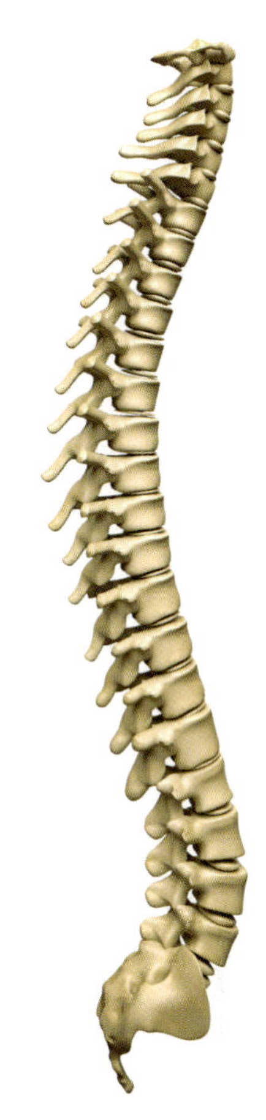

Like most drawings, this drawing of the vertebral column doesn't show the ligaments.

Wait a minute. Does that mean the vertebral column has joints in it? Yes! Each place one vertebra touches another is a joint, because it allows the vertebrae to move in relation to one another. What happened when you pulled on the top of the string in your contraption? The contraption bent, didn't it? What happened when you let go? It went back to standing straight up, didn't it? That's what happens to your vertebral column when you bend over or bend side-to-side. Your vertebrae slide against each other, allowing your vertebral column to bend! When you stop bending, your vertebrae stack right back together, because the elastic ligaments and the muscles connected to the vertebrae pull them back together.

Do you remember that I mentioned gliding joints in the previous lesson? That's what the joints between your vertebrae are. The ligaments keep the vertebrae from sliding too far, but they can slide a fair amount, which is good. Otherwise, you wouldn't be able to bend your vertebral column. Most drawings of the vertebral column don't include the ligaments, because they tend to cover up the bones. Without them, however, your vertebral column would not be stable.

LESSON REVIEW

Youngest students: Answer these questions:

1. What are the individual bones of the vertebral column called?

2. What does the vertebral column protect?

Older students: In your notebook, draw the contraption you made. Underneath the drawing, explain what each part of your contraption represents and how it is a model for the vertebral column.

Oldest students: Do what the older students are doing. In addition, explain why the shape of your contraption is not a good representation of the vertebral column. Explain what the real shape of the vertebral column does.

Lesson 83: Leonardo da Vinci and the Structure of Bones

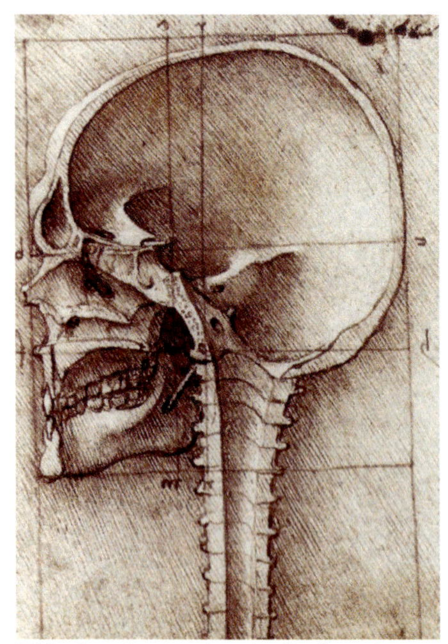

This drawing is a longitudinal section of the skull and part of the vertebral column.

As I have already told you, Leonardo da Vinci was very concerned about drawing human anatomy as accurately as possible. Based on his studies, he realized that it was not enough to simply draw what he saw on the outside of the bones. He knew that the body was so well designed that the *inner parts* of many bones were interesting and needed to be investigated. As a result, he cut bones open and drew what he saw on the inside.

Look, for example, at the illustration on the left. It's one of many that da Vinci drew of the skull and the top portion of the vertebral column. While it is not completely accurate (the skull and vertebral column don't connect exactly as they are drawn here), notice what it shows. Rather than showing what the outside of the skull looks like, he drew the skull as if he had chopped it in half. As a result, you can see the structures that exist inside the skull. You can see the open area where the brain sits in the skull, and you can clearly see the "tunnel" of the vertebral column that holds the spinal cord. This kind of view is called a **section**, and as far as we can tell, da Vinci was the first to draw such views.

Nowadays, many scientific illustrations show sections so that students can see inside all the wonderful things God has built in nature. In fact, sections are so important in scientific illustrations that they come in different varieties. I want you to learn about the two most common types of sections by doing the following activity.

Longitudinal Section and Cross Section

What you will need:

- Two bananas that are overripe (They should have several dark spots on their peels.)
- A sharp knife
- Your notebook (or a blank sheet of paper if you are doing the review exercises for the youngest students) and a pencil
- An adult to help you

What you should do:

1. In your notebook, start a new page and draw one of the bananas.
2. Have an adult help you use the knife to cut a small section out of one banana, as shown by the blue lines in the picture on the left.
3. Remove that little section and draw it right next to or below the first drawing you made.
4. Have an adult help you use the knife to cut the other banana in half the long way, as shown by the red curve in the picture on the left.
5. Separate the two halves and draw one of them right next to (or below) the second drawing you made.
6. Clean up your mess.

Have you ever looked at the inside of a banana that way? If you hadn't ever looked at a banana cut in half the long way, you probably didn't know that there are dark spots running down the center of the banana. Those are the seeds that will grow into a banana tree if they are planted in an area that has good weather for growing such trees. The other section you cut out of the banana probably had some dark spots in the middle. That shows you where the seeds are in the banana, but it doesn't show you that the seeds run pretty much all along the banana. To fully understand what is in a banana, you need to look at both views.

Even though both of the cuts you made produced sections of a banana, they showed you different things. Not surprisingly, then, they are given different names. When you cut the banana in half the long way, you made a **longitudinal** (lahn' juh tood' nul) **section**. The reason it's given that name should be pretty obvious – it involved cutting the long way. When you cut the small section out of the banana, you made a **cross section**, because you cut across the banana instead of down its length. If you just look at a banana without making any sections, you see only what is on the outside of the banana. If you look at only the longitudinal section, you see what is in the banana, but you get only one view. If you look at both the longitudinal section and the cross section, then you have a really good idea about what's on the inside of a banana.

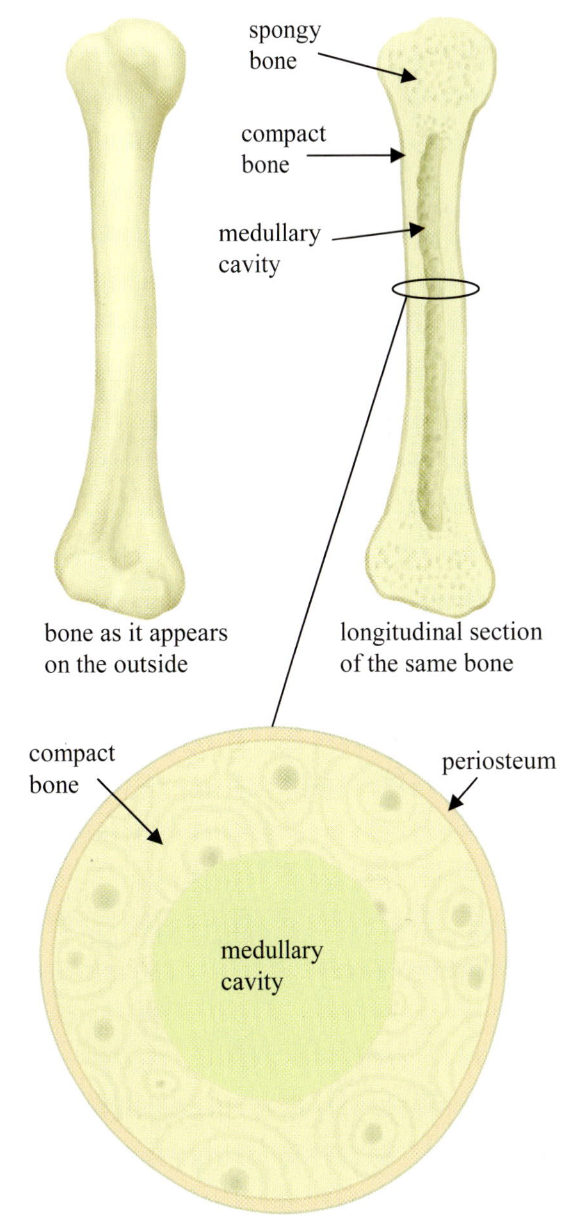

This drawing shows three different ways of viewing the same bone.

The same can be said for bones as well. There are a lot of things inside bones, and unless you look at both the longitudinal section and the cross section, you can't fully appreciate all that there is in a bone. Look at the three drawings on the right, for example. They are very similar to the drawings you made of the banana. The one on the top left shows a bone as it looks from the outside. The one on the top right shows you a longitudinal section of that same bone. Because of the longitudinal section, you can see that the inside of the bone looks very different from the outside. First, there are actually two different types of bone tissue. There is **spongy bone** and there is **compact bone**. As their names indicate, spongy bone has all sorts of spaces in it, just like a sponge. However compact bone is packed tightly together, and as a result, no spaces can be seen in it.

The other thing the longitudinal section shows is that in a long bone like this one, there is a "tunnel" running through the center of the bone. It's called the **medullary** (meh' duh lehr' ee) **cavity**, and it holds a substance called **bone marrow**. If you look at the cross section of the bone at the bottom of the illustration, you can see that the bone is actually covered by a thin layer called the **periosteum** (pehr ee os' tee uhm). You can also see that the medullary cavity is right in the center of the bone. Each view of the bone tells you something different, so each one is important in its own way.

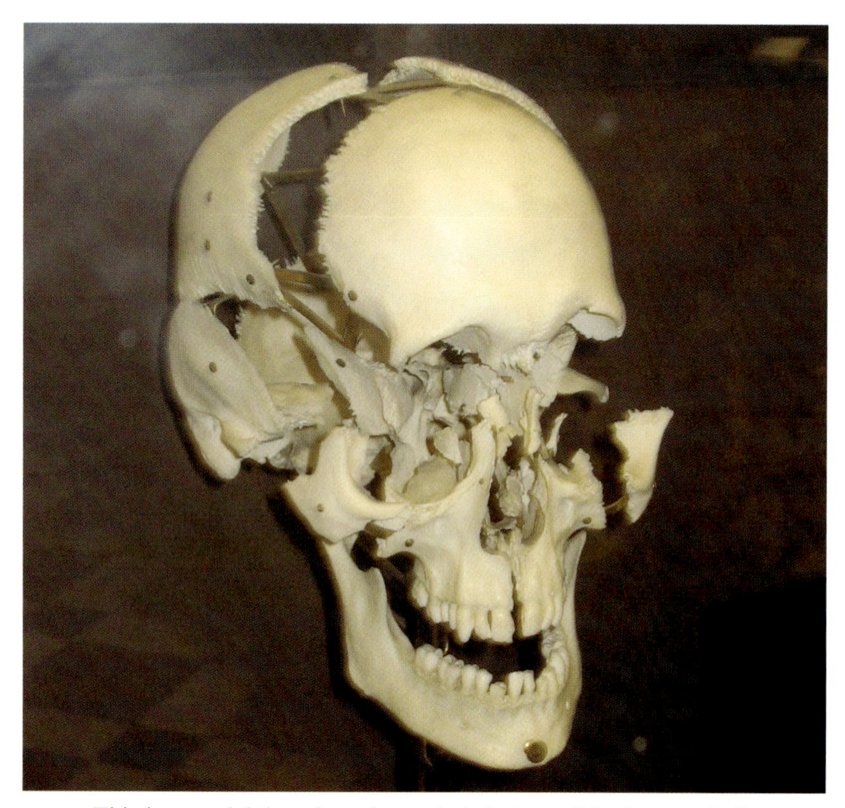

This is a model that gives the exploded view of the human skull.

Because the human body is so amazingly well designed, there are some parts of the body that require more than just a longitudinal section and a cross section in order to completely illustrate them. The skull is an example. It is actually made up of 28 separate bones. As a result, even a longitudinal section and a cross section of the skull isn't enough to show all of its detail. To fix this, some people like to use an **exploded view** of the skull. That's what is shown in the picture on the left. Can you see why it's called an exploded view? It looks like the skull exploded apart, doesn't it? This view not only allows you to see into the skull, but it also allows you to see how all the pieces fit together to make the skull. Did you have any idea that your skull was so complicated?

In the end, then, da Vinci was the first to realize that in order to illustrate all the amazing aspects of the human body, several different views had to be presented. Each view allows you to learn something different, so when you see them all, you have a much better understanding of the body part being illustrated.

LESSON REVIEW

Youngest students: Answer these questions:

1. When you draw something that is cut down its length, are you drawing a cross section or a longitudinal section?

2. What are the two types of bone tissue called?

Older students: In your notebook, label each of the sections you made as a longitudinal section or a cross section. Draw a longitudinal section of a long bone, labeling the spongy bone, compact bone, and medullary cavity.

Oldest students: Do what the older students are doing. In addition, draw a cross section of a long bone and label the periosteum, compact bone, and medullary cavity.

Lesson 84: Leonardo da Vinci and Muscles & Tendons

The joints of the skeleton *allow* the bones to move in relation to one another, but what actually *makes* the bones move? You should know the answer to that question. When you studied Galen's work, you learned a bit about skeletal muscles (Lesson 40). You learned that tendons attach the muscles to the bones of the skeleton, and when those muscles contract, they pull on one of the bones to which they are attached. Remember, muscles tend to work in antagonistic pairs so that when one of the muscles in the pair contracts, it flexes a joint, and when the other muscle in the pair contracts, it extends the joint.

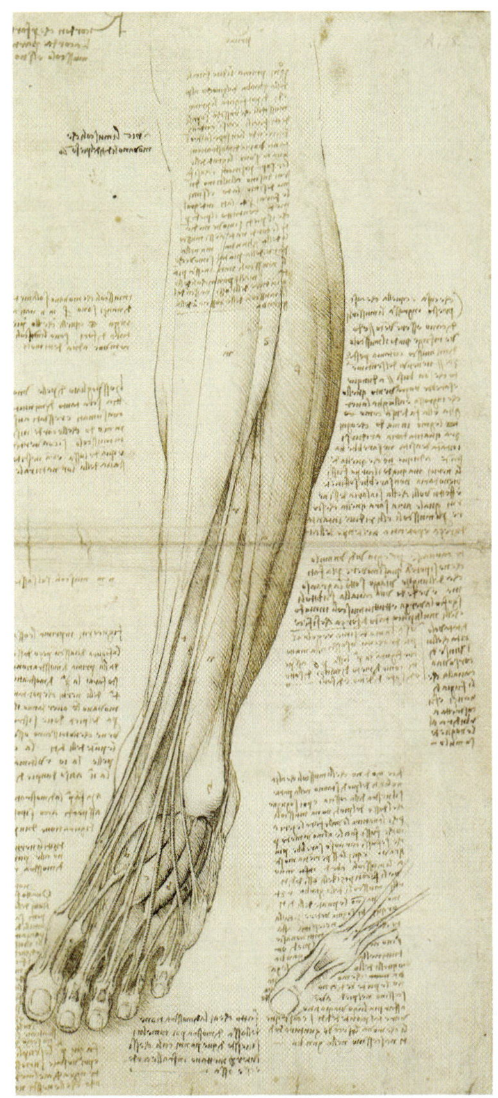

This is a portion of one of da Vinci's notebooks, showing some of the tendons in the foot.

Since Galen had already figured this out, Leonardo da Vinci was able to study the action of the muscles in detail. In particular, he was fascinated by a person's hands and feet, proclaiming, "The human foot is a masterpiece of engineering and a work of art." Why did he think that the foot was so amazing? Notice his drawing of the foot on the left. Do you see the tiny "strings" that travel from the leg to the toes of the foot? Can you guess what they are? Those are the *tendons* that attach many of the muscles that move the toes to the toes themselves. Think about what that means. Many of the muscles that move the toes aren't in the foot at all. Instead, they are in the legs.

Now why is that? Why are the muscles that control the toes in the legs? Think about it. Muscles are large. They have to be in order to have the strength they need to do their job. If those large muscles were in your feet, they would make your feet thick and bulky. However, in order to do their job, your feet have to be thin and nimble. Because of this, the big muscles are in your legs, which can be thick and bulky. Those muscles are attached to the toes by thin tendons that don't add much to the bulkiness of the toes. The same goes for your hands. Some of the muscles that control your fingers are actually in your arms, and they are attached to your fingers by thin tendons. See how this works by doing the following activity.

Controlling Your Fingers

What you will need:

- A sheet of construction paper
- A pencil or pen whose markings you can see on the construction paper
- Some string
- Scissors
- Tape

What you should do:

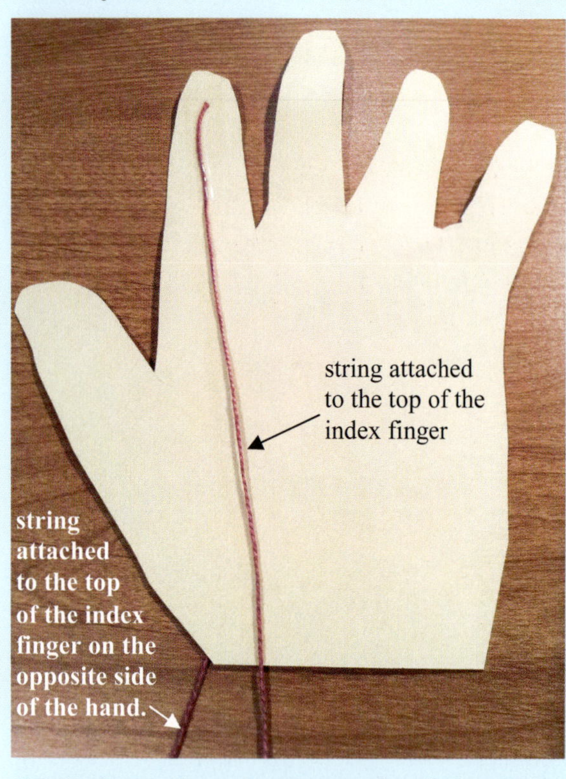

1. Lay your right hand flat on the piece of construction paper with your fingers spread apart.
2. Use the pencil or pen to draw an outline of your hand.
3. Use the scissors to cut the outline of your hand so that you have a construction-paper version of your hand.
4. Cut the string into two strings that are about twice as long as your construction-paper hand.
5. Tape one end of one of the strings to the top of the construction-paper hand's index finger.
6. Tape one end of the other string to the top of the construction-paper hand's index finger on the *opposite* side of the hand. The construction-paper hand should now look like the picture on the left.
7. Hold the construction-paper hand down on a table or desk by pressing down on it. Don't press down close to the strings; they need to be free.
8. Continuing to hold the construction-paper hand down, pull on the top string. Note what happens to the construction-paper finger.
9. Release the top string and pull down on the bottom string. Once again, note what happens to the construction-paper finger.
10. Play with the strings for a while so you see how they control the finger.
11. Clean up your mess.

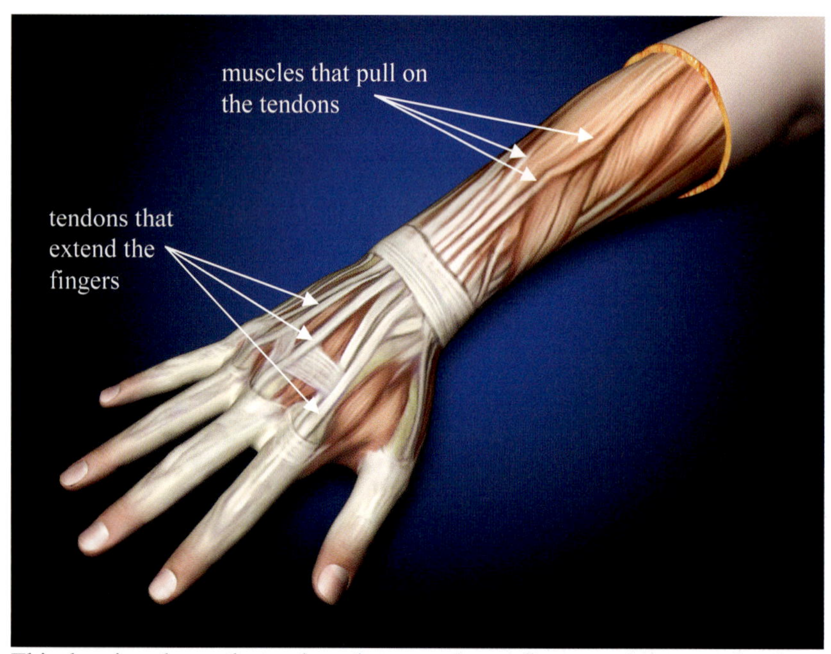

This drawing shows the tendons that extend your fingers and the muscles that pull on them.

What happened in the activity probably didn't surprise you, but it is actually a good illustration of how we flex and extend our fingers. Do you remember what "flex" and "extend" mean? When you straighten your finger out, you are extending it. When you pull it in so that it folds over your palm, you are flexing it. When you do that, you aren't really using any muscles that are in your hand. Instead, you are using the muscles that are in your forearm. However, those muscles are attached to your fingers with long, thin tendons, much like the strings in your activity. When one of the muscles contracts, it pulls on the tendon, which pulls on your finger.

When you extend your finger, then, a muscle in your forearm contracts. It pulls a tendon that runs down to the finger, and just like what happened when you pulled the top string in your activity, the finger extends. When you flex your finger, the other muscle in the antagonistic pair contracts. That pulls on another tendon, which pulls your finger in the other direction, flexing it. In other words, when it comes to these kinds of motion, your fingers are "cable operated!" This allows your fingers to be thin and nimble but at the same time fairly strong.

Now please don't think there aren't *any* muscles in your hand. There most certainly are! However, they are smaller muscles that control the finer movements of the hand. In the drawing below, for example, the red structures are muscles, while the white structures are tendons. These muscles allow you to spread your fingers apart and pull them back together. The big muscle at the base of the thumb allows you to fold your entire thumb across your palm.

These muscles are called the **intrinsic** (in trin' sik) **muscles** of your hand, because they are actually in your hand. The muscles that are in your forearm and pull on the tendons that are attached to your fingers are called the **extrinsic** (eks trin' sik) **muscles** of the hand. The extrinsic muscles allow you to grab something with strength, while the intrinsic muscles give you fine control so that you can grab it precisely in the way you want to grab it. Your foot is operated in exactly the same way. Extrinsic muscles in your leg pull on tendons that allow your toes to move with strength, while intrinsic muscles in your foot give you some fine control over your toes.

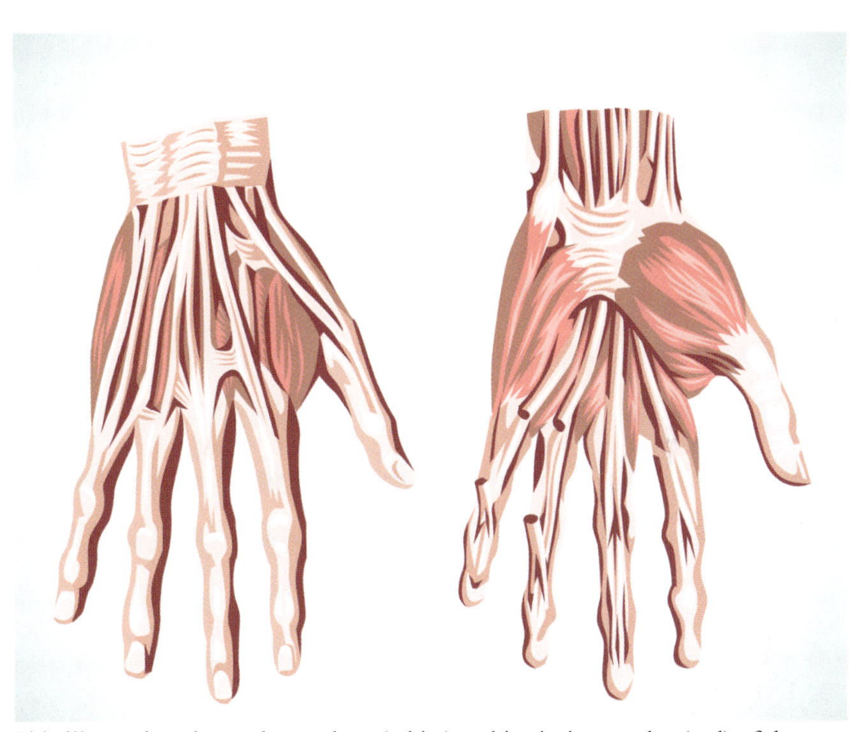

This illustration shows the tendons (white) and intrinsic muscles (red) of the hands. The drawing on the left shows the hand palm down, while the drawing on the right shows it palm up with some of the tendons cut.

LESSON REVIEW

Youngest students: Answer these questions:

1. Where are the intrinsic muscles of the hand found?

2. Where are the muscles that pull on the long tendons that are attached to your fingers?

Older students: In your notebook, explain what you did in your activity and why it is a good model for how you flex and extend your fingers. Also, explain the difference between the extrinsic and intrinsic muscles of the hand, including what kind of grip they provide.

Oldest students: Do what the older students are doing. In addition, look at the drawing above. Which side (left or right) shows the tendons that flex your fingers? Write your answer and your reasoning in your notebook and check to see if it is right. Correct it if it is wrong.

Lesson 85: Leonardo da Vinci and the Spinal Cord

As you learned back in Lesson 40, Galen had figured out how muscles work in groups (often pairs) of antagonists. In the previous lesson, you learned about how some of the muscles that control the fingers and toes aren't all in the hands and feet. Instead, the extrinsic muscles of the hands and feet are in the forearms and legs, and they operate the fingers and toes through long, cable-like tendons. But how are the muscles themselves controlled? When you decide you need to move your arm, what causes your muscles to contract in just the right way so that your arm actually moves?

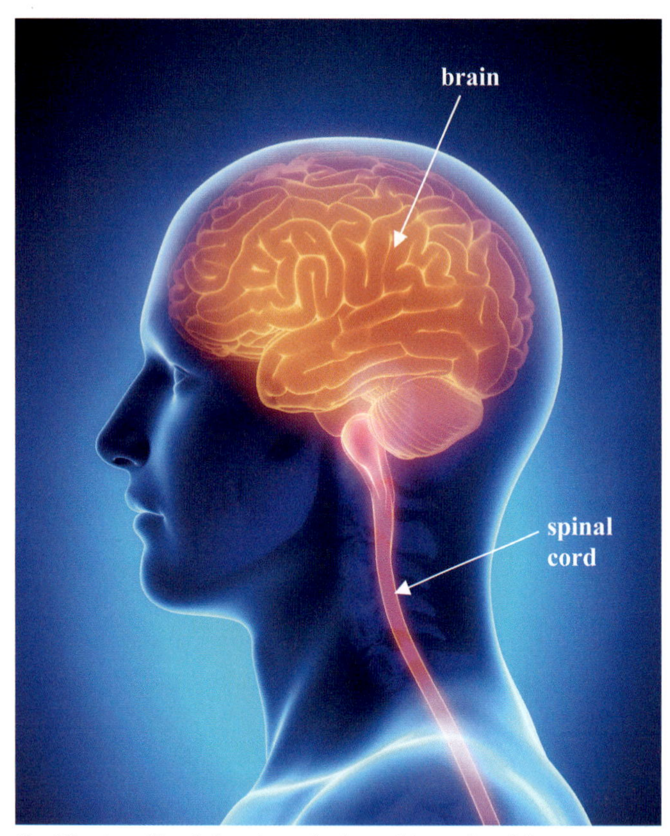

Da Vinci realized that the spinal cord is made of the same basic material that the brain is made of and can be thought of as an extension of the brain.

Galen had a partial answer to that. Remember, he showed that when he cut a specific nerve in a pig's body, the pig would no longer be able to squeal, even though it was clearly trying to. He traced that nerve back to the brain, and that made him realize that the brain controls how the pig makes noises. So Galen realized that nerves control muscles, and the brain controls the nerves. For most muscles, however, it's not quite that simple. Leonardo da Vinci showed why.

His study of dead bodies made him think that the spinal cord was really an extension of the brain. There's more to it than that, but it's not a terrible way to think of the spinal cord. Da Vinci specifically said that the spinal cord is made of the same basic material as the brain, and as a result, all the nerves of the body originate in the spinal cord. While that's not entirely true, it's close. There are some nerves that come directly from the brain. We call them **cranial nerves**. Most of the body's nerves, however, come from the spinal cord, and we call them **spinal nerves**.

Because da Vinci thought that all nerves started in the spinal cord, he thought that the spinal cord was very important to the movement of the muscles. Even though he was wrong about *all* the nerves starting in the spinal cord, he was right about the spinal cord being important to the movement of most muscles in the body. Perform the following experiment to get an idea of how the spinal cord helps to control the movement of muscles.

A Model of the Spinal Cord

What you will need:

- A working, hand-held flashlight
- Aluminum foil
- Scissors
- Tape

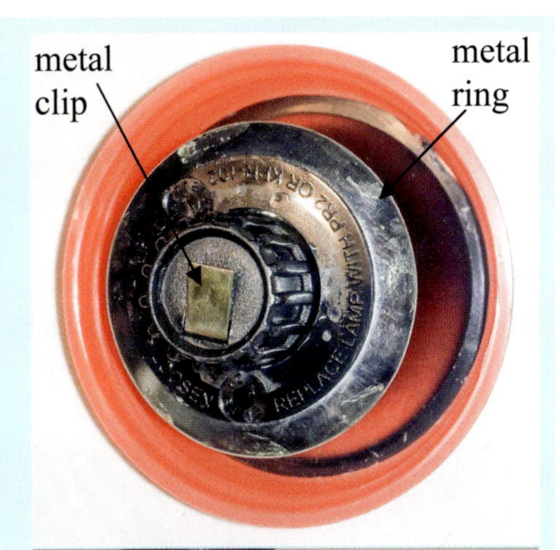

What you should do:

1. Make sure the flashlight works by turning it on and looking at the light it makes.
2. Unscrew the top of the flashlight and take it off.
3. Turn the top of the flashlight upside down so you see the parts that are normally inside the flashlight. You should see at least two pieces of metal – one that looks like a clip, and another that looks like a ring.
4. Use the scissors to cut two strips of aluminum foil. They should each be about 20 centimeters (about 8 inches) long and about 2 centimeters (about 1 inch) wide.
5. Fold the end of one strip on itself a few times to make a "bump," and push that bump against the metal clip.
6. Use the tape to hold the foil bump tightly against the metal clip. The foil needs to be in firm contact with the metal clip.
7. Tape one end of the other foil strip to the metal ring. Once again, the foil needs to be in firm contact with the metal ring. In the end, your flashlight top should look like the picture on the bottom right.
8. Remove one of the batteries from the flashlight.
9. Hold the battery in one hand, and hold the flashlight top in the other hand.
10. Bend the aluminum foil strips a bit so that they both point towards the battery. Be careful not to tear them or pull them off the flashlight top. Just bend them.
11. A battery has two sides. One side has a little bump in the center, and the other side is flat. Use the fingers on the hand holding the battery to hold one aluminum strip against one side of the battery and the other aluminum strip against the other side of the battery. It doesn't matter which strip is touching which end. Pinch the strips hard against the end of the battery. When you do that, you should see the top of the flashlight start shining. It probably won't shine as brightly as when it is in the flashlight, but it should shine.
12. Relax your hold on one of the foil strips so that it no longer touches the side of the battery. What happens to the light?
13. Tighten your grip so that both foils touch the sides of the battery like before. What happens to the light?
14. Loosen your grip on the other strip of foil so it no longer touches the side of the battery. What happens to the light?
15. Reassemble the flashlight and clean up your mess.

What does a flashlight and aluminum foil have to do with the spinal cord? Well, a flashlight works because electricity travels through the light bulb, causing it to light up. So in order to get the bulb in a flashlight to light up, electricity has to flow from the battery to the bulb, but it also has to flow from the bulb back to the battery. It turns out that in order for your muscles to move the way you want them to move, a similar thing must happen between your brain and your muscles, and the spinal cord makes this happen for most muscles in your body.

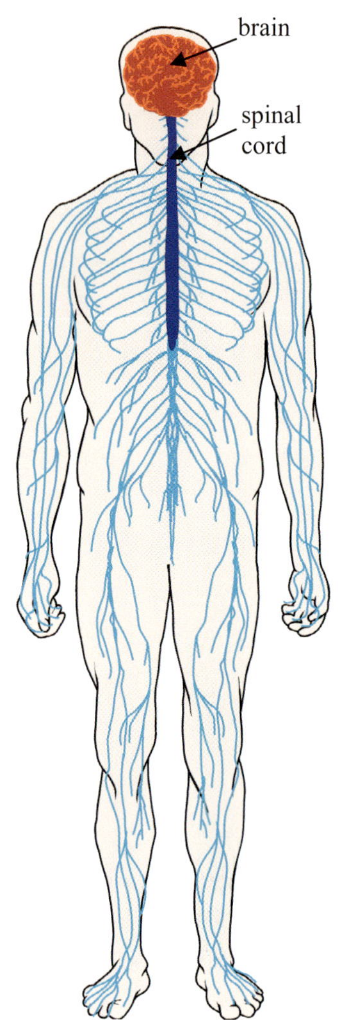

Most of the nerves in your body (the light blue lines in the drawing) send and receive signals through your spinal cord.

Your brain controls the muscles that move your skeleton, but since most nerves start in your spinal cord, the brain has to send its instructions through the spinal cord to get to those muscles. Guess how it sends its instructions? With electricity! Electrical signals flow from the brain, through the spinal cord, to the nerves that come out of the spinal cord. Those nerves then send those electrical signals to the muscle, causing the muscle to contract.

That's not the end of the story, however. In order for your brain to control the muscles properly, it needs to know how much they are contracted. As a result, nerves carry electrical signals from the muscles, through the spinal cord, and back to the brain. The brain uses those electrical signals to figure out how much the muscles you are moving are contracted so that it knows exactly what it needs them to do to produce the motion you want to make. Like the electricity in your experiment, then, electrical signals flow both ways in the spinal cord – some flow from the brain to the muscles, and some flow from the muscles to the brain. The key is that they all flow through the spinal cord, at least for most of the muscles in your body.

So in the end, da Vinci was mostly right. Your spinal cord isn't just an extension of your brain, but it is very important in producing movement, because your brain needs to send its instructions down the spinal cord to the nerves that control the muscles. In addition, it needs to get electrical signals from your muscles, and those come to it through the spinal cord. Without your spinal cord, then, you wouldn't be able to move most of the muscles in your body! Now remember, he was wrong about the spinal cord being the source of all nerves. In fact, there are some nerves that carry information directly from the brain and directly back to the brain. As a result, the spinal cord is not important in the movement of those muscles. However, most of the muscles in your body depend on the spinal cord for their movement.

LESSON REVIEW

Youngest students: Answer these questions:

1. (True or False?) The spinal cord is made of the same basic material as the brain.

2. What does the spinal cord do in order to allow the brain to control muscles in the body?

Older students: In your notebook, explain what you did in your experiment and why it is a good model for how the spinal cord helps the brain control muscles in the body. Also, explain why da Vinci concluded that the spinal cord is just an extension of the brain.

Oldest students: Do what the older students are doing. In addition, consider someone who is in a terrible accident and has his spinal cord cut apart halfway down. If the person survives, describe in your notebook how his ability to move would be affected. Check to see if you are right, and correct your answer if you are not.

Lesson 86: Leonardo da Vinci and the Heart

During Leonardo da Vinci's day, natural philosophers knew that every animal and person had an organ that was called the **heart**, but they didn't really know what it did. Many thought that it "heated" the blood with a "vital spirit" that was then carried through the body to keep a person alive. Da Vinci, however, had read of another idea from a Muslim natural philosopher named Avicenna (ah vuh see'nuh), who thought the heart was really a muscle, and it was related to a person's pulse. Da Vinci's studies led him to agree with Avicenna. Do you remember studying about your pulse in Lesson 39? Perform the following experiment to see how the heart and pulse are related.

The Heart and Pulse

What you will need:

- A cardboard tube from the center of a roll of paper towels
- A reasonably large funnel
- Wax paper or plastic wrap
- Tape
- Scissors
- Someone to help you
- A stopwatch or a watch with a second hand

What you should do:

1. Cut out a square of wax paper that is just slightly bigger than the wide opening of the funnel.
2. Cover the wide opening of the funnel with the wax paper, and use the tape to hold it tightly to the funnel. The paper needs to be stretched tightly over the funnel. It should be tight enough so that it makes a sound like a drum when you tap the paper.
3. Put the funnel in the tube so that its small opening is inside the tube, and the tube is pushed down as far as possible on the funnel.
4. Use the tape to hold the tube to the funnel. Make sure the tape wraps around the tube and funnel so that it forms a complete seal all around the tube.
5. Have your helper stand in front of you, and use your hand to find your helper's heartbeat in his or her chest.
6. Once you have found that spot, put the funnel there and press firmly but gently so the wax paper is right where you felt the heartbeat.
7. Put your ear to the open end of the tube. You should clearly hear your helper's heart beating.
8. Have your helper time you as you count the number of heartbeats you hear for 30 seconds.
9. Find your helper's pulse the way you did back in Lesson 39.
10. Once again, have your helper time you as you count the number of pulse beats you feel for 30 seconds. Compare that number to the number of heartbeats you heard.
11. Have your helper do 20 jumping jacks as fast as he or she can.
12. Repeat steps 5-10.
13. Compare the numbers you got this time to each other and to the numbers you got before your helper did the jumping jacks.
14. Clean up your mess.

Before I explain the results of the experiment to you, let me first say that in this experiment, you made a very simple **stethoscope** (steh' thuh skohp). When a doctor wants to hear your heart

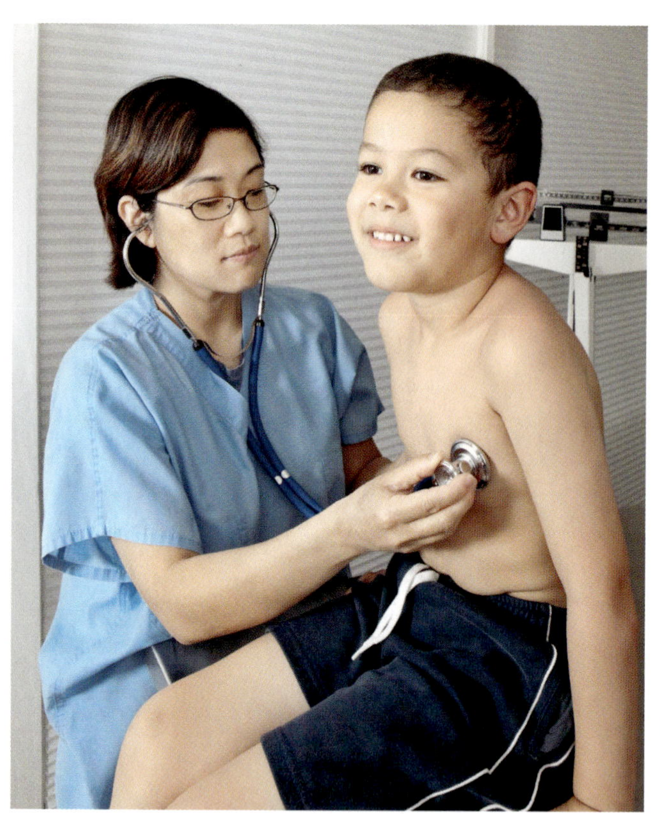
The doctor is using her stethoscope to listen to the boy's heart.

clearly, she uses a stethoscope. One end of the stethoscope goes where your heartbeat is strongest, and the other goes into the doctor's ears. As sound waves from your heartbeat hit one end of the stethoscope, a thin sheet of material makes the sound of the heartbeat louder, and that sound then travels through the tubes of the stethoscope and into the doctor's ears. Your stethoscope had the same basic elements: The wax paper was the thin sheet of material. The funnel held that material in place and also funneled the sounds to the cardboard tube, which carried the sounds to your ears. Because the stethoscope made it easy to hear your helper's heartbeats, you were able to count them easily.

Now, what did you find out when you compared the number of heartbeats you heard to the pulse that you counted? They should have been the same. Because you might not have counted every beat exactly right, however, they might have been off by a little. Now, what happened to both numbers after your helper did the jumping jacks? They both went up, didn't they? In fact, they should have gone up by the same amount. That's because the heartbeat and pulse are really the result of the same thing – the action of the heart as it pumps blood through the blood vessels of the body.

Da Vinci's experiments and his studies of dead people and dead animals led him to the correct conclusion that the heart is made of muscle. When you heard the heartbeat, what you actually heard were the sounds related to the heart pumping blood because of the way the muscles in the heart were relaxing and contracting. Da Vinci didn't figure it all out, but he realized that the heart must be pushing blood through the blood vessels in some way, because as you noticed in your experiment, the heartbeat and pulse were related. When one went up, the other went up. He reasoned that when the heart muscle contracted, blood was pumped through the blood vessels, and you could feel such pumping all the way down in your wrist.

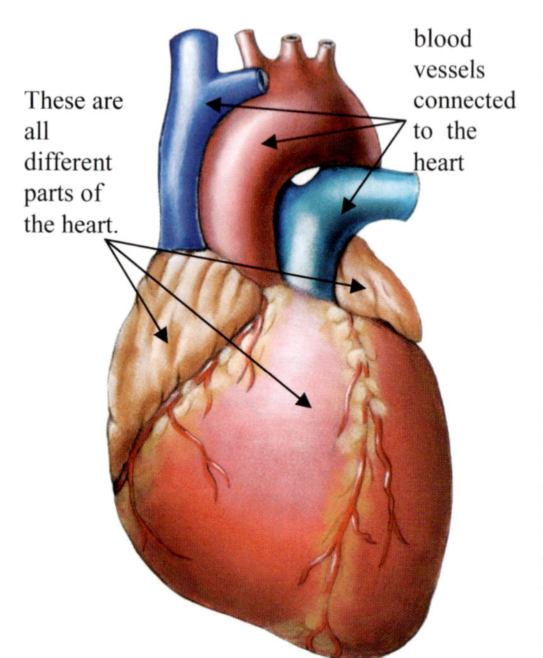

In this drawing, the heart and some of the blood vessels connected to it are shown.

Now, remember that da Vinci was not the first person to realize all this. As I mentioned at the beginning of this lesson, a Muslim natural philosopher named Avicenna was the first to think that the heartbeat was related to the pulse. Most historians think that da Vinci had read Avicenna's work and wanted to confirm what Avicenna suggested. So while da Vinci was important in helping us understand what the heart does, we can't overlook the person whose work led him to study the heart this way to begin with!

At the same time, however, da Vinci figured out more about the heart than what Avicenna had suggested. He showed that when you look inside of the heart, it is composed of four separate chambers. Different blood vessels flow into different chambers, which made da Vinci realize that there was a specific pattern in which blood flowed through the heart. He never figured out that pattern, at least as far as we can tell, but the fact that he recognized there was a pattern was a step in the right direction.

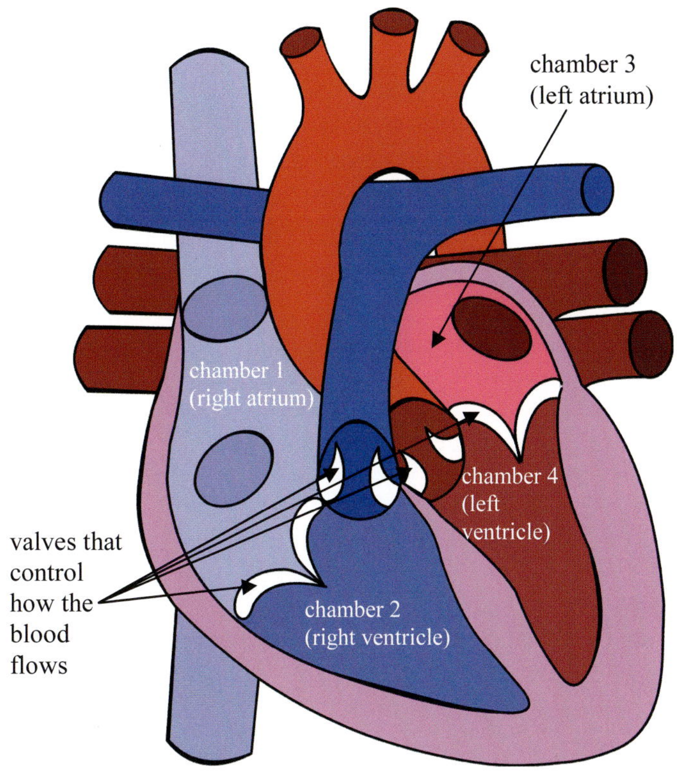

The interior of the heart is split into four chambers, and those chambers, as well as the openings of the blood vessels into those chambers, have valves that help control how blood flows.

He also noticed that there were valves in the heart and the blood vessels around the heart, and he realized that those valves helped control the way blood flowed through the heart. When the blood rushed in the right direction towards one of these valves, the valve would open. However, if blood started flowing the wrong way, he realized that the valves would close, stopping the incorrect motion. His previous studies of how water flowed helped him in this analysis. Interestingly enough, the sounds you head when you listened to your helper's heartbeat were actually the sounds of certain valves shutting when the heart muscles pumped in a certain way.

In the end, then, da Vinci's close study of the heart and how it works revealed that the heart did not add a "vital spirit" to the blood. They showed that it was a special muscle that controlled blood flow.

LESSON REVIEW

Youngest students: Answer these questions:

1. (True or False?) The heart is made of muscle.

2. What is the purpose of the valves in the heart?

Older students: In your notebook, explain how the heart is made of muscle and how it relates to the pulse. Also, explain what the valves inside the heart do.

Oldest students: Do what the older students are doing. In addition, explain what you were actually hearing when you listened to your helper's heartbeat.

Lesson 87: Leonardo da Vinci and Teeth

You would think that it would be easy to study a person's teeth, wouldn't you? After all, they are right there in a person's mouth. Just have someone open wide, and you can examine his teeth. Well, believe it or not, in Leonardo da Vinci's time, no one that we know of had ever studied teeth thoroughly enough to notice exactly how many and what kinds of teeth a typical person has. Part of the problem might have been that children have fewer teeth than adults. As a result, the number of teeth that you see in a person's mouth depends on his or her age.

Probably because da Vinci studied the bodies of dead adults, he realized that there was a specific pattern to a person's teeth. Nowadays, we call it the human **dental formula**, and da Vinci was the first to describe it accurately. Try to get a feel for your own dental formula and the dental formula of an adult by performing the following experiment.

Dental Impressions

What you will need:
- A block of hard cheese that is easy to bite into (A block of cheddar cheese is ideal.)
- A sharp knife
- An adult to help you

What you should do:
1. Have the adult help you cut a slice of cheese that is about ¾ of a centimeter (about ⅓ of an inch) thick and as wide as possible but still able to fit into your mouth. It should be long enough to reach the back of your mouth. In other words, you want a slice of cheese that you can push all of your teeth into in one bite.
2. Swallow any saliva that is in your mouth, then open wide and push the cheese in your mouth so that it goes all the way to the back. Be careful while you do this. You don't want to choke.
3. When the cheese is fully in your mouth, bite down on it slowly, sinking your teeth into the cheese. Don't bite all the way through the cheese. Just bite down so that each tooth has sunk into the cheese.
4. Open wide again and pull the cheese out of your mouth.
5. Use the knife to put a small mark on the top of the cheese so you know which side of the cheese was up when it was in your mouth.
6. Look at both sides of the cheese. You should see tooth marks. Notice how different they are. How many different kinds of tooth marks do you see?
7. Now repeat steps 1 – 6 with the adult who is helping you. In the end, you should have two pieces of cheese – one with your teeth marks in it, and one with the adult's teeth marks in it.
8. What differences do you notice between the adult's teeth marks and your teeth marks?
9. Keep the two pieces of cheese that have the teeth marks in them, but clean up the rest of your mess.

In the experiment, you made a partial **dental impression**, which showed how your teeth bite into food. What did you see in the partial dental impression you made with your teeth? You probably noticed that at the front of the slice of cheese, the marks that your teeth made were thin and slightly curved. However, near the back of the slice of cheese, the marks your teeth made were much wider. You might even have noticed that between the thin marks and the wide marks, there were slightly diagonal marks that were a bit wider than the thin marks but much thinner than the wide marks. These different marks demonstrate the different kinds of teeth you have.

What did you notice when you compared the adult's partial dental impression to yours? Most likely, you saw that the adult's cheese had more teeth marks in it than yours did. This might not have been the case, however, depending on how old you are. As I told you before, children have fewer teeth than adults. That's because the set of teeth with which you are born are not permanent. You probably refer to them as "baby teeth," but their scientific name is **deciduous** (dih sih' juh wus) **teeth.** Do you recognize the word "deciduous?" You should. You learned it in Lesson 72 when you were studying trees. Deciduous trees lose their leaves every winter. In other words, their leaves fall off. What happens to your baby teeth? They eventually fall out. That's why they are called deciduous teeth!

This young lady lost one of her deciduous teeth, which she is holding between her fingers. It will eventually be replaced by a permanent tooth.

When your deciduous teeth fall out, they are eventually replaced with **permanent teeth**. As long as you take good care of your permanent teeth and don't get hit too hard in the mouth, your permanent teeth will not fall out. That's why we call them permanent. Once they grow in your mouth, they should be there for the rest of your life.

Why are you born with one set of teeth only to have them replaced by another set of teeth? Because when you are born, your jaw just isn't big enough to hold all your teeth in their full-sized form. As a result, God gives you a "starter set" when you are born, and as your jaw grows, that set is slowly replaced by your full set. Most children get their permanent teeth by age 12 or so. That's why it's possible that you saw the same number of teeth marks in your partial dental impression as you did in the adult's. It's possible that you are old enough to have all your permanent teeth.

Now let's look more closely at the partial dental impressions that you made. Start by looking at the marks made by your front teeth on the top of the cheese. Those front teeth are called **incisors** (in sye' zurz). They have a particular job to do when it comes to eating. Can you tell what their job is by looking at the marks they made in the cheese? Their job is to cut the food you are eating. Now look at the same marks on the bottom of the cheese. They were made by your bottom front teeth. Unless you are missing one of your front teeth like the young lady in the picture above, you should see the same number of front teeth marks on the bottom of the cheese as you do on the top. That's because the bottom of your jaw holds the same kinds of teeth as the top of your jaw, and they are positioned so they can work opposite each other.

Look at the top of your partial dental impression again. You should see four thin, slightly curved marks in the front (two on the left and two on the right) made by your incisors. Behind them, you should see two marks (one on each side) that are still pretty thin, but they should not look exactly like the four marks in front. These are made by your two **canine** (kay' nine) teeth. They have a different job from your incisors. Their job is to hold on to a piece of food that you need to tear apart.

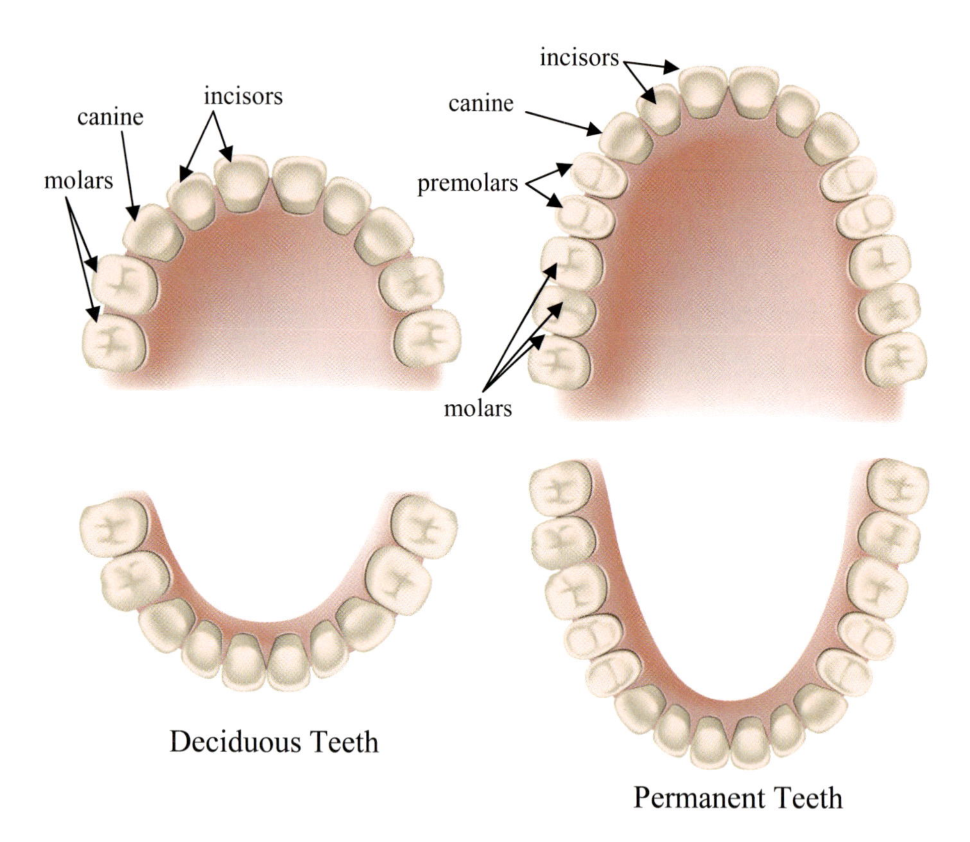

This illustration shows a child's dental formula on the left and an adult's dental formula on the right.

Behind the incisor and canine marks, you should see wide tooth marks. This is probably where you see the most difference between your partial dental impression and the adult's. An adult has two kinds of wide teeth: **premolars** (pree moh' lurz) and **molars** (moh' lurz). They are both used for grinding food into tiny bits so that it is easy to swallow, which is why they are so wide. An adult has two premolars on each side of his or her mouth and three molars. Most children start out with only two molars on each side of the mouth, and then as the permanent teeth come in, the premolars and the extra molars are added.

In the end, then, scientists say that the dental formula (the number and kinds of teeth) of a child is two incisors, one canine, and two molars. Now please understand what this means. It is listing the teeth in one half of the mouth (upper or lower) on one side (left or right). So if you have two incisors, one canine, and two molars in one side of one half of your mouth, you have eight incisors, four canines, and eight molars in your entire mouth, for a total of twenty teeth. The dental formula of an adult is two incisors, one canine, two premolars and three molars. Can you tell me how many teeth an adult has? Most adults have 16 teeth on top and 16 teeth on the bottom.

LESSON REVIEW

Youngest students: Answer these questions:

1. What is a dental formula?

2. Who has more teeth, a child or an adult?

Older students: In your notebook, draw a diagram like you see above and label each kind of tooth, as in the diagram above. Explain what each type of tooth does, and list the number of teeth in a child and in an adult.

Oldest students: Do what the older students are doing. In addition, explain what deciduous teeth are and why they are called by that name.

Lesson 88: Leonardo da Vinci's "Inventions"

In addition to all of his other accomplishments, Leonardo da Vinci was an inventor of sorts. In his notebooks, you can find drawings of all kinds of incredible inventions, including flying machines, carts that move without someone pushing or pulling on them, machine guns, and many other things. There's just one problem. As far as we know, da Vinci didn't try to build most of them. He *thought* they would work as drawn, but he didn't actually build most of the inventions he drew. That's why I say he was an inventor "of sorts." He designed a lot of inventions. It's just not clear how many of them he actually made.

Now even though we don't think da Vinci built many of his inventions, we know that some of them did, indeed, work. For example, da Vinci used a lot of **ball bearings** in his inventions. While ball bearings had probably been used in the past, da Vinci's drawings were the first real indication of what they could do. What are ball bearings? Perform the following experiment to find out.

Ball Bearings

What you will need:
- Two flat pieces of wood, wooden boards, or rectangular wooden blocks (They don't have to be very big at all.)
- A few marbles that are all the same size

What you should do:
1. Put one piece of wood on the floor.
2. Put the other piece of wood on top of the first piece of wood.
3. Holding the bottom piece of wood with one hand so that it doesn't move, push on the top piece of wood to get it to move. Try to use as little force as possible to get it moving. Experiment with this a few times, getting used to how much you have to push in order to get the top piece of wood sliding along the bottom piece of wood.
4. Remove the top piece of wood.
5. Put a single marble on the piece of wood that is still on the floor, and before it can roll off, put the other piece of wood on top of it to hold it in place. You now have a single marble sandwiched between two pieces of wood.
6. Push a few more marbles in between the two pieces of wood so that they are spread somewhat evenly between the pieces of wood.
7. Repeat step 3. Do you notice a difference in terms of how hard it is to move the top board?
8. Play with the setup a little more, getting an idea of how the marbles allow the top board to be moved.
9. Clean up your mess.

What did you find out in your experiment? You should have discovered that moving the top board was *a lot* easier with the marbles than without them. In fact, you might have had a hard time getting the top board to sit still when the marbles were underneath it. Were you able to figure out how the marbles were able to do this? It's because they rolled when the top board was moved. Their rolling motion made it very easy to move the top board. That's what a ball bearing is. It is a ball that sits between two surfaces that have to move against each other. Ball bearings are used because it makes that kind of motion much, much easier.

Now, of course, the real question is, "*Why* do ball bearings make it so much easier for two surfaces to move against each other?" The answer has to do with something called **friction** (frik' shun). When two surfaces come into contact, they tend to "hold on" to each other. As a result, there is some resistance when you try to get them moving, and that's friction – the resistance two surfaces experience when they are moving against one another.

Friction applies to any two surfaces that move against one another. For example, put your hands together so that your palms are touching. Now rub your palms back and forth against each other as fast as you can for at least ten seconds. Do you feel anything? You should. Your palms should have gotten warm. Why did they get warm? Because of friction. Your palms were "holding on" to each other a bit, and in order to move them, you had to overcome the force with which they were holding on to each other. That took energy. Do you know that heat is a form of energy? The heat that you felt in your palms was the result of friction converting some of the energy you were using to move your hands into heat.

When you want to move something, then, you have to overcome the friction that is holding it in place. Even when you do that, however, you still aren't done fighting friction. It still continues after the surfaces are moving. Think about what you just did with your hands. The heat built up as time went on, didn't it? That's because the friction was still there, even though your palms were moving against one another. In general, however, I can say this: Once you get something moving, there is less friction involved than before you got it moving. In other words, you have to use more force to overcome friction to get something moving. Once you have done that, there is still friction, but it is less than the friction that was holding it in place.

The ball bearings in this engine allow the two brown discs to move easily against one another by reducing friction.

What does all this have to do with ball bearings? They make it easier to move surfaces against one another. They don't *get rid of* friction. That's not possible. They just reduce its effect on moving surfaces. How do they do that? Only a small part of the ball bearing touches each surface. There is friction between the ball bearing and the surfaces, but when the surfaces begin to move, that friction causes the ball bearing to roll, which encourages the surfaces to continue to move. So ball bearings don't get rid of friction; they *use* the friction to help the motion.

Because friction exists between pretty much any two surfaces that are trying to move against one another, most machines try to reduce friction so that they can operate more efficiently. Your automobile, for example, has lots of ball bearings in it. So do many other machines. In fact, you can find biological versions of ball bearings in living organisms as well. For example, some of the smallest forms of life on the planet are called **bacteria** (bak tear' ee uh). They are so small that you have to magnify them about 1,000 times just to be able to see them. To see them in any detail, you have to magnify them even more than that. You can see a drawing of a bacterium (singular of "bacteria") after it has been magnified several thousand times in the upper part of the illustration on the right.

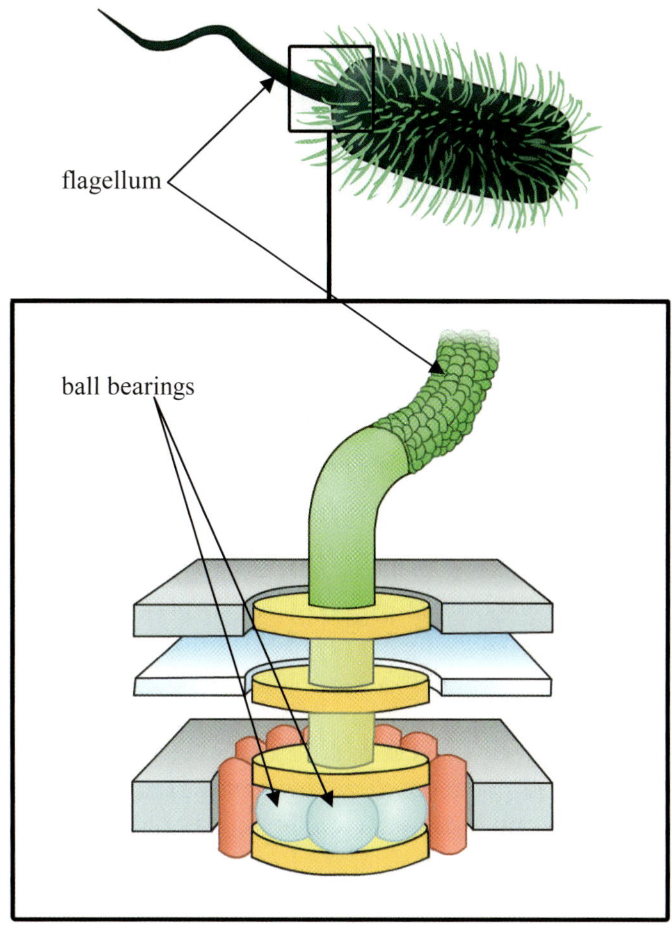

The top of this illustration shows a bacterium, magnified much larger than normal. The bottom shows just the flagellum where it attaches to the bacterium. That part is magnified even more!

Do you see the bacterium's "tail?" It's called a called **flagellum** (fluh jel' uhm). The bacterium doesn't wag its flagellum. Instead, the bacterium spins the flagellum like the propeller on a plane. This moves the bacterium around in the water.

The rest of the illustration is a drawing of the flagellum, right where it attaches to the rest of the bacterium. It's very complicated, so don't get lost in the details. I just want you to notice what's pointed out in the figure. Those are natural ball bearings! They aren't made of metal, of course. The bacterium makes them out of chemicals called proteins, and they allow the two disks at the bottom of the flagellum to spin much more easily than they would if the ball bearings weren't there at all. In other words, even though scientists and engineers (with the help of Leonardo da Vinci) figured out how to use ball bearings, God had already used them in designing the bacterium!

LESSON REVIEW

Youngest students: Answer these questions:

1. What is friction?

2. What is a ball bearing?

Older students: In your notebook, draw a diagram like you see above. It doesn't have to contain nearly as much detail. All you have to do is show the two disks and the ball bearings in between. Explain what the ball bearings do and how they work.

Oldest students: Do what the older students are doing. In addition, try to explain why there is less friction when two surfaces are already moving against one another than when the two surfaces are at rest. Check the answers to see if your explanation is correct, and change it if it is not.

Lesson 89: More About Friction

Since Leonardo da Vinci worked with ball bearings, he obviously knew about friction. He studied friction quite a bit, because he really wanted to know about this thing that resists motion. One way he studied friction was to see how it changed from situation to situation. For example, he did some very simple experiments that allowed him to learn some very important things about friction. To see what he learned, perform the following experiment.

Friction

What you will need:

- A smooth countertop that has an edge off of which you can hang something
- An empty CD case (A small box will work as well.)
- At least 40 pennies (Any coin will work, as long as it is all the same kind of coin.)
- String
- Scissors
- Tape
- A Ziploc bag

What you should do:

1. Put the empty CD case on the counter, at least 60 centimeters (24 inches) from the edge.
2. Cut the string so that it is long enough to reach from the CD case to the edge of the counter and hang several centimeters (a few inches) off the edge.

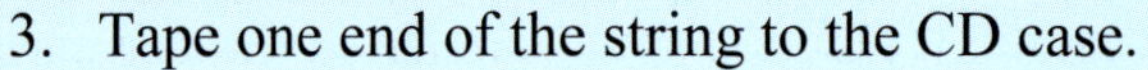

3. Tape one end of the string to the CD case.
4. Tape the other end of the string to the top of the Ziploc bag, near the center.
5. You should now have a CD case sitting on the countertop with a string attached to it. The other end of the string should be attached to a Ziploc bag that is hanging off the edge of the countertop, as shown in the picture on the left.
6. Hold the CD case so that it doesn't move, and put a penny in the Ziploc bag.
7. If the penny causes the bag to move around, steady it with your other hand. Once the system is completely steady, release the CD case and see if it moves.
8. Repeat steps 6 and 7 until the CD case moves when you release it.
9. Put ten pennies on the CD case. Distribute them evenly.
10. Reset the system so that the CD case is back where it was in step 6 and the Ziploc bag with the pennies in it is hanging over the edge of the countertop. Does the CD case move? It shouldn't.
11. You are going to start adding pennies to the bag until the CD case moves again. Before you do that, however, guess how many pennies it will take to get the CD case moving again. Say the number out loud.
12. Repeat steps 6 and 7 until the CD case moves again. Count the pennies you are adding as you repeat those steps.
13. How many pennies did it take to get the CD case moving again?
14. Add ten more pennies to the CD case. Once again, distribute them evenly on the case.

15. Reset the system so that the CD case is back where it was in step 6 and the Ziploc bag with the pennies in it is hanging over the edge of the countertop. Does the CD case move? It shouldn't.
16. Repeat steps 6 and 7 until the CD case moves again. Count the pennies you are adding as you repeat those steps.
17. How many pennies did it take to get the CD case moving again?
18. Clean up your mess.

What did you see in the experiment? First, it should have taken a few pennies to get the empty CD case moving. You probably expected that. When you added ten pennies to the case, you probably weren't surprised to see that the case didn't move until more pennies were added. But how did your guess turn out in step 11? Most students guess that it will take ten more pennies to get the CD case moving again. In the end, it took a lot less than ten pennies, didn't it? Why?

Let me ask you a simpler question first: Why did it take a few pennies to get the empty CD case moving? Most people think it's because the CD case has some weight to it, and you have to put *more* weight in the Ziploc bag to get it moving. Surprisingly, that's not true! Suppose the countertop was made of smooth, slippery ice. Would it take more pennies or fewer pennies to get the CD case moving again? It would take fewer pennies to get the CD case moving. That's because the weight of the Ziploc bag is not fighting against the weight of the CD case. It is fighting against *friction*.

Remember, when two surfaces rub up against each other, they tend to "grab" onto each other, and that makes it hard for the surfaces to move. So in your experiment, the countertop and CD case were "grabbing" onto each other, and that's why you had to add pennies to the Ziploc bag to get it to move. You weren't fighting the weight of the CD case. You were fighting the friction that was caused by the countertop and CD case "grabbing" each other.

When you added more weight to the CD case, you were increasing the friction. After all, with more weight, the CD case was being pushed onto the countertop with more strength. That made it easier for the countertop and CD case to "grab" each other, so that increased the friction. That's why you had to add more pennies to the Ziploc bag to get the heavier CD case to move. However, you didn't have to add the same amount of weight to the bag, because it's not the added weight you had to overcome to get the CD case to move again. It was just the added friction you needed to overcome.

This looks like an impossible task, but if the friction between the rock and the ground is small enough, the man could push this rock, because he is not fighting the weight of the rock to get it to move. He is fighting the friction between the rock and the ground.

Now don't get confused here. If you were *picking up* the CD case, then you would have to overcome its weight to get it to move. After all, an object's weight pulls it *down*. If you were trying to lift it *up*, you would have to overcome the weight in order to get it to move. However, that's not what you were doing. You weren't trying to lift the CD case up. You were trying to slide it along the countertop. The only thing the CD case's weight does in that situation is to increase the friction. Thus, you weren't really fighting the CD case's weight. You were just fighting friction.

This is one of the many things Leonardo da Vinci learned about friction. He realized that friction results from two surfaces "grabbing" onto one another. In order to get one of those surfaces to slide against the other one, then, you had to overcome the strength with which they were grabbing onto each other. He also learned that different surfaces grab onto each other differently. Smooth ice, for example, doesn't grab onto another surface nearly as well as rough wood. In the end, then, the amount of friction you have to overcome depends on the two surfaces that are rubbing against each other. Finally, he learned that the heavier something is, the harder its surface can grab onto the other surface, so the more friction there will be. These facts can be summarized as follows:

1. **In order to get an object to slide on a surface, you have to overcome the friction that holds the surface to the object.**
2. **The amount of friction between the surface and the object depends on their characteristics. Some surfaces and objects produce a lot of friction, while other surfaces and objects produce little friction.**
3. **The amount of friction depends on the weight of the object resting on the surface. The heavier the object, the greater the friction.**

Believe it or not, da Vinci learned all of this by doing experiments like the one you did. Simply by seeing how much weight he needed to slide objects across a surface, he was able to learn a lot about friction, which is obviously an important thing to learn about if you want to understand how things move.

LESSON REVIEW

Youngest students: Answer these questions:

1. Fill in the blank: In the experiment, the weight of the pennies in the Ziploc bag was used to overcome the ________ between the countertop and the CD case.

2. (Is this statement True or False?) The only thing that determines the friction between an object and the surface it is sliding on is the nature of the surface.

Older students: In your notebook, draw a diagram of your experiment. Explain what the weight of the pennies in the Ziploc bag had to do in order to get the CD case moving. Also, explain why the CD case with ten pennies added to it didn't take ten extra pennies in the Ziploc bag to get it moving.

Oldest students: Do what the older students are doing. In addition, write down (in your own words) the three things that da Vinci learned about friction.

NOTE: The activity in the next lesson requires time for glue to dry. You might want to start your day with science when you do that lesson so you can do other things while the glue dries.

Lesson 90: More on da Vinci's "Inventions"

Even though Leonardo da Vinci didn't seem to test many of his inventions, others have. For example, da Vinci had a great idea for how to build a bridge in sections. The sections could be carried easily, and then the bridge could be put together wherever it was needed. While we don't know whether or not da Vinci ever built such a bridge, others have done so, and it works quite well. See for yourself by building a model of his bridge.

Leonardo da Vinci's Bridge

What you will need:
- 16 craft sticks (or popsicle sticks)
- Glue

What you should do:
1. Glue four craft sticks together as shown in the picture labeled "A." Note that for the two horizontal sticks, one is glued on *top* of the vertical sticks, while the other is glued on the *bottom*.
2. Glue three craft sticks together as shown in the picture labeled "B. "
3. Repeat step 2 three more times, so you have four sets of sticks that look like the ones in the picture labeled "B."
4. Let the sticks sit so the glue can dry.
5. Once the glue is dry, hold the set of sticks that look like picture "A" in one hand and one of the other sets of sticks in the other hand.
6. Weave the bottom of the sticks that look like picture "B" between the horizontal sticks on the "A" set as shown in the picture labeled "C." Note how the bottoms of the "B" sticks are above the lower crossbar in the "A" sticks but below the top crossbar.
7. Add the next set of sticks by doing the same thing, as shown in the picture labeled "D."
8. Continue the process, threading each set of sticks like the one before, as shown in the pictures labeled "E" and "F."
9. Once you are done, notice how the bridge stands on its own.
10. Use your hand to press down on the top of the bridge. Notice how sturdy it is.
11. Clean up your mess.

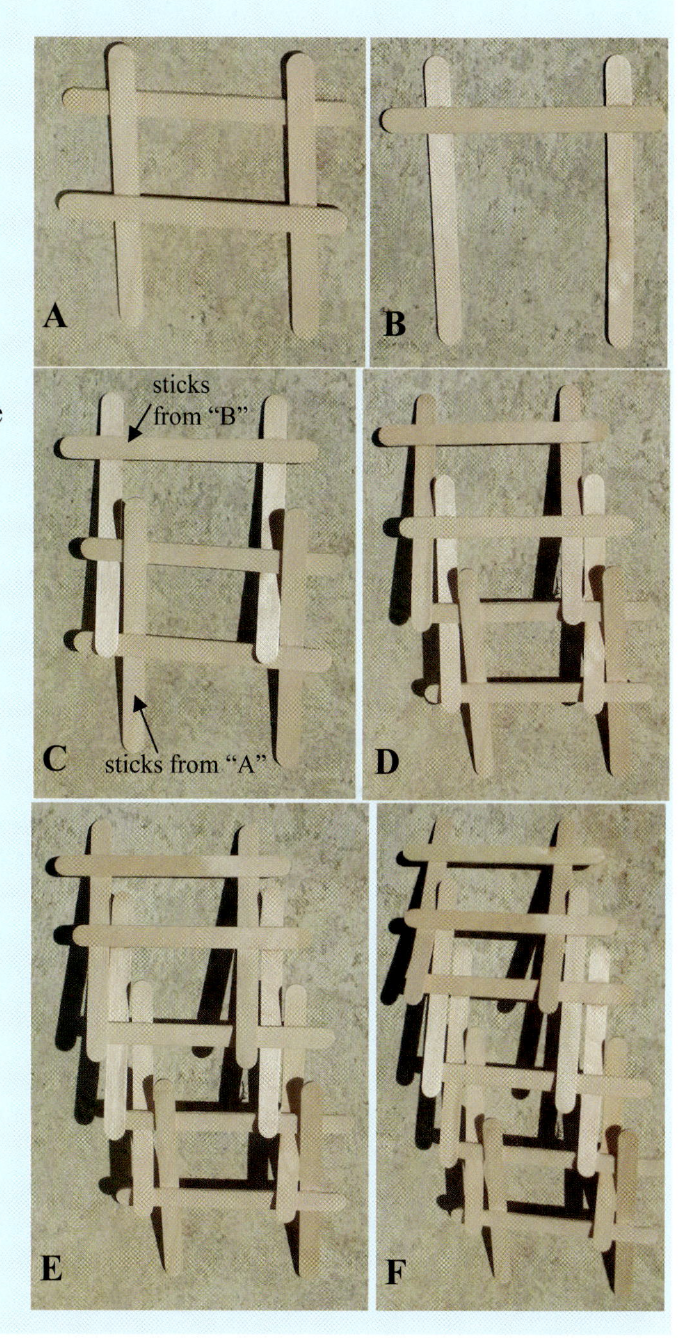

Leonardo da Vinci's bridge design is quite clever, isn't it? Not only would the sections be easy to carry around, but notice that the bridge is so sturdy once the sections are put together properly that you don't have to glue them together in any way. They just "hold on" to each other using friction.

That way, the bridge can be put together and taken apart very quickly. It is just ideal for travelers (and armies) who are going a long distance and don't know what kind of obstacles they might encounter on the road.

This is a model of Leonardo da Vinci's parachute.

Another interesting invention that da Vinci drew was a parachute shaped like a pyramid. Most people say that the parachute was invented by Louis-Sébastien Lenormand in 1783. While it is true he was the first to design and then successfully use a parachute, da Vinci beat him by about 300 years when it came to designing one. Since the parachute designed and built by Lenormand was so effective, most people were not interested in trying to make one according to da Vinci's design. As a result, no one knew for a long time whether or not da Vinci's design was any good. Some experts argued that it was far too bulky to work properly. However, in the year 2000, a British man by the name of Adrian Nicholas built and used a parachute according to da Vinci's specifications. It worked very well. In fact, while the parachute was harder to handle than a modern parachute, he said that it produced a smoother ride.

Da Vinci also designed the first diving suit. He thought that if ships came to invade a port, the defenders could use diving suits to go underwater and attack the ships from there. His diving suit involved two tubes that were attached to a float. One end of the tubes would stay above water because of the float, and the other end would be attached to a leather mask that the diver would wear. The diver could breathe through the tubes as if they were long snorkels. Once again, this invention was tested in modern times by a filmmaker named Jacquie Cozens. She said that it worked well in shallow water, though it surely got harder and harder to breathe the deeper she went.

Do you remember learning about water pressure in Lesson 63? Do you remember that the deeper you go in water, the more pressure you feel from the water? That's why da Vinci's diving suit would work well only in shallow waters. As a diver goes deeper, the water presses on him with more and more pressure. With da Vinci's diving suit, however, the air that the diver is breathing is at a lower pressure, because there is no water pushing on it where the float is. So the diver is being pushed on with more pressure than the air the diver is breathing. That means the diver has to work hard to suck the air into his lungs. The more pressure he experiences, the harder it is for him to do that, so eventually, he just isn't strong enough to suck in the air that is at a lower pressure.

Modern divers use scuba gear, and they bring their air down with them. The pressure of the air is controlled by the mouthpiece through which the diver breathes. That mouthpiece is called a "regulator," and it puts the air that the diver is breathing at the same pressure as the water around the diver. That way, there is no pressure difference between the diver and his air supply, so he can breathe air easily.

Da Vinci was also very interested in the idea of people being able to fly, and he studied birds carefully in order to try to learn the secrets of flight. He quickly realized that people are too heavy to fly by simply attaching wings to their arms, but he thought a well-designed machine might be able to accomplish the task. A model of one of his designs is shown below. In this design, a man would put his feet in the stirrups that hang from the bottom and would hold on to the bar in front. As the man moved his feet, it would cause the wings to flap. Da Vinci thought that the large wings would make up for the fact that a person is so heavy.

None of da Vinci's flying machines would have worked had they been tested. As you probably already know, God designed birds for flight, and one way He did that was to give them bones that were mostly hollow. As a result, they are very light. In addition, they can adjust their wings so that they attack the air differently when they flap down compared to when they are pulled up. That's very hard to design into a machine. As a result, man figured out how to fly by designing airplanes with wings that didn't flap. Instead, the shape of the wing gives the airplane lift, as long as the airplane is traveling quickly enough, which is accomplished with powerful engines.

This is a model that was built based on one of da Vinci's designs for a flying machine. It would not have worked, but it is a beautiful design.

LESSON REVIEW

Youngest students: Answer these questions:

1. What were the benefits of Leonardo da Vinci's bridge?

2. Did da Vinci's design for a parachute work?

Older students: In your notebook, draw a diagram of the bridge model you built, and explain why it would have been a very convenient thing for a group of people taking a long trip into unknown terrain. In addition, explain how da Vinci's parachute was different from a modern parachute and how we know it is a good design.

Oldest students: Do what the older students are doing. In addition, explain da Vinci's diving suit and why it would only worked when the diver was in shallow water.

Some Final Thoughts

You have reached the end of learning about the science that was learned from the time before Christ through about AD 1500. If you continue this series, the next book will cover some of the science that was learned after Leonardo da Vinci. For now, however, I want you to sit back and think about what you have learned. Some people have the idea that during the times you studied in this book, people were not very intelligent. Is that the impression you got? I hope not! In fact, people who lived in these times were, as far as I can tell, just as intellectual as you and me. They had less technology to work with, but the things they were able to accomplish were truly astounding!

Now think about what you have learned in the last four sections of this course (Lessons 31-90). Did you notice that most of the natural philosophers you learned about in those lessons were Christians? There were some who were not, but by and large, most of the natural philosophers who contributed to our understanding of the natural world after Christ was born were Christians. That's not a coincidence. You see, science is a natural consequence of a Christian worldview. When someone understands that a logical, rational Creator made this world and set laws in motion that would govern how the world works, he realizes that it is possible to use reason and actually understand nature.

This is why so many of the great scientists of the past were Christians. Science makes perfect sense in a Christian worldview, and it makes less sense in most other worldviews. Lots of scientists and science historians understand this. For example, Dr. Loren Eiseley was a professor of Anthropology and History of Science at the University of Pennsylvania. He was not a Christian, but he knew enough about the history of science to say:

> … we must also observe that in one of those strange permutations of which history yields occasional rare examples, it is the Christian world which finally gave birth in a clear articulated fashion to the experimental method of science itself…It began its discoveries and made use of its method in the faith, not the knowledge, that it was dealing with a rational universe controlled by a Creator who did not act upon whim nor interfere with the forces He had set in operation. The experimental method succeeded beyond man's wildest dreams but the faith that brought it into being owes something to the Christian conception of the nature of God. It is surely one of the curious paradoxes of history that science, which professionally has little to do with faith, owes its origins to an act of faith that the universe can be rationally interpreted, and that science today is sustained by that assumption. [Loren Eiseley, *Darwin's Century: Evolution and the Men Who Discovered It*, (Doubleday Anchor Books: New York, 1961), p, 62].

As Dr. Eiseley correctly points out, it was an act of faith that brought us science. Since early Christians believed that God created the universe and that God is a Lawgiver, it only made sense to them that the universe would operate under certain laws. They realized that if they could understand those laws, they could better understand the universe. Without such a creationist view, science would have never developed into what we see today.

It's no coincidence, then, that the most talented natural philosopher I have discussed so far, Leonardo da Vinci, was the man who painted *The Last Supper*, *The Virgin of the Rocks*, and *The Virgin and Child with Saint Anne*. The same faith that led him to paint those beautiful pictures was the faith that told him he could understand the universe by studying it.

Glossary

Acid – A type of chemical that acts opposite of a base

AD – Abbreviation for *Anno Domini*, a Latin phrase that means "Year of our Lord," used to date things that happened after Christ was born

Aeolipile (ee ol' uh pahyl) – A steam engine made from a round container and two bent tubes

Aether (ee' thur) – One of the elements Aristotle thought made up everything in nature. It was supposed to be associated with the planets and the stars

Air pressure – The strength with which the air pushes on everything that it touches

Air resistance – The molecules in air must be moved out of the way in order to move through it, but that requires work, which means that air is hard to move through

Anatomy (uh nah' tuh me) – The branch of science focused on identifying the organs and their placement in a living thing

Ancient – Belonging to the very distant past

Angle – The space between two intersecting lines

Anno Domini – Latin phrase that means "Year of our Lord," used to date things that happened after Christ was born

Apogee (ap' uh jee) – The point at which the moon is farthest from the earth

Archimedes' principle – Water pushes up on an object with a force equal to the weight of water that the object displaces

Arteries (in the body) – Blood vessels that carry blood away from the heart

Astrology (uh strah' luh jee) – The study of the stars, moon, and planets assuming that they affect the lives of people

Astronomer – A scientist who studies the sun, moon, and stars

Astronomy (uh strah' nuh mee) – The scientific study of the objects in the sky, such as the stars, moon, planets, and sun, as well as the universe as a whole

Atom – The smallest chemical unit of matter

Bacteria (bak' tear ee uh) – Microscopic creatures that are not plants or animals. They are considered the simplest form of life on earth

Baleen (bay' leen) **Whales** – Whales that have rows of feathery projections (baleen) instead of teeth

Base – A type of chemical that acts opposite of an acid

Basic Law of Magnetism – In magnets, opposite poles attract each other, but like poles repel each other.

Bathometer – A device that measures the depth of water

BC – Abbreviation for "Before Christ," used to date things that happened before Christ was born

Bile – A liquid produced in the liver that aids in the digestion of fat

Block and Tackle – A system of multiple pulleys that allows a heavy weight to be lifted with a small force

Blood – A liquid in the body that transports oxygen and nutrients to the tissues and picks up waste from the tissues

Blood Vessels – Tubes in the body that carry blood

Bone marrow – Soft tissue found inside bones that is important for making blood and components of the immune system

Bronchial (brahn' kee uhl) **tubes** – The tubes in the body that carry air to and from the lungs

Camera obscura (ob skur' uh) – A device that consists of an enclosed chamber with a hole on one end. It is used to project an image of its surrounding on a screen.

Cardinal Directions – North, South, East, and West

Center of Gravity – The point on an object from which all of its weight can be assumed to act

Chemical Energy – The energy stored in chemical bonds, such as food or oil

Cilia (sil' ee uh) – Tiny hairlike projections that produce movement

Circulatory (sur' kyoo luh tor' ee) **System** – The body system that includes the blood, blood vessels, and heart. It transports materials throughout the body.

Circumference (sur kum' fuh rents) – The distance around a circle or sphere

Classification – The act of grouping things together based on their similarities

Combustion (kum bust' yun) – A chemical reaction that occurs when oxygen combines with other substances to produce heat and usually light

Compass – A device with a magnetic pointer that always points north

Condensation (kon' den say' shun) – The process by which a gas turns into a liquid

Cranial (kray' nee uhl) **Nerves** – Nerves that come from the brain

Crescent moon – The phase of the moon when only a small sliver of the surface is lit

Crest – The part of a wave that is highest

Deciduous (dih sih' juh wus) **Teeth** – The first set of teeth with which a person or animal is born. They are eventually replaced by permanent teeth.

Deciduous Tree – A tree that loses its leaves in the fall

Decompose – To break a substance down into simpler substances

Density – A measure of how much matter exists within a given volume. It can be calculated by dividing the mass by the volume.

Dental Formula – A list of the number and kinds of teeth in an animal or person

Detect – To discover or become aware of

Diagnose (dye' uhg nohs) – The process by which a doctor tries to determine what is wrong with a patient

Diameter (dye 'am uh tur) – The length of a line running from one edge of a circle or sphere to the other while passing through the circle or sphere's center

Digestion – The process of breaking down food so that it can be used by the body for energy

Displace – To cause something to move from where it was

Dissection (dy sek' shun) – The process by which dead things are cut so that their various components (including internal organs) can be studied

Dynamics (dy na' miks) – The branch of science that studies the motion of objects and what causes that motion

Earthshine – Light from the sun that reflects off the earth and shines onto the moon

Electrons (ih lek' trahn) – One of the components of an atom. It orbits the nucleus and is negatively charged.

Element (el' uh muhnt) – The simplest component of a substance that cannot be broken down

Enamel – The hardest substance in the body. It covers the teeth.

Energy – The ability to do work

Enzyme (en' zeyem) – A chemical (specifically, a protein) produced by living organisms used to speed up chemical reactions in the body

Epicycle – A small circle whose center travels in a larger circle. Ptolemy used it to explain retrograde motion.

Erosion (ih roh' zhun) – The process by which rocks and soil are broken up and taken away

Evaporation (ih vap' uh ray' shun) – The process by which a liquid changes into a gas

Evergreen Tree – A tree that keeps its leaves (needles) year round

Exhale – To breathe out

Experimental Error – The error associated with an experiment

Factor – A number used to divide or multiply another number

Fahrenheit – A temperature scale in which water freezes at 32 degrees and boils at 212 degrees

Fat – A greasy or oily deposit found in many animals that is an excellent thermal insulator

Focal point – The point to which horizontal light rays are focused in a lens or curved mirror

Frequency (free' kwuhn see) – The number of times a crest or trough of a wave hits a specific point per second

Friction (frik' shun) – The resistance that two surfaces experience when they move against one another

Fulcrum – The part of a lever (or balance) that doesn't move

Full moon – The phase of the moon when the entire surface that is facing earth is lit

Gall – Another word for bile

Gas – One of the three phases of matter. This phase occupies the most volume.

Geocentric (jee' oh sen' trik) – A view in which the earth is at the center of the universe or solar system

Graph – A method that allows you to visually illustrate mathematical relationships

Gregorian (grih gor' ee uhn) **calendar** – A calendar that uses the earth's orbit around the sun to track the passage of the seasons

Heart – The organ in the body that pumps blood through the blood vessels

Heliocentric (hee' lee oh sen' trik) – A view in which the sun is at the center of the universe or solar system

Heliotropism (he' lee uh troh' piz um) – The process by which a plant changes the positions of its leaves so that they follow the sun as it travels across the sky

Hippocratic (hip' ih krat' ik) **Oath** – An oath taken by physicians that contains the major ethics that guide medical practices

Humidity – A measure of how much water vapor is in the air

Hydrometer (hi' drah muh tur) – A device that measures the density of a liquid

Hydroponics (hi druh' pahn iks) – The process of using a water-based solution to grow plants without soil

Hygrometer (hi grah' muh tur) – A device that measures humidity

Indicator – A chemical that changes color depending on the amount of acid or base present

Indivisible (in' duh vih' zuh buhl) – Cannot be broken down into smaller parts

Infinite – Limitless or endless

Inhale – To breathe in

Invertebrate (in ver' tuh brut) – An animal that does not have a backbone

Involuntary Motion – A motion you cannot control with your brain

Ion (eye' on) – An atom that has lost or gained electrons and now has an electrical charge

Irrigation (ihr uh gay' shun) – The process by which water is taken from one area and given to crops in another area.

Joint – A point at which two bones in the skeleton are joined so they can move relative to one another

Kinematics (kih' nuh ma' tiks) – The branch of science that studies the motion of objects without reference to the cause of the motion

Law of Continuity – In a system where water is being fed in at a constant rate, the smaller the opening through which the water leaves, the faster the water must move through that opening.

Law of Reflection - When light reflects from a smooth surface, the angle of reflection will be equal to the angle at which the light hit the surface.

Leeches – Bloodsucking animals that were used by ancient physicians who thought their patients had too much blood

Lens – A device that focuses light

Lever – A rigid bar resting on a fulcrum, usually used to lift heavy objects

Liquid – One of the three phases of matter. For most substances, this phase occupies more volume than the solid phase but less than the gas phase.

Loadstone (also Lodestone) – A rock that is naturally magnetic

Lungs – The organs that bring in air and allow the blood to absorb oxygen and release waste gases

Matter – A general term for any physical substance

Medium – A substance through which something travels or in which something is contained

Melting – The process by which a solid turns into a liquid

Mirror Writing – Writing words backwards so they appear normal in a mirror

Molecule (mol' ih kyool) – Two or more atoms linked together to make a substance with unique properties

Mucus (myoo' kus) – A slimy substance produced by the body to trap foreign particles and lubricate organs

Neutrons – (new' trahn) One of the components of an atom. It resides in the nucleus and has no charge.

Nucleus (new' klee us) – The center of an atom (which holds the neutrons and protons), or the portion of the cell that holds the DNA

Octave – A musical interval consisting of eight notes

Organ – A structure in a living creature that performs a specific function

Organism (or' guh nih zuhm) – A general term that refers to anything that is alive, be it plant, animal, or something else

Oxygen – A gas that people and animals breathe in and is necessary for most living things. Plants produce it through photosynthesis.

Perigee (pehr' uh jee) – The point at which the moon is closest to the earth

Phase – (in matter) One of the three ways matter appears: solid, liquid, or gas; (for the moon) A distinctive shape of the moon as seen in the night sky

Philosopher (fuh lah' suh fur) – A person who seeks understanding through reason

Phlegm (flem) – Mucus that is expelled from the body

Photosynthesis (foh' toh sinth' uh sis) – The process by which a plant makes food for itself from water, carbon dioxide, and sunlight

Phototropism (foh' toh troh' piz uhm) – The process by which a plant alters its growth so that it can be exposed to more light

Physiology (fih' zee ah' luh jee) – The branch of science focused on the activities and functions of the organs in a living thing

Pi – The ratio of a circle's circumference to its diameter, usually abbreviated as "π"

Pitch - The highness or lowness of a musical note

Pointillism (poyn' tuh lih' zuhm) – The practice of using tiny dots to construct an image, particularly in painting

Pore – A small opening or space

Projectile (pruh jek tyl') – An object that is launched but has no way to propel itself after it is launched

Proton (pro' tahn) – One of the components of an atom. It resides in the nucleus and is positively charged.

Ptolemaic (tah' luh may' ik) **system** – Another way to describe the geocentric view of the universe. It refers to the view's best-known champion, Ptolemy.

Pulley – A wheel with a grooved rim around which a cord passes. It is typically used to change the direction of a force.

Pulse – The throbbing of blood vessels as blood is pumped through them

Pupil – The circular opening of the eye that changes in size to regulate how much light enters the eye

Quarter moon – The phase of the moon when half the surface that is facing earth is lit, which means only a quarter of it is easily visible to observers on earth

Radiant (ray' dee unt) **Energy** – The energy of light

Range – The horizontal distance traveled by a projectile

Reflect – To bounce off an object and change direction

Refraction (rih frak' shun) – A process by which light bends when it encounters a new substance through which it must travel

Respiration (res' puh ray' shun) – The act of breathing

Respiratory (res' puh ruh tor' ee) **system** – The body system that contains the mouth, nose, bronchial tubes, and lungs. It brings oxygen into the body and gets rid of waste gases.

Retina (ret' nuh) - The layer of tissue in the eye that detects light

Retrograde Motion – Motion in which a planet initially moves one direction in the sky and then changes direction for a while before returning back to its original direction

Scientific Method – A process of doing science in which you make observations, form a hypothesis, and test the hypothesis

Shorthand – A method of writing that abbreviates words or uses symbols to replace words in order to reduce what has to be written

Siphon – A tube that moves water from a higher to a lower position due to air pressure and gravity

Skeletal Muscles – The muscles that attach to your skeleton, allowing you to move it

Skeleton – The system of support in an animal. For vertebrates, a skeleton is made of bones. For invertebrates, it is a hard outer covering.

Solar system – The system in which the earth, the planets, their moons, and asteroids orbit the sun

Solid – One of the three phases of matter. For most substances, this phase occupies the least volume.

Solute – A substance being dissolved in a solvent

Solution – A mixture of at least one solute in a solvent

Solvent – The substance a solute is being dissolved into

Soot – A black powder made of unburned carbon left over from burning something

Sound wave – A vibration in the air that consists of crests and troughs

Sphere – A round solid object in which all points on its surface are the same distance from the center

Spinal Nerves – Nerves that come from the spinal cord

Star – A large body of gas that produces its own light through thermonuclear reactions

Stethoscope (steh' thuh skohp) – A medical device used to hear a person's heartbeat or the sounds of the person's breathing

Summer Solstice (sohl' stis) – The day on which the sun reaches its highest noontime point in the sky

Tapeworm – A flat worm that lives in the intestines of people and other vertebrates

The Academy – The school of higher learning founded by Plato in Athens, Greece

Theologian – A person whose career is devoted to studying and understanding Scripture

Thermal (thur' muhl) **Energy** – The energy associated with heat

Thermometer – A device that measures temperature

Tissue – The material that makes up different organs in a living thing

Toothed Whales – Whales that have teeth in their mouth, such as sperm whales

Trial – One repetition of an experiment

Trough – The part of a wave that is lowest

Universe – The sum total of all that God created

Veins – (in the body) Blood vessels that carry blood back to the heart

Vertebrate (ver' tuh brut) – An animal that has a backbone

Vivisection (vih' vuh sek' shun) – The process by which living things are cut so that their various components (including internal organs) can be studied

Void – An empty space

Volume – A measure of how much room something takes up

Voluntary Motion – A motion you can control with your brain

Winter Solstice (sohl' stis) – The day on which the sun reaches its lowest noontime point in the sky

Work – Exertion or effort directed to produce or accomplish something

Photo and Illustration Credits

Photos by Kathleen J. Wile:
5, 7 (bottom), 10, 13 – top, 27, 35, 51, 53, 56 (all), 62 (both), 69 (both), 71 (bottom), 75, 77 (bottom), 78 (both), 81, 83, 105 (top), 109 (both), 117, 121, 126 (bottom), 137 (all), 146, 157, 158 (top), 163, 178, 197-198, 200 (bottom), 201 (both), 224, 232, 237 (bottom), 243 (bottom), 246, 250 (top), 256 (top), 259, 270, 273

Photos and illustrations by Dr. Jay L. Wile:
13 – bottom, 14, 25, 34, 57, 68 (bottom), 75 (bottom), 77 (top), 80, 88 (Earth and moon are public domain images.), 105 (bottom), 110, 112, 113, 149 (flat mirrors), 159 (bottom), 161 (everything but the man pointing), 188-190, 199, 215 (top), 216 (right), 225, 236

Illustrations by Julia Marie Ciferno:
24, 63, 66, 122, 124, 144, 170, 179 (both), 245, 253, 263, 269 (flagellum)

Photos and illustrations from www.shutterstock.com (Copyright holder in parentheses): xviii (Dimitrios), 1 (Jubal Harshaw), 4 (Marino Bocelli), 6 (pr2is), 7 (Offscreen – top), 8 (Zadorozhnyi Viktor), 11 (Lavitrei – just the face, Sunsetman – bottom), 15 (Asaf Eliason), 19 (Potapov Alexander), 29 (Craig Wactor), 36 (Leonello Calvetti), 38 (ZouZou), 40 (Margouillat Photo), 44 (Andrea Danti), 46 (Portokalis), 47 (Anastasios71), 50 (Panos Karapanagiotis), 52 (Hartphotography), 55 (Arkady), 58 (mmutlu), 59 (rng – top left, Dmitrijs Mihejevs – top center, Thomas Zobl – top right, Sean Nel – bottom left, Bill Mack – bottom center, Josh T. – bottom right), 61 (Olkhovsky Nikolay – left, Jo Crebbin – right), 73 (Krzysztof Odziomek), 74 (Ermak Vadim – top), 79 (Aragami12345s), 89 (Birute Vijeikiene), 90 (PerseoMedusa), 94 (montage includes illustrations by Dennis Cox and Hudyma Natallia), 101 (Pete Pahham), 104 (PHB.cz [Richard Semik]), 106 (Aceshot1), 115 (URRRA), 118 (Lisa F. Young), 120 (Digital Storm – top, hkannn – bottom), 123 (Sebastian Kaulitzki), 126 (Matthew Cole – top), 128 (hkannn), 130 (Alex Mit), 131 (rob3000), 133 (somchaij), 134 (adziohiciek), 137 (mareandmare), 141 (pryzmat), 149 (MilanB – curved mirrors), 150 (MilanB), 152 (Jan Mika), 153 (irabel8), 154 (Africa Studio), 155 (T.W.), 156 (Dan Holm), 158 (Reinhold Leitner), 164 (Laborant), 167 (montage of images by Magcom, Roland Glukhov, and Dirk Ercken), 168 (Blue Lemon Photo), 169 (Michael Hare), 171 (Dusan Po), 173 (John Wynn), 174 (Ensuper), 176 (Yayayoyo), 177 (Paul B. Moore), 180 (diez artwork), 184 (Dja65), 185 (Brad Collett), 187 (Kiev.Victor), 192 (Ronald Caswell), 193 (JaySi), 195 (Alex Staroseltsev), 196 (Jorg Hackemann), 200 (montage of Alekcey and marilyn barbone – top), 202 (Donald Gargano), 204 (marekuliasz), 207 (Diego Barucco), 211 (OtnaYdur), 214 (12_Tribes), 215 (SlavaK – bottom), 216 (rodho – left, SlavaK – middle), 217 (Zastolskiy Victor Leonidovich), 218 (Mr Doomits), 219 (Yuriy Boyko), 220 (Anmor Photography), 221 (Anest), 222 (Piotr Tomicki), 223 (Pakhnyushcha), 226 (montage of Kletr, vita khorzhevska, and paulrommer), 227 (Africa Studio), 231 (Tonis Pan – top, Lightspring – bottom), 234 (claudio zaccherini), 235 (Boris Sosnovyy), 238 (Nazzu), 241 (Kirsty Pargeter), 242 (kokkodrillo), 243 (Gemenacom – top), 244 (Skyhawk), 247 (Alila Sao Mai), 248 (sharethnegative), 250 (CLIPAREA l Custom media – bottom), 251 (malinx), 252 (WimL – bottom), 256 (BioMedical – bottom), 257 (URRRA), 258 (CLIPAREA l Custom media), 260 (natasha58), 262 (forestpath – top, Andrea Danti – bottom), 265 (Paul Michael Hughes), 266 (Alila Sao Mai), 268 (Oleg Kozlov), 269 (whatie –bacterium) 271 (gualtiero boffi), 275 (Leo Blanchette)

Illustrations from www.clipart.com:
70, 107, 161 (man pointing)

Photos and illustrations from the public domain:
2, 9, 16, 18, 31, 49, 67, 68 (top), 71 (top), 92, 108, 111, 114, 135, 138, 139, 145, 165, 166, 181, 191, 203, 206, 208, 209, 212, 213, 230, 237 (top), 239, 240, 249, 252 (top), 254, 255

Photo from http://www.istockphoto.com:
76 (homermi)

Photo from Eon Images:
86

Photo by John Skipper
42

Illustration published under the Creative Commons Attribution 3.0 Unported license (http://creativecommons.org/licenses/by/3.0/):
274 (Nevit - http://simple.wikipedia.org/wiki/File:Leonardo_da_Vinci_parachute_04659a.jpg)

Index

-C-

-D-

-E-

-F-

-G-

-H-

-I-

-Q-

-R-

-S-

-T-

-U-

-V-

-W-

-X, Y, Z-